Library of
Davidson College
VOID

DEVELOPMENTAL AND CELL BIOLOGY SERIES

EDITORS
D.R.NEWTH J.G.TORREY

PRIMORDIAL GERM CELLS
IN THE INVERTEBRATES

PRIMORDIAL GERM CELLS IN THE INVERTEBRATES
From epigenesis to preformation

PIETER D. NIEUWKOOP
Hubrecht Laboratory, Utrecht, The Netherlands

LIEN A. SUTASURYA
Dept of Biology, ITB, Bandung, Indonesia

CAMBRIDGE UNIVERSITY PRESS

CAMBRIDGE

LONDON · NEW YORK · NEW ROCHELLE

MELBOURNE · SYDNEY

Published by the Press Syndicate of the University of Cambridge
The Pitt Building, Trumpington Street, Cambridge CB2 1RP
32 East 57th Street, New York, NY 10022, USA
296 Beaconsfield Parade, Middle Park, Melbourne 3206, Australia

© Cambridge University Press 1981

First published 1981

Printed in Great Britain at the
University Press, Cambridge

British Library cataloguing in publication data

Primordial germ cells in the invertebrates. –
(Developmental and cell biology).
1. Germ cells
2. Embryology – Invertebrates
I. Nieuwkoop, Pieter D   II. Sutasurya, Lien A
III. Series
592'.03'2   QL364   80–41651
ISBN 0 521 22189 7

*Dedicated with gratitude to
Professor Dr D. A. Tisna Amidjaja,
the former teacher of L.A.S.*

Contents

Preface		xi
Acknowledgements		xiii
1	**General introduction**	1
2	**Protista, in particular some colonial flagellates**	9
	Introduction	9
	General characterisation	9
	Colonial forms	10
	Conclusion	13
3	**Porifera (sponges)**	14
	General characterisation	14
	Regeneration and reconstitution	14
	Asexual reproduction	15
	Sexual reproduction	16
	Origin of the germ cells	18
	Conclusion	18
4	**Coelenterata (Cnidaria)**	21
	General characterisation	21
	HYDROZOA	21
	Adult organisation	22
	Regeneration	23
	Interstitial cells in the adult animal	24
	Asexual reproduction	27
	Embryonic development and origin of the I-cells	27
	Sexual reproduction and origin of the germ cells	29
	SCYPHOZOA	34
	ANTHOZOA	35
	General conclusion	35
5	**Platyhelminthes**	37
	TURBELLARIA	37
	General characterisation	37

	Origin of the regeneration blastema	37
	Neoblasts	40
	Asexual reproduction	42
	Embryonic development and origin of the neoblasts	44
	Origin of the germ cells	45
	Sexual differentiation	46
	Conclusion	46
	TREMATODA	46
	General characterisation	46
	Life cycle and origin of the germ cells	47
	Conclusion	50
	CESTODA	50
	General characterisation	50
	Life cycles	51
	Embryonic development	51
	Adult organisation and origin of the germinative cells and the germ cells	52
	Conclusion	54
6	**Nematoda**	55
	General characterisation	55
	Embryonic development and origin of the germ cells	55
	Chromatin diminution	59
	Conclusion	61
7	**Ectoprocta (Bryozoa)**	62
	General characterisation	62
	Asexual reproduction	62
	Sexual reproduction	64
	Embryonic development and origin of the germ cells	64
	Conclusion	65
8	**Annelida**	66
	General characterisation	66
	Asexual reproduction and regenerative capacity	66
	Embryonic development	69
	Origin of the germ cells	73
	Sexual differentiation	74
	Conclusion	75
9	**Echinodermata**	76
	General characterisation	76
	Regenerative capacity	78
	Embryonic development	78

	Origin of the germ cells	83
	Conclusion	85
10	**Mollusca**	86
	General characterisation	86
	Embryonic development of the spiralian molluscs	88
	Polar lobe formation and its possible role in development	91
	Further experimental analysis	95
	Embryonic development of the Cephalopoda	97
	Origin of the germ cells and gonad formation in the molluscs	101
	Conclusion	103
11	**Onychophora**	104
	General characterisation	104
	Embryonic development	104
	Origin of the germ cells	106
	Conclusion	106
12	**Arthropoda**	108
	Introduction and general characterisation	108
	MYRIAPODA	109
	General characterisation	109
	Embryonic development	109
	Origin of the germ cells	111
	HEXAPODA (INSECTA)	112
	General characterisation	112
	APTERYGOTA	112
	Embryonic development	113
	Origin of the germ cells	115
	PTERYGOTA	116
	Embryonic development of the Hemimetabola: descriptive data	117
	Embryonic development of the Holometabola: descriptive data	120
	Embryonic development of Hemi- and Holo-metabola: experimental data	123
	Origin of the germ cells in Hemi- and Holo-metabola	127
	CRUSTACEA	148
	General characterisation	148
	Embryonic development	149
	Origin of the germ cells and development of the gonads	151
	CHELICERATA (ARACHNOIDEA)	152
	General characterisation	152
	Embryonic development	154
	Origin of the germ cells	155

General conclusion   155

13 Remaining phyla   158
Ctenophora   158
Nemertini (Rhynchocoela)   159
Mesozoa   162
Rotifera   163
Gastrotricha   165
Entoprocta   165
Echiurida   168
Chaetognatha   168
Conclusions   171

14 General discussion   172
Embryonic development in the animal kingdom   172
Weismann's *Keimplasma* theory and the phenomenon of chromosome elimination   173
The germ line concept   174
Epigenesis and preformation in germ cell development; a continuous variable in the animal kingdom   177
Stability of cellular differentiation in epigenetic and preformistic development   180
Germ cell determination   182
Determinancy in cleavage, embryonic development and germ cell formation   183
Possible role of the germinal granules in germ cell formation   184
Fundamental features of germ cells   186
Suggestions for further analysis   187

References   189

Author index   233

Taxonomic index   239

Subject index   244

Preface

In his book *L'Origine des Cellules Reproductrices et le Problème de la Lignée Germinale* Bounoure (1939) summarised the knowledge of the origin of the germ cells in the animal kingdom which was available at the time. Much work has been done in the succeeding four decades, particularly on the ultrastructure of the germ cells in the insects and the anuran amphibians. The request of the editors of the Developmental and Cell Biology Series of Cambridge University Press to write a new, up-to-date monograph on the origin of germ cells in the animal kingdom therefore seemed to us well justified. However, we did not feel it right to treat the very extensive subject in a single volume. As we mentioned in the companion volume to this *Primordial Germ Cells in the Chordates* (Nieuwkoop & Sutasurya, 1979), we feared that a single volume would not only be difficult to write, but also hard to read.

Our personal contributions to the subject have dealt with the origin of the germ cells in the vertebrates, in particular in the urodele amphibians, where the germ cells show both a site and a mode of origin that are essentially different from those in the anuran amphibians. This discrepancy, in our opinion, points towards a different phylogenetic origin of the two groups of Amphibia. The chordates, of which the vertebrates are the main representatives, form a rather homogeneous group, the phylogeny of which has been widely studied due to the availability of an extensive fossil record. In the companion volume we therefore treated the original of the germ cells in the chordates against the background of early embryogenesis as well as phylogeny.

Unfortunately, in the invertebrates, the origin of the germ cells cannot be treated in the same way, since too little is known about the phylogenetic relationships among the various invertebrate phyla. Germ cell origin in the invertebrates must therefore, of necessity, be discussed against the background of their embryonic development only. However, in most instances, this knowledge is of a purely descriptive nature. Fortunately, there are a number of groups where experimental analysis has given us some insight into the mode of origin of the germ cells. Emphasis will be placed on these groups.

We will discuss primarily how far the origin and development of the germ cells in the various invertebrate groups can be characterised as *epigenetic*, and how far as *preformistic* (for definitions of these concepts see Hertwig, 1900 (reprinted in 1977), Raven, 1958 and Maresquelle, 1978). In the introduction we try to formulate some basic questions concerning different mechanisms which may act in germ cell formation. After discussing germ cell formation in the various phyla, we will try to formulate some common principles of germ cell development in the concluding chapter. We hope that such a treatment will stimulate interest as well as lead to active participation of developmental biologists in this fascinating field of research.

Acknowledgements

First of all we would like to thank the editors of the Developmental and Cell Biology Series of Cambridge University Press for allowing us to treat the subject of the origin of the germ cells in the animal kingdom in two separate volumes, the first on the chordates and the second on the invertebrates.

We are very grateful to the Department of Biology of the Institut Teknologi Bandung for granting leave to L.A.S. in order to come to Holland for the extensive literature study for the second book and the final preparation of the manuscript, and for the hospitality extended to P.D.N. during the preparation of the definitive text.

We also extend our thanks to the Governing Board of the Hubrecht Foundation for repeatedly providing the necessary funds for the travel of P.D.N. to Indonesia and of L.A.S. to Holland.

We want to thank the staff of the Library of the Hubrecht Laboratory for their assistance in compiling the extensive bibliography and Mrs C. L. Kroon for preparing the majority of the illustrations. We express our deep gratitude to Dr J. Faber for his valuable suggestions and for the correction of the English text.

Finally, we are indebted to several colleagues for allowing us to reproduce figures or sending us original illustrative material.

1
General introduction

Towards the end of the nineteenth century Weismann formulated his theory of the functioning of heredity in development, laid down in his classical works *Die Continuität des Keimplasmas als Grundlage einer Theorie der Vererbung* (1885) and *Das Keimplasma. Eine Theorie der Vererbung* (1892). According to this theory pluricellular organisms consist of two main components, the somatic cells, constituting the body of the individual of a particular generation, and the germ cells, representing the forerunners of the next generation. The two components were called *soma* and *germen* respectively.

The analysis of the origin of the germ cells has been strongly influenced by the theoretical concepts of *soma* and *germen*. Weismann postulated that the development of the unicellular egg into a complex organism with different cells, tissues and organs was based upon the differential distribution of the various *Determinanten* (later called genes) amongst the different cell types, allowing each cell type a specific but restricted mode of development. The only exception to this rule would be the germ cells, which would retain the full complement of *Determinanten* originally present in the fertilised egg, thus forming the so-called 'germ line'.

The distinction between *soma* and *germen* as formulated above is necessarily an essential and permanent one, since the somatic cells, which have received only part of the genetic complement of the egg, would no longer be able to form *totipotent* germ cells. Conversely, the germ cells would at any time be able to form somatic cells of the organism.

In the years that followed many investigators have shown that, in both the vertebrates and the invertebrates, the germ cells either segregate from the somatic cells very early in embryonic development, or are discernible only at much later stages of development. In the latter situation, during a rather long initial period of development, no clear distinction between *germen* and *soma* can be made. Bounoure (1939) reviewed the literature up to the late 1930s on the early versus late appearance of the germ cells in the various groups of the animal kingdom in his book *L'Origine des Cellules Reproductrices et le Problème de la Lignée Germinale*. Referring primarily to the vertebrates, Cambar (1956) called the early segregation of the germ cells the *preformistic* mode of germ cell formation; this is often characterised by the presence of a special cytoplasmic structure, the 'germ plasm', which acts as a germ cell determinant. He called the late appearance of the germ cells the *epigenetic* mode of germ cell formation; this often seems to occur under inductive influences. The first question to be answered is whether this distinction also holds for the invertebrates, and if so, whether it has the same or a broader significance there.

What is our present insight into the potentialities of germ cells and somatic cells? We know that during the subdivision of the egg into a large number of cells forming different cell types, the division of the nucleus is *not* characterised by a differential distribution of genes among the daughter nuclei, but that the accurate replication of the full complement of nuclear genes and the subsequent distribution of the two identical sets of genes among the daughter nuclei renders them potentially isopotent. This has been demonstrated convincingly by, among other methods, transplantation of nuclei from differentiated somatic cells into enucleated eggs (see reviews by Gurdon & Woodland, 1968 and Gurdon, 1974a on vertebrates, and by Gurdon, 1974b and Illmensee, 1976 on insects). In addition, it has been shown that different genes are active in different phases of development and in different cell types (see review by W. Beermann, 1967). This has led to the conclusion that the differentiation of cells is due to differential gene activation or derepression and not to differential distribution of genes, so that Weismann's theory has been essentially refuted. We must therefore ask ourselves whether the distinction between *soma* and *germen*, which is a direct consequence of Weismann's theory, should not be abandoned as being inadequate and obsolete. Since we will argue that it should, the 'germ line' concept will be avoided as much as possible in the following chapters.

On the basis of the above considerations Davidson (1976) has advanced the hypothesis that in the nuclei of different cells different genes are repressed or derepressed due to the different cytoplasmic composition of the cells. This hypothesis, for which Davidson and others brought together extensive evidence, places the distinction between *soma* and *germen* in an entirely different light. Germ cell and somatic cell nuclei are *potentially* equal, since they both contain the full complement of genetic information. Due to their different cytoplasmic composition, however, germ cells and somatic cells may express different potentialities due to the activation of different parts of the genome.

An interesting question arises here. Are the more restricted expressions of the somatic cells permanent or only temporary? In other words, can the differential inactivation of genes in somatic cell nuclei be reversed under certain conditions? Transplantation of nuclei from differentiated somatic cells into enucleated eggs has demonstrated that, in a low percentage of cases, such nuclei *can* support normal development, so that somatic cell nuclei *can* reacquire their full potentialities. However, transplanted germ cell nuclei do so in an appreciably higher percentage of cases. This relative difference is probably due to the fact that germ cell nuclei can adapt more easily to the special requirements of the egg cytoplasm than can the nuclei of differentiated somatic cells.

Germ cells are considered to be *totipotent* since they can give rise to complete new individuals. The question should, however, be raised of whether totipotency is actually an adequate concept. Embryonic development, including germ cell formation, requires at any time a very accurately programmed, sequential release of the potentially present, but functionally repressed, genetic information. Moreover, it should be realised that, although the nuclei of egg and sperm may be more or less equivalent – in androgenesis the sperm nucleus can support normal development of the enucleated egg almost equally as well as the female pronucleus of a parthenogenetically activated egg – they belong to entirely different, specialised cell types. They must therefore have different sets of derepressed genes. One must conclude that the concept of totipotency of the *germen* is inadequate, and that the distinction between *germen* and *soma* on the basis of fundamental differences in potentialities is erroneous. The distinction between *germen* and *soma* is only a relative one. This makes it theoretically possible that somatic cells can be converted into germ cells.

According to Davidson's hypothesis, the cell-type-specific machinery keeps the somatic cell nucleus engaged in releasing only that genetic information which is relevant for a particular differentiated state. Conversion of a somatic cell into a germ cell is possible, therefore, only after the specific cytoplasmic differentiation of the somatic cell has been 'erased'. In other words, the somatic cell must have dedifferentiated to such an extent that the information for other types of differentiation can be released; in its most extreme form this is the information for germ cell formation. Consequently, the more highly differentiated a somatic cell is, the more unlikely it is that it can be converted into a germ cell.

A clear distinction must be made between nuclear potentialities and cellular expression. Although nuclei of differentiated somatic cells *can* support normal development of enucleated eggs, in normal development somatic cells are rarely transformed into germ cells. This holds particularly for the more highly evolved forms, such as holometabolous insects and vertebrates. These are generally characterised by a relatively early segregation of the germ cells during embryonic development, as well as by the non-replaceability of the germ cells. On the other hand, the non-convertibility of *soma* into *germen* and vice versa certainly does not hold for several lower animal phyla, such as the sponges and the coelenterates. As well as a sexual form of reproduction with gamete formation, the majority of these animals show an asexual form of reproduction by bud formation or schizogenesis. The two forms of reproduction often alternate.

The germ cells of highly evolved forms, which are usually segregated during early embryonic development, are often characterised by the presence of a special cytoplasmic structure which is supposed to act as a germ cell determinant (see Hegner, 1914 and Gehring, 1976*b*), giving germ cell development in these forms a strongly *determinate* or *preformistic* character. This aspect has been studied extensively in several holometabolous insects as well as in the anuran amphibians. How general is this phenomenon among the invertebrates? Do the special organelles encountered in the germ plasm actually act as a germ cell determinant? In our previous book, dealing with primordial germ cells in the chordates (Nieuwkoop & Sutasurya, 1979), this notion has been seriously questioned. What is the evidence for it in the invertebrates?

Weismann's germ plasm theory has been rejected on the grounds of the essential equipotentiality of all the nuclei in a developing organism. What, then, is the significance in this context of the phenomena of chromosome elimination and chromatin diminution encountered in the somatic cells of certain nematodes, crustaceans and insects? Do these phenomena support Wiesmann's theory? Is it really a unique part of the genetic information that is eliminated, or only reduplicated chromosomes or amplified genes?

Contrary to the highly evolved forms, where germ cell development seems to be strongly preformistic, many lower invertebrate forms show a more or less typical *epigenetic* mode, where the germ cells are formed from cells of one or the other 'germ layer' under inductive influences from adjacent organ anlagen, or even under the influence of external environmental factors. Among the vertebrates a typically epigenetic mode of germ cell development has been demonstrated in the urodele amphibians. Is this also encountered in invertebrate groups? Is the mechanism in invertebrates and vertebrates the same or different? Can a more-or-less continuous transition between the typically epigenetic and the typically preformistic modes be found among the invertebrates, or are the two modes of germ cell formation mutually exclusive?

Whatever the answer to these questions, the essential problem with which we are faced is whether we can find a denominator common to the different modes of germ cell formation, the end-product being the same in all cases. The solution to this problem may lead to a deeper understanding of germ cell formation generally. We must ask ourselves whether indications can be found that germ cells show a particular form of activity or inactivity of the genome which is conditioned by a specific composition or functioning of their cytoplasm. If we cannot yet define such requirements, how should we proceed towards this goal?

We now come to some more practical points which have to be settled before we can start the survey of the various groups of invertebrates.

As mentioned in the Preface, we know hardly anything about the phylogenetic relationships among the different invertebrate phyla. Consequently, the invertebrates can be arranged only according to a purely *taxonomic* classification. It is evident that classification into separate phyla and their arrangement in a hierarchical order from 'lower' to 'higher' forms of organisation is subjective. Opinions differ rather strongly among the taxonomists themselves; consequently, several schemes have been proposed. We have, on the whole, followed the classification of the Dutch Leyden School, which is based mainly on the work of Hyman (1940, 1951a,b, 1955, 1959, 1967), Barnes (1963) and Karstner (1965/7). The primary subdivision of the Metazoa into Radiata and (acoelomate, pseudocoelomate and eucoelomate) Bilateralia makes for easy surveyability. It should, however, be regarded as a purely practical one. The classification used in this book is given in table 1.1. The well-known phyla and classes will be treated in this sequential order. However, there are a number of metazoan phyla in which the mode of germ cell formation is almost completely unknown. These will be discussed briefly in a separate chapter at the end of the systematic part.

Table 1.1. *Classification of the animal kingdom used by the authors based mainly on Hyman (1940, 1951a,b, 1955, 1959, 1967), Barnes (1963) and Karstner (1965)*

Kingdom ANIMALIA
  Subkingdom Protista
      Phylum Protista
  Subkingdom Parazoa
      Phylum Porifera
  Subkingdom Metazoa
    Division Radiata
      Phylum Coelenterata or Cnidaria
      Phylum Ctenophora
    Division Bilateralia
     Subdivision Acoelomata
                                               ⎧ Class Turbellaria
                Phylum Platyhelminthes ⎨ Class Trematoda
                                               ⎩ Class Cestodes
        Phylum Nemertini
        Phylum Mesozoa
     Subdivision Pseudocoelomata
        Phylum Acanthocephala
        Phylum Rotifera
        Phylum Gastrotricha
        Phylum Kinorhyncha
        Phylum Nematomorpha
        Phylum Nematoda
        Phylum Entoprocta

Table 1.1 continued

Subdivision Eucoelomata
Phylum Brachiopoda
Phylum Ectoprocta
Phylum Phoronida
Phylum Annelida
Phylum Echiurida
Phylum Sipunculida
Phylum Priapulida
Phylum Echinodermata
Phylum Mollusca
Phylum Tardigrada
Phylum Pentastomida
Phylum Onychophora
Phylum Arthropoda { Class Myriapoda; Class Hexapoda { Apterygota, Pterygota }; Class Crustacea; Class Chelicerata }
Phylum Chaetognatha
Phylum Pogonophora
Phylum Hemichordata
Phylum Chordata

In each of the various groups germ cell formation will be classified (often tentatively) into one of the following three main modes: the typically *epigenetic* mode, an *intermediate* mode with relatively late appearance of the germ cells, and the typically *preformistic* mode. In the concluding chapter we shall try to arrange the various phyla and classes into a more-or-less continuous series. We realise that such an arrangement is, again, a subjective one. We hope, however, that it will make the discussion more adequate because, on the one hand, it emphasises the great diversity in modes of germ cell formation, even in related groups, while, on the other hand, it points towards the possible existence of a common mechanism of germ cell formation.

For each of the phyla or classes, early embryonic development (as well as asexual reproduction and regeneration) will be discussed first, wherever appropriate, and will be followed by a discussion of germ cell origin serving to place germ cell formation against the background of embryonic development.

It would be desirable to view germ cell formation against the background of both descriptive and experimental analysis of early development, so that not only the *site* but also the actual *mode* of germ cell formation could be elucidated. It must, however, be realised that in many groups hardly any experimental analysis has been performed, so that, in these cases, nothing definite can be said about the actual mode of germ cell formation. In other groups, even normal development is only fragmentarily known, and germ cells are recognised at only rather late stages of development, making even their site of origin questionable. Groups of which we know so little must necessarily be treated briefly. In the more thoroughly studied groups, normal development is discussed in broad outline only, since this is not a textbook of invertebrate development. A similar restriction holds for the references on normal development, since in certain groups, such as the insects, the literature on normal development is far too extensive for complete coverage. We shall, therefore, refer mainly to review articles, citing the original publications only for aspects of particular interest. For further information the reader will find the full titles in the reference lists of the reviews cited here.

As in our previous book (Nieuwkoop & Sutasurya, 1979), we shall restrict ourselves to the initial development of the germ cells, leaving oogenesis and spermatogenesis out of consideration. This means that the book will deal only with the *primordial germ cells* (PGCs). We shall treat the subject as comprehensively as possible, but in a concise form which we hope will ensure easy reading. Given the excellent review of the older literature by Bounoure (1939), we shall restrict ourselves mainly to the literature of the last four decades, referring to older literature only where no recent literature is available or where reference to important older investigations cannot be omitted.

Some of the relevant terminology must be discussed briefly here. The development of the germ cells from their first detectable origin until their release as mature gametes, called gametogenesis, is subdivided into two major periods, that preceding and that succeeding sexual differentiation. During the first period the germ cells are called 'primordial germ cells' (PGCs). The forerunners of the germ cells, while they are segregating from the somatic cells of the embryo, will be called 'presumptive primordial germ cells' (pPGCs), in contradistinction to the fully segregated or 'true' primordial germ cells. For more detailed information the reader is referred to the chapter on terminology in the companion volume, *Primordial Germ Cells in the Chordates* (Nieuwkoop & Sutasurya, 1979, pp. 5–7).

In the invertebrates the gonads may either be formed from cells split off from the oogonia and spermatogonia, which come to surround the germ cells, or may develop independently of the PGCs, often in another part of the embryo. In the latter case the PGCs must move actively or passively from their site of origin to the 'gonadal anlagen'. The latter may differentiate into an 'ovary', a 'testis' or an 'ovotestis'. In some forms the sexes are separate, and male and female animals can be distinguished. In so-called 'hermaphrodites' male and female gonads may be found in the same or in different parts of the body, or they may be fused into an 'ovotestis'. In other forms again the animal functions initially as a male and subsequently as a female, or vice versa (called 'protandry' and 'protogyny', respectively).

2

Protista, in particular some colonial flagellates

Introduction

It does not seem obvious to include the Protista in a monograph which deals essentially with the segregation of germ cells from somatic cells in multicellular organisms. There are, however, a number of reasons for discussing the Protista briefly at the beginning of this book.

The Protista possess essentially the same cytoplasmic and nuclear organisation as multicellular organisms. Their protoplasm may be more or less undifferentiated or may contain special cell organelles for locomotion, food capture and perception, as well as conductile and contractile fibrils and water-regulatory vacuoles.

It is very likely that the higher multicellular organisms have evolved from unicellular organisms, in particular from the flagellates, which are considered to have a position at the bottom of both the animal and plant kingdoms.

Certain colonial flagellates, the Phytomonadina or Volvocales, may, for all practical purposes, be considered as multicellular organisms, since they show some degree of cellular differentiation and spatial organisation. These features make them important for understanding the complex organisation of the higher organisms.

Besides asexual reproduction, the Protista exhibit sexual reproduction with all possible degrees of diversity of the fusing gametes. Some forms have complicated life histories showing alternation of sexual and asexual generations. Asexual reproduction is always preceded by nuclear multiplication, while sexual reproduction is always characterised by reduction of the number of chromosomes and meiosis, as in the Metazoa. Gamete types range from isogametes through anisogametes to flagellated microgametes and immobile macrogametes, called spermatozoa and eggs respectively.

General characterisation

The Protista comprise four classes: the Flagellata, the Rhizopoda or Sarcodina, the Sporozoa and the Ciliata. The Flagellata may show either plant- or animal-like characteristics, are often endowed with yellow, brown or green chromoplasts, and carry a varying number of flagella for locomotion. They reproduce asexually as well as sexually. The Rhizopoda possess pseudopodia for locomotion and food catching, and usually reproduce asexually by binary fission. The Sporozoa are parasitic forms with contractile vacuoles. They reproduce asexually by multiple fission as well as sexually through zygote formation, and may form cysts. The Ciliata constitute a highly specialised group showing ciliary locomotion, and are endowed with both somatic and generative nuclei. They usually reproduce asexually by binary fission but sometimes sexually by conjugation. It is highly probable that the Flagellata represent the most primitive group, from which the other groups have evolved (see Hyman, 1940, vol. 1, pp. 44–232).

Colonial forms

Colony formation occurs among the Rhizopoda, Ciliata and Flagellata. Among the Rhizopoda, some Radiolaria form spherical or cylindrical colonies surrounded by a gelatinous capsule. The ciliate *Zoothamnion* forms regular-branching colonies which consist of an axial stem terminating in a single large cell, with alternating left and right branches arranged in a two-dimensional branching pattern. New colonies are formed when a zooid breaks off from the colony (asexual reproduction). After attachment of the new zooid to a substrate, a non-contractile peduncle with a thickened cuticle develops, after which a neuromuscular cord grows out. Fifteen hours after attachment an unequal division occurs, by which a terminal macrozooid and a smaller daughter zooid are formed. The colony is a true symplastic one without being syncytial (see Berrill, 1961).

Special attention is paid here to various *Volvox* species since, among the Flagellata, the Phytomonadina or Volvocales show an interesting supra-cellular organisation. *Volvox* forms green, multicellular, hollow mucilaginous spheres of up to 50 000 cells (fig. 2.1). In several species the individual cells are permanently connected to neighbouring cells in a hexagonal pattern by strands of protoplasm (fig. 2.2). Each cell has two flagella. The colony is polarised in that the protoplasts at the anterior pole have better-developed eye-spots and are more widely spaced (fig. 2.1a). The colony generally rotates counterclockwise about the antero-posterior axis and moves with the anterior pole in front.

In asexual reproduction a restricted number of larger cells – from one to twenty, but usually eight – are formed in the posterior half of the colony by unequal division; they are arranged alternately in two perpendicular planes (fig. 2.1a). These so-called 'gonidia' sink in and lose their flagella. They form small new colonies (fig. 2.1b), which may finally escape through the original pore of the mother colony (see below), or break through its wall. The number of cell divisions in the new colony is genetically determined. A

Colonial forms 11

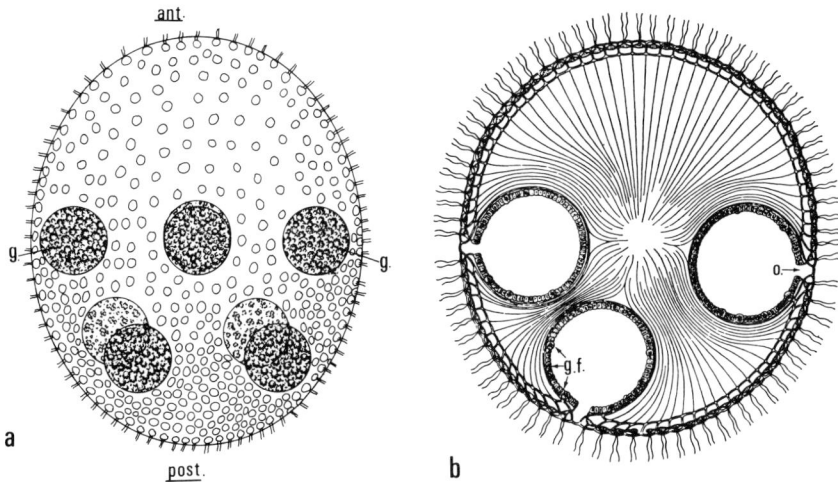

Fig. 2.1. (a) External view of an asexual *Volvox* colony with eight uncleaved gonidia (g.) in posterior half. <u>ant.</u> = anterior; <u>post.</u> = posterior. (After Kochert, 1975.) (b) Diagrammatic cross-section through an asexual *Volvox* colony, showing daughter colonies having completed embryonic development, with gonidia formation in their walls (g.f.) and ostiole (o.). (After Berrill, 1961.)

Fig 2.2. Surface view of an hermaphroditic colony of *Volvox aureus*, showing protoplasmic connections between individual protoplasts as well as between them and male (m.) and female (f.) initial cells. (After Janet, 1923.)

daughter colony is not a complete and closed sphere but has an ostiole or original pore at the anterior pole. A complete inversion of the daughter colony occurs through this ostiole (fig. 2.3).

Many *Volvox* species are protandrous hermaphrodites (see fig. 2.2). In the dioecious form of *Volvox rousseletti* several hundred male initial cells are scattered over the surface of the colony except for the anterior pole region. Other species have a smaller number of male initial cells. In *V. spermatosphaera* every cell of a developing gonidium becomes a male germ cell. Male initial cells look similar to gonidia but are much more numerous. After sinking in they orient themselves towards the anterior pole, begin to divide, and form a hollow sphere of about 512 cells (nine divisions). Inversion occurs as in a developing asexual colony, but the individual male cells have a single long flagellum and are ultimately set free as spermia. Female initial

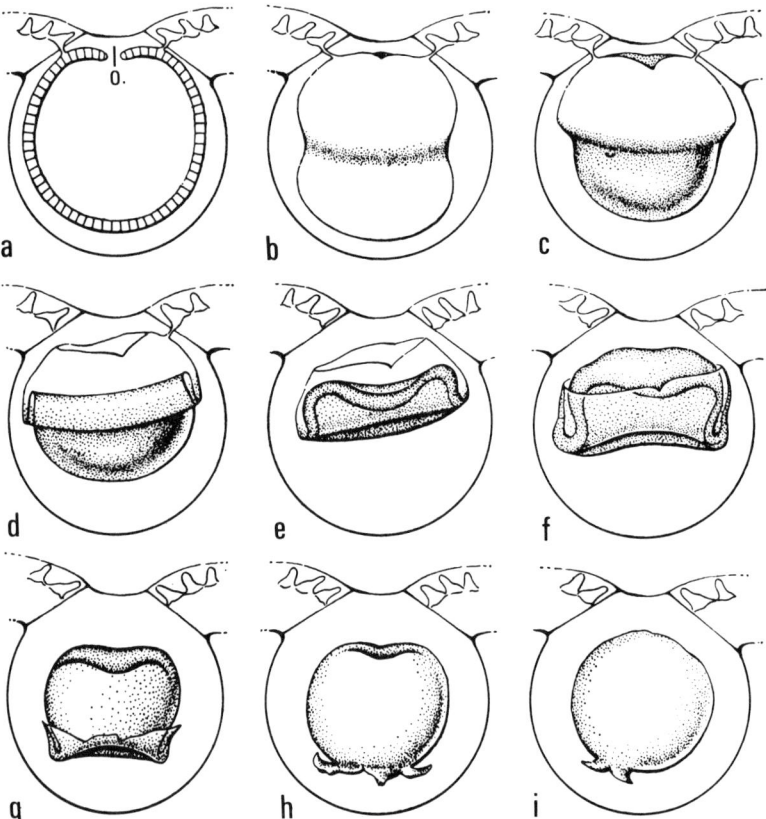

Fig. 2.3. Successive stages (a–i) of inversion in a developing *Volvox* colony. o. = ostiole. (After Pocock, 1933.)

cells also look similar to gonidia but become intensely dark green and begin to increase in size. They occur in much smaller numbers than male initial cells. When they attain a diameter of about 30 μm the flagella disappear. When the egg is maturing the green colour is replaced by red haematochrome. Eggs are fertilised *in situ* and then sink in. Although the zygote is diploid, the cells of a *Volvox* colony are haploid since only one product of meiosis survives. The zygote undergoes seven to nine synchronous divisions, enters a resting phase, and then goes through the process of inversion (see the review by Berrill, 1961).

In certain *Volvox* species or strains the shift from asexual to sexual development occurs under the influence of a species-specific inducer. Male spheroids produce a large amount of inducer; but in some strains both male and female spheroids do so. To explain the spontaneous appearance of male spheroids among asexually reproducing colonies it must be assumed that asexual spheroids also produce a small amount of inducer. Female spheroids are apparently formed only under inductive influence (Kochert, 1977). The inducer, which has been identified as a glycoprotein and is released into the surrounding medium, acts in extremely low concentrations (10^{-14} to 10^{-16} M). It must act during the initial growth phase of the gonidium, prior to first cleavage. Its first visible effect is a delay in the appearance of the unequal divisions which leads to gonidium formation. (See the reviews by Starr, 1970, Darden, 1973 and Kochert, 1975.)

Conclusion

In the colonial Protista sexual reproduction is initiated by changes in environmental conditions, demonstrating that gamete formation is a typically *epigenetic* process. In many species it occurs periodically, so that asexual and sexual generations alternate. Since, as far as we know, every cell can, in principle, be transformed into a gamete, there is no question of any preformation in germ cell development, so that the germ line concept is not applicable.

3
Porifera (sponges)

General characterisation

The majority of the sponges have no definite form but some show radial symmetry. The surface of the sponge is perforated by numerous small apertures serving for water ingress and leading into a complicated system of canals and chambers. The internal canal system is again connected with one or a few large apertures serving for water egression.

The sponges consist essentially of an outer epidermal layer of pinacocytes, a mesenchymal layer of varying thickness containing several cell types and minute spicules and fibrils, and an inner epithelium of choanocytes and endopinacocytes. The chambers are lined with choanocytes, the canals with pinacocytes. The sponge maintains its form by means of an internal skeleton consisting of small skeletal elements (spicules or fibrils) secreted by scleroblasts. Most sponges are hermaphroditic, but some dioecious forms exist. The great majority of sponges are marine animals, though some live in fresh water (Hyman, 1940, vol. 1, pp. 284–364).

The sponges are primarily classified according to the nature of their skeletal elements: the Calcarea have spicules of calcium carbonate ($CaCO_3$), the Hyalospongia spicules of silica (SiO_2), and the Desmospongia siliceous spicules and spongin fibrils or spongin fibrils only.

Regeneration and reconstitution

Sponges have great powers of reconstitution and regeneration. Reconstitution also occurs during the normal growth of the sponge, when old flagellated chambers are transformed into canals, and larger cavities and new chambers are formed in the growth zone. In this process the majority of the choanocytes of the old chambers are transformed into amoeboid lophocytes, which migrate to the growth zone where they reform into choanocytes. The remaining choanocytes evolve into endopinacocytes lining the newly formed canals (Diaz, 1974). Similar processes occur in regeneration after parts of the sponge are damaged or lost. Upon mechanical dissociation into a cell suspension a new sponge can be formed by reaggregation and subsequent reorganisation. The new sponge arises mainly from two or three cell types, viz. amoeboid archaeocytes, pinacocytes or epidermal cells (which probably represent a special form of archaeocytes), and choanocytes. (See review by Stolte, 1936.)

Asexual reproduction

Dis- and re-organisation occur during seasonal asexual reproduction through the formation of reduction bodies without and gemmules with a resistant envelope. In the Calcarea both are formed by massive emigration of choanocytes from the flagellated chambers and by their transformation into amoeboid archaeocytes, which accumulate at certain places in the old sponge (Duboscq & Tuzet, 1939). Gemmule formation in the freshwater sponge *Ephydatia fluviatilis* starts with a local aggregation of trophocytes accumulating reserve food, archaeocytes and spongoblasts. The archaeocytes phagocytose the trophocytes, which become transformed into complex vitelline corpuscles (De Vos, 1971). In the marine desmosponge *Suberites domuncula* similar processes occur, finally leading to a typical vitellogenesis (Connes, 1977).

Gemmule formation leads directly to the formation of a new sponge, all its tissues deriving from the yolk-containing archaeocytes. When development starts the archaeocytes become temporarily polynucleated. They later fragment into mononucleated cells which behave like embryonic blastomeres. After their division into numerous histioblasts, which leave the gemmular envelope through the micropyle (fig. 3.1), a mesenchymatous

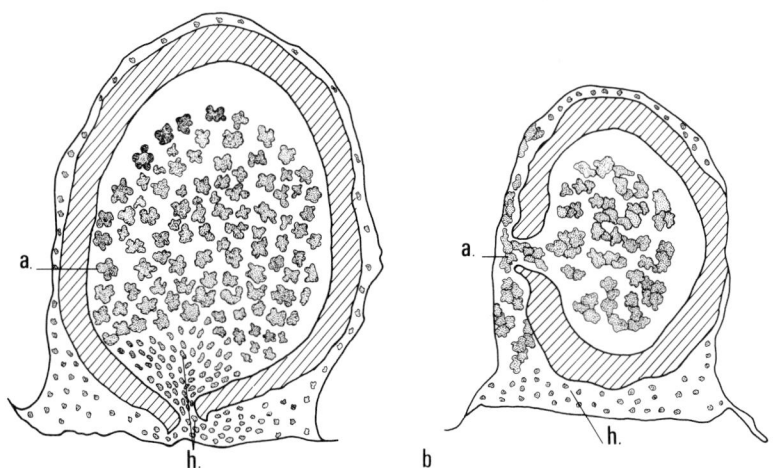

Fig. 3.1. (a) Section through gemmule of *Spongilla lacustris* during emergence of histioblasts (h.) through micropyle. (b) Section through gemmule of *Ephydatia fluviatilis* during subsequent emergence of archaeocytes (a.). (After Brien, 1932.)

16   *Porifera (sponges)*

mass is formed. Its outer cells form epidermis while the inner cells form a flat epithelium surrounding intercellular lacunae. The majority of the epidermal cells become perforated and constitute the epidermal pores. The nutritive chambers form from small accumulations of histioblasts. Spicules are secreted by so-called scleroblasts, which represent transformed archaeocytes. In the newly constituted sponge, archaeocytes may transform into oogonia and spermatogonia, the remaining archaeocytes representing an embryonic reserve of pluripotent cells (Brien, 1932; Stolte, 1936). In the freshwater sponges *Spongilla lacustris* and *Tubella pensylvanica* massive gemmule formation leads to the destruction of the old sponge, of which only a skeletal framework remains (Simpson & Gilbert, 1973).

Sexual reproduction

In the sponges normal meiosis occurs during oogenesis and spermatogenesis. In contrast to spermatogenesis, oogenesis is characterised at first by slow and later by rapid cellular growth due to extensive vitellogenesis. In the Calcarea the ovoid egg, after a period of amoeboid activity, becomes embedded in the mesenchyme underneath the wall of a flagellated chamber. The egg is typically polarised, its long axis being oriented parallel to the chamber wall. Its short axis is the animal–vegetal axis, the main axis of the future embryo, the animal pole pointing towards the chamber wall (Duboscq & Tuzet, 1937*a*; Tuzet, 1970).

Fertilisation takes place *in situ* but follows a peculiar course: the sperm does not enter the egg directly but penetrates the nearest choanocyte, which encapsulates the sperm; the choanocyte then acts as a carrier cell and transfers the spermiocyst to the underlying egg. Zygote formation occurs after the disappearance of the spermiocyst wall, which may take place only after a long stay in the egg cytoplasm (Duboscq & Tuzet, 1942*b*; Tuzet, 1964, 1970; Sarà & Orsi, 1975).

Embryogenesis resembles gemmule formation but development is indirect and first gives rise to a free-living larval stage. The sponge only secondarily acquires its definitive morphology, after attachment to a substrate and subsequent metamorphosis (Brien, 1932). During larval metamorphosis the choanocytes originate from archaeocytes and not from ciliated cells (Bergquist & Green, 1977).

Depending on the quantity of yolk, the fertilised egg undergoes a number of equal or unequal holoblastic divisions. In the Calcarea the 16-cell stage, called the diploblastic placula, consists of eight superior and eight inferior blastomeres. While the eight superior ectoblast cells do not divide, the eight inferior cells form a large number of small, flagellated endoblast cells. This so-called stomoblastula is characterised by a large opening among the eight ectoblasts. Subsequently, the stomoblastula turns inside out – as in *Volvox* –

so that the flagellated endoblast cells come to face outward, leading to a free-swimming amphiblastula (Duboscq & Tuzet, 1937a, 1942b) (fig. 3.2).

The amphiblastulae of *Sycon* and *Grantia* contain four special cells situated symmetrically in cross-like position in the third or fourth tier of flagellated cells. These cells lose their flagella and contain chromatic bodies at their lower pole, conferring a tetraradial symmetry on the larva (fig. 3.3). The chromatic bodies seem to derive from remnants of the mitotic apparatus of the first two divisions. They later degenerate and are extruded. Duboscq & Tuzet (1937b, 1942a) suggest that these cells function as temporary larval sensorial elements. The chromatic body is certainly not identical or similar

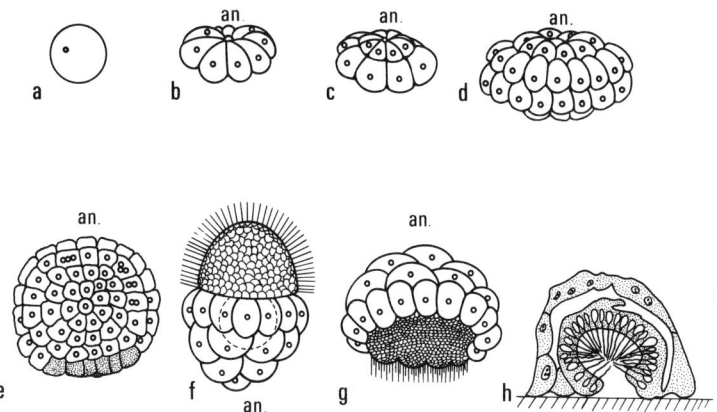

Fig. 3.2. Diagrammatic representation of cleavage, blastula formation and subsequent inversion of larva in *Sycandra*. (a) Uncleaved egg, (b) 8-cell stage, (c) 16-cell stage, (d) 48-cell stage, (e) hatching stage, (f) free-swimming amphiblastula, (g) inversion of germ layers, (h) fixation to substrate. an. = animal side. (Redrawn after Schulze, 1875.)

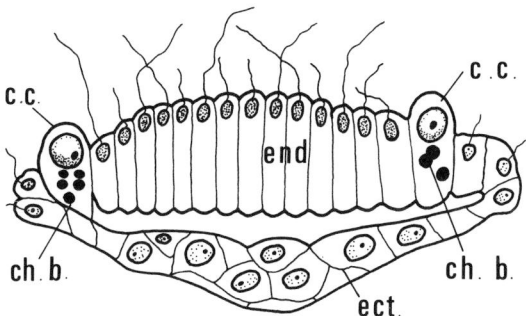

Fig, 3.3. Cross-section through amphiblastula of *Sycon elegans*, showing two of the four cross-wise positioned cells (c.c.), which lack flagella but contain chromatic bodies (ch.b.). ect. = ectoblasts; end. = flagellated endoblasts. (After Duboscq & Tuzet, 1942a.)

to the 'germinal determinant' of other invertebrates and has nothing to do with germ cell formation.

In the Calcarea gastrulation seems to occur by invagination of the small flagellated endodermal cells and overgrowth by the large, granular, non-flagellated cells, leading to a typical gastrula or a massive parenchymula. Attachment occurs on the blastoporal side (see fig. 3.2). In the Desmospongia embryonic development leads first to a solid blastula with a polar axis, then to a stereogastrula with a flagellated outer layer and a non-flagellated inner cell mass. Here inversion of layers also occurs, the flagellated cells becoming the choanocytes and the inner cell mass forming epidermis and mesenchyme. In some freshwater sponges no inversion of the stereoblastula occurs and the choanocytes develop directly from the inner cell mass.

Origin of the germ cells

It is generally agreed that in the sponges the PGCs originate from amoeboid archaeocytes. They constitute one of the four different cell types in the mesenchyme (Fincher, 1940).

In the desmosponge *Suberites massa* choanocytes sink into the mesenchyme and transform first into amoeboid archaeocytes, then, at the onset of sexual reproduction, into oogonia and spermatogonia (Diaz, Connes & Paris, 1973, 1975; Diaz, 1974). In *Sycon* Duboscq & Tuzet (1942b) have the PGCs deriving from amoeboid archaeocytes, which in turn originate from endodermal choanocytes, while Sarà & Orsi (1975) have both the PGCs and the nutritional cells or 'dolly cells' deriving from choanocytes. In *Grantia* the PGCs derive from amoebocytes representing transformed choanocytes which have lost their flagella and have migrated into the mesenchyme. In *Sycon* there seems to be a direct transformation of choanocytes into oogonia (fig. 3.4a) (Duboscq & Tuzet, 1937a).

The same holds for the spermatogonia, which form large aggregates of small cells (fig.3.4b) (Duboscq & Tuzet, 1939). In the desmosponge *Aplysella rosea* all the choanocytes of a flagellated chamber transform simultaneously into spermatogonia, extruding most of their cytoplasmic inclusions but retaining their flagellum and centriole (Tuzet, Garronne & Pavans de Ceccatty, 1970a,b). Therefore, there seem to be two possible origins of the germ cells in the sponges, either from amoeboid archaeocytes or from choanocytes. The two are closely related, since the amoeboid archaeocytes themselves derive from flagellated endodermal choanocytes.

Conclusion

A survey of sexual and asexual reproduction in the sponges makes it evident that the PGCs develop *epigenetically* from amoeboid archaeocytes rep-

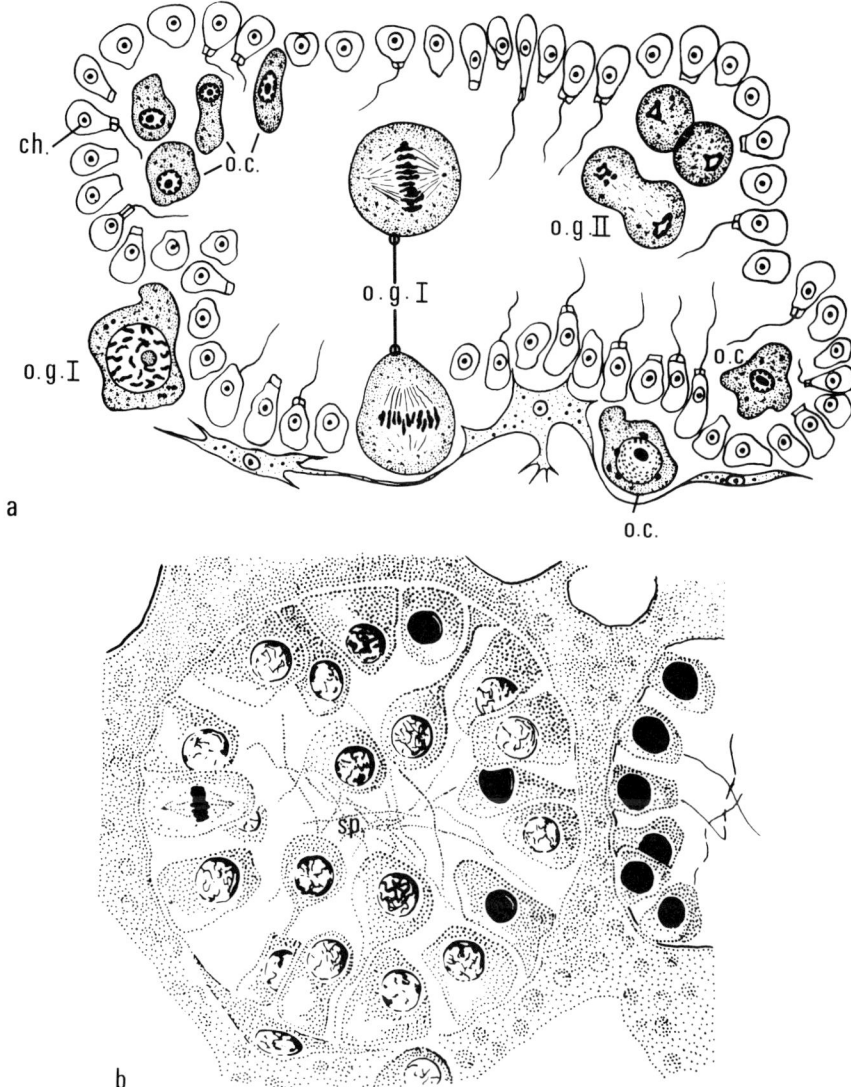

Fig. 3.4. (a) *Sycon raphanus*, cross-section through chamber, with choanocytes (ch.) and various stages of oogonia (o.g., I and II) and oocytes (o.c.). (b) Cross-section through chamber of *Hippospongia communis* with spermatogonia (sp.). (After Tuzet, 1964.)

resenting transformed choanocytes, the latter being functionally differentiated cells. Seasonal environmental factors determine the transition from the asexual to the sexual state. The PGCs may appear in very different phases of the life cycle, so that there is no question of a continuous germ line.

It is not known, however, whether germ cells can be reconverted into archaeocytes during the reverse transition from the sexual to the asexual state. Therefore, Brien's (1932) statement that 'only the archaeocytes represent immortal, totipotent cells, whereas all somatic cells as well as the germ cells are mortal, differentiated cells' still does not seem entirely proven. (See the evidence for his corresponding statement with regard to the coelenterates, p. 36.) The reader is further referred to Duboscq & Tuzet (1937a) and to the general reviews by Brien (1964, 1966) and Tuzet (1964).

4
Coelenterata (Cnidaria)

General characterisation

Coelenterates have a definite form with a radial to radio-bilateral symmetry and a distinct proximo-distal polarity. Their bodies consist essentially of two epithelial layers, an outer epidermal layer consisting of epidermal and epithelio-muscular cells and an inner gastrodermal layer. The two layers are connected by the mesoglea, which may be acellular or cellular. The digestive cavity has a single aperture functioning as both mouth and anus. The body bears tentacles provided with nettle cells, called cnidoblasts or nematocysts, which constitute a chief characteristic of the phylum. The coelenterates show typical polymorphism, taking essentially two different, often alternating forms, the sessile *polyp* form and the usually free-swimming *medusa* form, representing the asexual and sexual generations respectively.

The Coelenterata are subdivided into three classes: the Hydrozoa, the Scyphozoa and the Anthozoa. (See Hyman, 1940, vol. 1, pp. 365–661.)

There is a much more extensive literature on hydrozoan than on scyphozoan or anthozoan development. This is due mainly to the fact that some hydrozoan species are easily kept under laboratory conditions and are particularly suitable for experimental analysis. This holds, for example, for the solitary form *Hydra* and for several colonial forms such as *Tubularia* and *Campanularia*. Since the problems in which we are interested have been most thoroughly studied in the hydrozoans we shall concentrate mainly on this group. Moreover, we shall restrict ourselves to the most important facts and refer only to the most relevant literature. (For further information, see the recent general reviews on coelenterate development by Uchida & Yamada, 1968*a*, Campbell, 1974 and Tardent, 1975, 1978.)

For a proper understanding of the origin of the germ cells in the coelenterates a knowledge of their regeneration is indispensable. However, regeneration can be properly understood only on the basis of the normal life cycle.

HYDROZOA

In the Hydrozoa the asexual polyp often alternates with the sexual medusa. There are, however, also sexually reproducing polyps and asexually

22 Coelenterata/Hydrozoa

reproducing medusae. There are solitary as well as colonial forms; in the latter the individual polyps are connected by stolons. The Siphonophora are polymorphic, swimming or floating colonies with modified polypoid and medusoid individuals, which have acquired different functions. Since nothing is known about the origin of the germ cells in the Siphonophora, they will not be discussed separately. The mesoglea of the Hydrozoa is essentially acellular.

Adult organisation

Although in this monograph we cannot enter into the nature of polarity, a few words must be said about the proximo-distal polarity of the hydroid body. Polarity is firmly established and can be reversed only under rather extreme experimental conditions (see Tardent, 1954; Wilby & Webster, 1970).

The solitary form *Hydra* consists of a proximal disc by which it is attached to the substrate, an elongated peduncle or stalk region, a large gastric region including a gonadal or budding zone, and a hypostome region with a ring of tentacles (fig. 4.1). In *Hydra* polarity seems to be maintained by the hypostome and possibly by the basal disc region (Webster, 1971).

The colonial forms, such as *Tubularia, Campanularia* and *Pennaria*, have a stolon connecting the individual polyps, here called hydranths. The hydranths of *Pennaria* are attached to the stolon by a stalk or caulus and consist of a gastric region with a basal ring of tentacles, the gonophores situated slightly more apically, and an apical hypostome region with a ring of small tentacles (fig. 4.2). The stolon and caulus are stiffened by an external perisarc secreted by the epidermis.

The growth and maintenance of the polyp have been studied extensively in *Hydra*. Transplantation of a vitally stained graft into an unstained host by

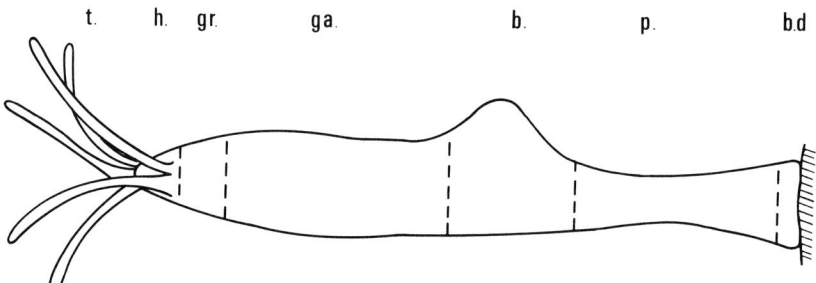

Fig. 4.1. Diagrammatic representation of successive body regions of *Hydra*. t. = tentacle region; h. = hypostome; gr. = main growth zone; ga. = gastric region; b. = budding region; p. = peduncle; b.d. = basal disc. (After Burnett, 1961.)

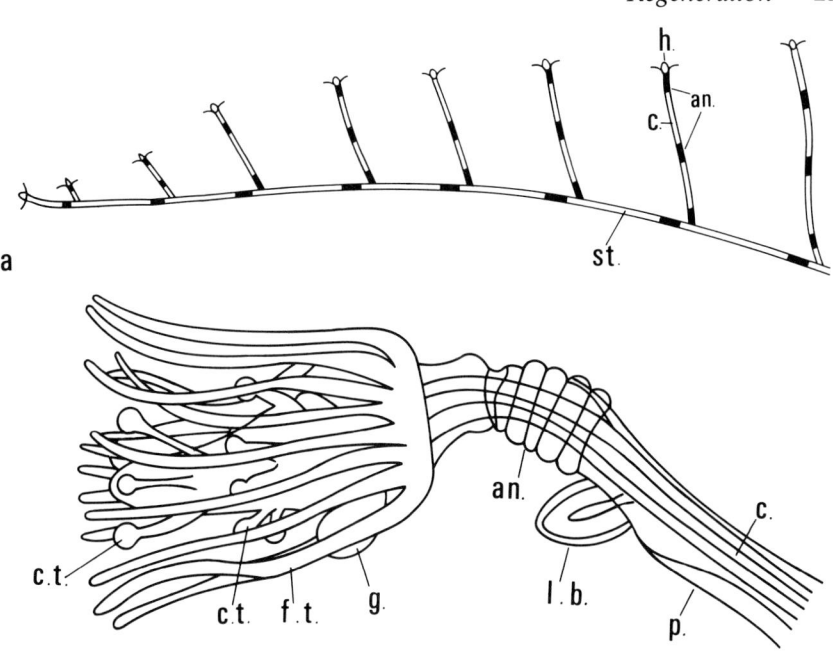

Fig 4.2. (a) Diagram of branching colony of *Pennaria tiarella* with stolon (st.), caulus (c.), annulated regions (an.) and terminal hydranths (h.). (b) Terminal part of caulus (c.) with surrounding perisarc (p.), young lateral bud (i.b.) and annulated region (an.), and hydranth with two rings of capitate tentacles (c.t.), proximal ring of filiform tentacles (f.t.) and early stages of gonophore formation (g.). (After Berrill, 1952a.)

Brien & Reniers-Decoen (1949) demonstrated that cells move away from the subhypostomal region, either in a distal direction towards the tip of the tentacles, or in a proximal direction towards the basal disc. At both extremities cells are sloughed off. This growth pattern used to be ascribed to the existence of a well-defined growth zone. Although the general pattern of tissue movement was confirmed by Campbell (1967a,b,c), he could not demonstrate a localised growth zone; mitoses were found throughout the length of the polyp and in all tissues. Cells in the epidermis and gastrodermis marked with ^{3}H-thymidine or with colloidal carbon move as coherent sheets; only the nematocysts are able to move individually by amoeboid activity (Campbell, 1967c, 1973).

Regeneration

All coelenterates show a very high regenerative capacity after wounding.

Virtually any fragment of a *Hydra* can form a new individual. Zwilling (1963) and Diehl (1969) found that isolated epidermis of *Cordylophora* can form an entire animal with epidermis and gastrodermis, but X-irradiated epidermis is unable to do so. Neither normal nor irradiated gastrodermis alone can undergo morphogenesis, but X-irradiated epidermis plus normal gastrodermis form a normal animal. These potentialities of the isolated epidermis have recently been confirmed by Marcum & Campbell (1978). A new individual can even be reconstituted upon dissociation and subsequent reaggregation, a phenomenon described, among others, by Föyn (1927) in *Clava squamata* and more recently by Diehl (1969) in *Cordylophora*. The high regenerative capacity is also expressed in the phenomenon of asexual reproduction by budding. Both regeneration and budding capacity are attributed primarily to the presence of so-called 'interstitial cells' (I-cells), since both processes come to a standstill after destruction of the I-cells by X-irradiation (see review by Stolte, 1936). Brien (1953, 1961, 1965) considers the I-cells as the 'reserve cells' capable of differentiating into a large variety of cell types. (See also the general review of Brien, 1964.)

Interstitial cells in the adult animal

I-cells are small, darkly stained basophilic cells with prominent nuclei and nucleoli. Ultrastructurally, they look like small undifferentiated cells, with an abundance of free ribosomes but only a few other organelles such as endoplasmic reticulum vesicles, mitochondria and Golgi complexes (fig. 4.3). In *Hydra* I-cells are most numerous in the hypostome and the budding and gastric regions, particularly at the distal end of the latter. They are rare in the tentacles and the basal disc. They are predominantly localised in the epidermis at the base of the epithelio-muscular cells, only a few being found in the gastrodermis (Lentz, 1965, 1966). In the polyps of *Eleutheria*, *Cladonema* and *Campanularia* I-cells are found in the epidermis of the hydrocaulus and hydrorhiza, while in the medusa they lie in the epidermis at the base of the tentacles and the velum and in the gastrodermis of the manubrium (Weiler-Stolt, 1960). Tardent & Morgenthaler (1966), by making chimaeras of ^{3}H-thymidine-labelled proximal halves and X-irradiated distal halves of *H. attenuata*, showed that I-cells as well as nematocysts are migratory elements.

Tardent (1954) was able to stain the I-cells of *Hydra* selectively with toluidine blue after separation of the epi- and gastro-dermal layers. He observed an axial gradient in I-cell distribution along the proximo-distal axis of the polyp, with a peak in the hypostome region. Moreover, he found a quantitative correlation between the number of I-cells and the rate of regeneration at various levels along the proximo-distal axis. Burnett (1961) showed that all regions of the polyp possess regenerative capacity except the

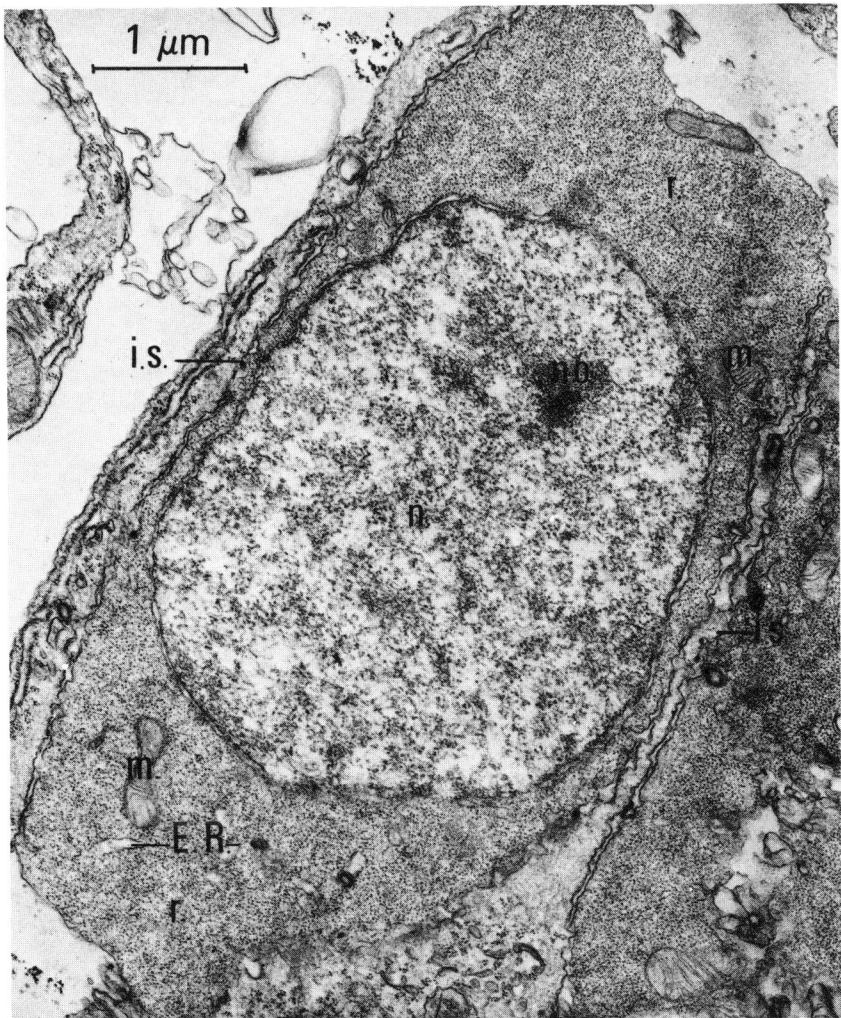

Fig. 4.3. Electron micrograph of interstitial cell of *Hydra* with abundant ribosomes (r.), other organelles scarce (endoplasmic reticulum (E.R.) and mitochondria (m.)). i.s. = intercellular spaces; n. = nucleus; n.o. = nucleolus. (Courtesy T. L. Lentz, 1965.)

tentacles, the only parts which seem to be devoid of I-cells. Brien & Reniers-Decoen (1955) stated that, in *Hydra*, the I-cells are responsible for the normal replacement of epi-and gastro-dermal cells and for the formation of nematocysts, as well as for growth, regeneration, budding and gametogenesis, giving rise to all the cellular elements of the polyp. Belousov (1963) observed periodic bursts of I-cells multiplication in *Campanularia* and *Opercularella*, while *Obelia* and *Dynamena* showed a more constant rate of cell division in the I-cells of the hydranth. In regenerating and budding hydras I-cell multiplication seems to be preceded by an increase in the number of nerve cells (Bode *et al.*, 1973). Stem cell division and differentiation are, moreover, affected by the local density of I-cells (Bode & David, 1978).

Diehl (1969) found that, in *Cordylophora*, no new I-cell formation occurred after previous destruction of all I-cells by X-irradiation. This observation has recently been confirmed by Marcum & Campbell (1978) in *Hydra* treated with colchicine. Such animals soon lack nerve cells, nematocysts, gametes and endodermal gland cells and consist strictly of epi- and gastro-dermal cells. This is partially in agreement with the following observations. Lentz (1965, 1966) found that in *Hydra* I-cells differentiate into epithelio-muscular cells, digestive cells, glandular cells, nerve cells and nematocysts. David (1975) found that I-cells can form nematocysts, nerve cells and germ cells, but not epithelial cells or gland cells.

Campbell & David (1974) and David & Challoner (1974) studied particularly nematocyst and nerve cell formation in *Hydra attenuata*. Small I-cells with a short cell cycle and forming nests of 4 to 32 cells differentiated into nematocysts, whereas single or paired larger I-cells with a long and variable cell cycle formed nerve cells and pluripotent stem cells. David & Murphy (1977) found that clones of I-cells, obtained by introducing viable I-cells into tissue aggregates inactivated with nitrogen mustard, usually differentiate into either nematocysts or nerve cells, but that some clones can give rise to nematocysts as well as nerve cells, demonstrating their intrinsic pluripotentiality.

What is the origin of the I-cells in the adult animal? Tardent (1954) and Brien & Reniers-Decoen (1955), working on *Hydra*, and Bouillon (1957), working on *Limnocnida* (Limnomedusae), consider the epithelio-muscular elements of the epidermis as the source of the I-cells. According to these authors there exists no permanent 'embryonic reserve'. Haynes & Burnett (1963), Burnett, Davis & Ruffing (1966) and Davis *et al.* (1966) observed in *Hydra viridis* that I-cells were formed by dedifferentiation of specialised gastrodermal cells, i.e. mucus and gland cells, epithelio-muscular cells probably being formed directly from gastrodermal cells. Diehl & Burnett (1965a,b), after I-cell destruction with nitrogen mustard, found mitoses in epithelio-muscular cells but not in gastrodermal cells. In embryonic development (see below, p. 27) the I-cells seem to be either of ectodermal or of endodermal origin; in the latter case they migrate from the endoderm through the mesoglea into the ectoderm. This somewhat contradictory evidence actually pleads in favour of a multiple origin of the I-cells.

How far are the I-cells indispensable for maintenance, growth, regeneration and budding? Brien & Reniers-Decoen (1955) observed a continuation of regeneration and bud formation in *Hydra* after destruction of all I-cells by X-irradiation. This was confirmed by Diehl & Burnett (1965a,b) in *Hydra* after I-cell destruction with nitrogen mustard, by Müller (1967) in *Hydractinia* colonies after treatment with alkylating agents, and recently by Marcum & Campbell (1978) in *Hydra* treated with colchicine. An I-cell-free hydroid remains capable of tissue renewal as well as bud formation, so that I-cells are apparently not indispensable for these functions. However, nematocyst formation and proliferation of sex cells cease completely. It must be emphasised, however, that after such treatments the animals gradually degenerate and ultimately die. They can be prevented from dying only by supplying them with new I-cells. Apparently, cells other than I-cells can also be involved in cell replacement, regeneration and budding, probably through partial or complete dedifferentiation and subsequent mitosis, but these vital functions are primarily performed by the I-cells (see review by Diehl, 1973).

Asexual reproduction

Berrill (1949a, 1952a,b) described polymorphic development in gymnoblastic hydroids and distinguished various forms of growth at the growing points of a colony: stolon growth, hydranth formation, medusa formation and gonophore formation, the latter representing an 'abortive' type of medusa formation. In *Limnocnida* Bouillon (1957) distinguished four types of blastogenesis: polyp budding, medusa budding, formation of frustules (elongated mobile elements), and formation of spherical, resistant winter buds. All these forms of asexual reproduction begin as a local outbulging of both the epidermal and the gastrodermal layers.

In *Hydra* bud formation is preceded by an overall increase in cell multiplication and is accompanied by cell migration towards the budding region (Webster & Hamilton, 1972). Otto & Campbell (1977) prepared a fate map of the polypoid bud anlage in *Hydra*. (See further Brien's studies on the influence of temperature on asexual and sexual reproduction, pp. 31–2.)

Embryonic development and origin of the I-cells

Embryonic development shows rather pronounced variations depending on the amount of yolk in the egg and the time of onset of metamorphosis. Bodo & Bouillon (1968) described the early development of the Lepto-, Antho- and Limno-medusae. The eggs show total, usually unequal, cleavage and form a hollow coeloblastula or a massive stereoblastula. Gastrulation occurs through unipolar ingression, delamination or cellular infiltration and gives rise to a planula larva. I-cells and nematocysts are usually of endodermal origin when formed early, but of ectodermal origin

Table 4.1. *Data on egg type, cleavage, blastula, gastrulation, type of larva, embryonic origin of I-cells and time of metamorphosis in a number of different groups and genera*

Group or genus	Egg type/ amount of yolk	Cleavage	Blastula	Gastrulation	Type of larva	Origin of I-cells	Metamorphosis	References
Capitata	Poor in yolk	Total	Coeloblastula	Ingression or delamination	Parenchymula planula	Endodermal	Gradual, late	Van de Vyver, 1967, 1968a,b
Filifera	Fairly rich in yolk	Total, equal, radiate	Stereoblastula	Delamination	Parenchymula planula	Endodermal	Fairly early	Van de Vyver, 1967, 1968a,b
Limnocnida	Fairly poor in yolk	Total	?	?	Ciliated larva planula	?	Early	Bouillon, 1957
Corydendrium	Fairly poor in yolk	Total, equal	Stereoblastula	Delamination	Planula	Endodermal	Fairly late	Glätzer, 1971
Eleutheria	Fairly poor in yolk	Total, equal	Coeloblastula	Unipolar ingression	Parenchymula planula	Endodermal, transformation into nematocysts	Early	Weiler-Stolt, 1960; Van de Vyver & Bouillon, 1969
Cladonema	Fairly poor in yolk	Total, equal	Coeloblastula	Unipolar ingression	Parenchymula planula	Endodermal, transformation into nematocysts	Early	Weiler-Stolt, 1960
Campanularia	Fairly poor in yolk	Total, equal	Coeloblastula	Unipolar ingression	Parenchymula planula	Endodermal, transformation into nematocysts	Early	Weiler-Stolt, 1960
Pennaria	Fairly rich in yolk	Total, unequal or irregular	Stereoblastula	Delamination	Planula	Endodermal	Late	Cowden, 1965; Summers & Haynes, 1969; Martin & Thomas, 1977
Tubularia	Fairly rich to rich in yolk	Total, irregular or initially syncytial	Modified coeloblastula	Multiple ingression or delamination	Planula	Endodermal	Gradual, late (actinula)	Nagao, 1965; Fennhoff, 1978
Eudendrium	Very rich in yolk	Syncytial	Syncytial blastula	Gradual delamination	Planula	Ectodermal	Gradual, late	Mergner, 1957

when formed late in development. Table 4.1 gives data on egg type, cleavage, blastula and gastrula formation and larval development, as well as on the embryonic origin of the I-cells in a number of different genera. Summarising all these data, it may be stated that the majority of authors observed an endodermal origin of the I-cells, although an ectodermal origin has also been observed. Nematocysts usually form from undifferentiated I-cells but may also arise more or less directly from endodermal cells of the planula.

Sexual reproduction and origin of the germ cells

In many Hydrozoa asexual and sexual generations alternate. In the latter the germ cells are formed in the free-living medusa or in the 'reduced' gonophores. According to Bouillon (1957) medusa formation in Limnomedusae is characterised by a proliferation of the ectodermal layer at the tip of the gonophoral bud. This leads to the formation of an ectodermal vesicle, the entocodon or bell nucleus, which represents the primordium of the subumbrellar cavity of the medusa. The gastrocoelic cavity of the bud extends into four radial canals which fuse sideways, thus forming the ring canal. The gastrocoelic cavity also grows out centrally, giving rise to the manubrium. The epidermis and entocodon together form the velar plate, which ruptures in the centre. Through this opening the manubrium may extend. Tentacles and special sense organs are formed at the margin of the umbrella (fig. 4.4). Finally, the stalk by which the medusa is attached to the mother animal narrows and breaks off, setting the medusa free. In the medusa gonads are formed in the wall of the manubrium, but the latter may also function as a budding region for new medusae. Many hydroids do not form free-swimming medusae, however, but only develop sessile gonophores representing intermediate stages in medusa formation with accelerated gonadal development. Berrill & Liu (1948) and others confirmed the original observations of Weismann (1883), who distinguished various degrees of regression of the sexual generation, from the fully developed free-living medusa to the rudimentary sporosac. Gonad formation in the wall of the gastric region of *Hydra* may be considered as the complete suppression of the medusal generation.

Föyn (1929) observed male and female germ cell formation in the endoderm of the young gonophore of *Clava squamata* at the time of bell nucleus formation. The male germ cells penetrate the bell nucleus and differentiate into spermatozoa. In *Tubularia*, Dupont (1942) observed formation of germ cells in the ectoderm of the evaginating bud. These migrate to the boundary of the inner layer of the subumbrellar vesicle. Liu & Berrill (1948) and Berrill (1952b) claimed that in *Tubularia crocea* the entocodon as well as the germ cells which develop from it are of endodermal origin, and that the I-cells are not involved in germ cell formation. The

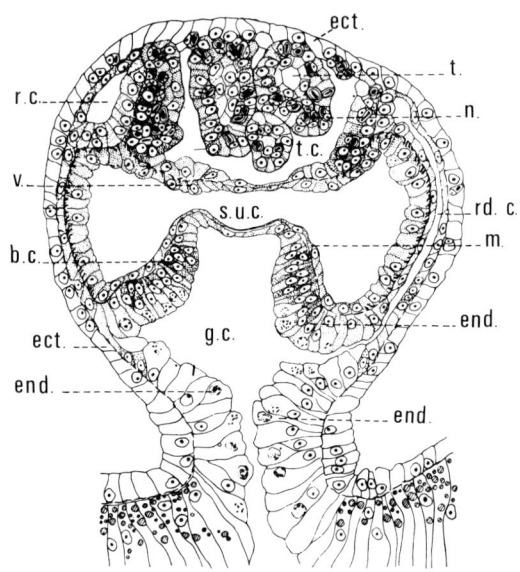

Fig. 4.4. Longitudinal section through developing medusal bud of a Limnomedusa. b.c. = basophilic cells involved in manubrial bud formation; end. = endodermal epithelium; ect. = ectodermal epithelium; g.c. = gastric cavity; m. = manubrium; n. = nematocysts; r.c. = ring canal; rd.c. = radial canal; s.u.c. = subumbrellar cavity; t. = tentacles; t.c. = tentacle cavity; v. = velum. (After Bouillon, 1957.)

former statement seems to be at variance with observations on many species where the entocodon is formed as a proliferation of the ectodermal layer (e.g. Dupont, 1942). Fennhoff (1978) claims that the entocodon is formed by a few I-cells coming from the ectoderm. The latter statement is contradicted by, among others, Boelsterli (1975), who found that the oogonia derive from clusters of I-cells lying on the outside of the spadix of the female gonophore and by Fennhoff (1978), who states that the sexual cells are former I-cells which have migrated into the endoderm at the base of the gonophore. Brien (1954) observed precocious germ cell formation in the hydranths of *Tubularia* and *Clava* near the site of later gonophore formation. In *Bougainvillia* and *Aselomaris*, Berrill (1949a) observed that oogonia develop when the gonophoral entocodon is formed early and develops rapidly, whereas spermatogonia are formed when entocodon formation is late and slow. Germ cells and gonads arise in the ectodermal wall of the manubrium of the medusa of *Rathkea octopunctata*, but in the gonophoral endoderm in *Pennaria*, *Acaulis* and *Eudendrium* (Berrill, 1952a). In *Eudendrium racemosum* oogonia appear in the ectoderm of the upper branches below the budding zone, where they may induce gonophores, although gonophores may also develop independently (Mergner, 1957). Cowden (1964) observed that, in *Pennaria tiarella*, the majority of the entocodon cells differentiate into oogonia. In the medusa of *Limnocnida* the germ cells arise in the ectodermal layer of the proximal half of the manubrium (Bouillon, 1957). In *Corydendrium parasiticum* Glätzer (1971) found that ectodermal I-cells migrate through the mesoglea into the endoderm of the bud, differentiate into germ cells and, in the case of early migration, induce gonophore formation. In the anthomedusa *Podocoryne carnea* the germ cells derive from undifferentiated cells of the endodermal layer which migrate through the mesoglea into the ectoderm and differentiate into oogonia (Boelsterli, 1977).

In *Hydra* Brien (1954) observed that the germ cells are formed from embryonic cells belonging to the epidermal layer of the polyp. They may temporarily migrate to the endodermal layer for nutritional purposes. Brien & Reniers-Decoen (1955) and Brien (1961, 1962, 1963, 1965, 1966) found that, in several *Hydra* species, the switch from asexual to sexual reproduction is effected by environmental factors. At 19°C reproduction is strictly asexual (fig. 4.5b) and the animal does not contain any germ cells. The I-cells are engaged in somatocyte and nematocyst formation. At 8°C germ cells suddenly appear as a result of massive transformation of I-cells into germ cells in the now sexually active animals (fig. 4.5a,c). This switch from asexual to sexual reproduction leads to a cessation of vegetative functions;

Fig. 4.5. (a) Female *Hydra* with growing oocytes. (b) Asexual *Hydra* with completed bud. (c) Male *Hydra* with spermatic ampullae. (After Brien, 1961 and 1966.)

continued sexuality leads to the complete exhaustion of the I-cell population and to ultimate death. Return of the animal to higher temperatures leads to the converse switch, i.e. the cessation of sexuality. I-cells are formed anew either from still undifferentiated oogonia and spermatogonia or by dedifferentiation of epithelio-muscular cells; another source is the redistribution of remaining I-cells from the subhypostomal region of the polyp, which is never involved in gametogenesis. This temperature effect occurs in *Hydra pirardi* and *H. fusca* but not in *H. viridis* or *H. littoralis*. In the latter two species nerve cells are less numerous and the number of I-cells is about half that in the former species (Burnett & Diehl, 1964). Brien concluded that somatogenesis and blastogenesis on the one hand, and gametogenesis on the other, represent two complementary physiological states which can be switched one into the other.

At low temperatures some hydras are apparently able to resist degradation by exhaustion of the I-cell population and can form a new asexual line. Longitudinal parabiosis of such an asexual animal with a normal sexual one leads to extension of gametogenesis into the asexual partner (Brien, 1963). In dioecious species Brien (1962) observed masculinisation of the female animal in parabiosis with a male animal, demonstrating that the male sex is dominant. Sexual differentiation is apparently controlled by male gonadal influences. This may lead to intersexuality or to complete reversal of genetic females into functional males. A similar situation was found by Tardent (1968) in heterosexual chimaeras of *Hydra*, which always become males. This masculinisation is blocked by X-irradiation, the male germ cells apparently being more sensitive than the female germ cells. Weiler-Stolt (1960) observed that, in *Eleutheria*, sexuality can be induced in asexual medusae (which only reproduce by budding) by implantation of tissue from the normal sexual form, containing I-cells with undiminished potency for germ cell formation. Hydras deprived of their I-cells by a short treatment with alkylating agents, and subsequently inoculated with I-cells from a heteroclonal donor, acquire the genetic features and the sex of the donor, demonstrating the dominant role of the I-cells in sexual and asexual reproduction (Müller, 1967).

The regional distribution of 'ovaries' and 'testes' along the longitudinal axis of hermaphroditic *Hydra* species is attributed by Vannini (1963) to so-called sex gradients. (See also chapter 5, section Turbellaria, p. 46.)

We now come to a recent observation which, in our opinion, places the problem of the origin of the germ cells in the Coelenterata, and possibly in other lower invertebrates as well, in a new light. Noda & Kanai (1977) observed that, in *Pelmatohydra robusta*, the I-cells and early nematocysts of the asexual form, as well as the oogonia of the sexual form, are characterised by electron-dense bodies which strongly resemble the germinal granules found in the pole cells of higher insects. The bodies consist of loosely interwoven 10 nm thick fibrils, lack a limiting membrane and are situated near or around the nucleus (fig. 4.6). They may be produced in the nucleus and exported through the nuclear pores. In the cytoplasm they are often found in close association with mitochondria. The three cell types have different numbers of electron-dense bodies, however: the oogonia contain a large amount of electron-dense material, the I-cells a fair but lower amount,

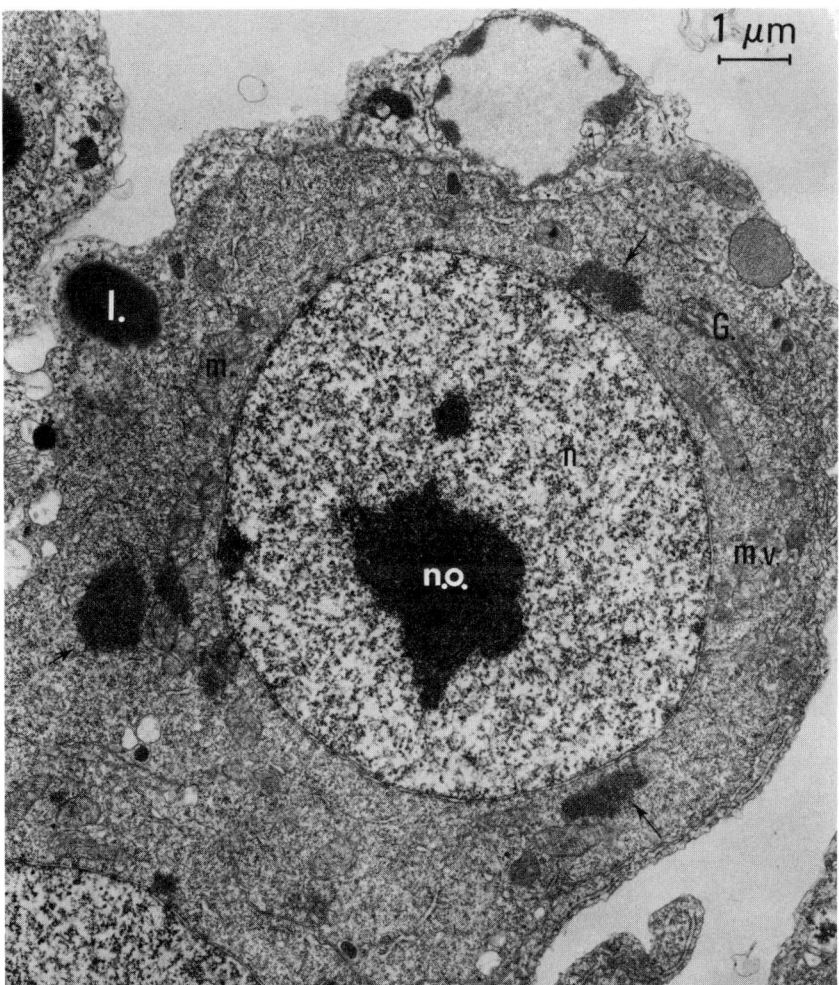

Fig. 4.6. Electron micrograph of young oogonium of *Pelmatohydra robusta* showing electron-dense bodies in cytoplasm (arrows), some of them near nuclear membrane and some associated with mitochondria. G.= Golgi apparatus; l.= lipid droplet; m.v.= multivesicular bodies; n.= nucleus; n.o.= nucleolus. (Courtesy K. Noda & C. Kanai, 1977.)

and the young nematocysts a small amount which drops to zero as cellular differentiation proceeds. The reversibility of the I-cell/germ cell transformation, being the essential fact behind the alternation of asexual and sexual reproduction, is more easily understood in the light of these observations. It is not known whether somatic cells other than nematocysts also contain 'germ plasm' during their initial stages of differentiation from pluripotent I-cells. While on the one hand the presence of 'germ plasm' represents a typically *preformistic* feature of germ cell formation, on the other hand their formation from I-cells under the influence of environmental factors is a typical example of *epigenetic* development. In fact, a major key to our understanding of germ cell development generally may lie in a deeper understanding of 'germ plasm' formation and breakdown.

SCYPHOZOA

The mainly descriptive literature on the development of the Scyphozoa or Scyphomedusae has been reviewed by Berrill (1949b) and more recently by Uchida & Yamada (1968a), Campbell (1974) and Tardent (1975, 1978).

In the Scyphozoa the polyp generation is insignificant in comparison with the medusoid generation. The 'scyphistoma' polyp is characterised by four longitudinal septa which arise as folds of the endodermal layer covering extensions of the cellular mesoglea. Medusae are formed by budding or by strobilation.

Development is rather strongly influenced by egg size and yolk content. Small eggs which are relatively poor in yolk develop into a free-swimming 'planula' larva after unipolar ingression of the endoderm of the coeloblastula. After attachment to a substrate, a 'scyphistoma' polyp is formed, which may produce 'ephyra' medusoids by mono- or poly-disc strobilation. Eggs intermediate in size and yolk content have only a very transient planula stage and develop more or less directly into 'scyphistoma' polyps by unipolar ingression or by invagination. Large eggs with abundant yolk may develop directly into an 'ephyra' medusoid.

The 'scyphistoma' of *Cassiopeia* exhibits two forms of asexual reproduction, strobilation of 'ephyra' medusoids or formation of a long chain of ciliated buds, from which individual buds break off terminally and develop into polyps (Curtis & Cowden, 1971).

The 'scyphistoma' has a rather high regenerative capacity. Basal fragments form complete polyps, while apical fragments only form head structures (Neumann, 1977). Steinberg (1963) and Curtis & Cowden (1972) showed that isolated epidermis of the 'scyphistoma' larva can form a complete larva. They could not identify typical I-cells but found large amoeboid cells arising from epidermal cells and transforming themselves into gastrodermal cells. According to Campbell (1974) gametogenic cells have the character of undifferentiated cells but are slightly larger than ordinary I-cells.

In the Scyphozoa the PGCs are considered to be of endodermal origin. They are found initially at the base of the gonophore anlage and later move to the tip, mainly as a result of gonophore growth, which occurs by recruitment of cells from the surrounding parent tissues. The PGCs then enter the entocodon and are finally found in the manubrium of the medusoid. The passage from one layer to the other must be due to active migration.

ANTHOZOA

The purely descriptive literature on the development of the Anthozoa has been reviewed by Uchida & Yamada (1968a), Campbell (1974) and Tardent (1975, 1978).

The Anthozoa are represented only by the polyp form. They reproduce both sexually (by gamete formation) and asexually (by budding or longitudinal fission). Polyp formation is characterised by invagination of an ectodermal stomodaeum – erroneously called pharynx – and by the extensive formation of septa by longitudinal folding of the endodermal layer with involvement of the cellular mesoglea. The number of complete and incomplete septa is six or a multiple of six in the Hexacorallia, and eight or a multiple of eight in the Octocorallia.

The yolky eggs often show intralecithal cleavage initially, the first three nuclear divisions not being followed by cytokinesis; after this total cleavage takes over. This leads to the formation of a ciliated coeloblastula, which transforms into a didermic planula by invagination, multipolar ingression of the endoderm, or delamination of ecto- and endo-derm (Nyholm, 1943, 1959; Chia & Crawford, 1973). The PGCs are said to be of endodermal origin and appear in the septal filaments of the young polyp (Jennison, 1979).

General conclusion

It may be concluded that, in the Coelenterata, the PGCs derive from pluripotent I-cells, which, in turn, may be of either ectodermal or endodermal origin. I-cells can change into young oogonia and spermatogonia, and vice versa, under the influence of external environmental factors. That the germ cells are formed *de novo* from existing I-cells is at complete variance with Weismann's (1883) interpretation of germ cell development in the coelenterates. (See also Berrill & Liu's (1948) criticism of Weismann's theoretical interpretations.)

It is perfectly evident that, in the coelenterates, no fundamental distinction exists between germ cells and somatic cells – a situation similar to that in the sponges and colonial protists – since both germ cells and somatic cells can derive from the same pluripotent stem cells. In 1932 Brien stated that, in the coelenterates, the I-cells are the only pluripotent cells of the animal, while somatic cells and germ cells represent differentiated elements destined for particular functions. Noda & Kanai's (1977) observations plead in favour of a germ-cell-determining role for the 'germ plasm', but this would be a *quantitative*, rather than a qualitative role, since 'germ plasm' is not unique to germ cells and germ cell formation is achieved by a marked increase (synthesis) of 'germ plasm'.

5

Platyhelminthes

TURBELLARIA

General characterisation

Turbellaria are dorso-ventrally flattened, acoelomate Bilateralia without segmentation, but with a distinct antero-posterior polarity and bilateral symmetry. The antero-posterior polarity is expressed in distinguishable head, trunk and tail regions. Except for the Acoela, which have no digestive tract, the other Turbellaria have a well-developed digestive system with mouth, stomodaeum or 'pharynx', and a blind-ending intestine. The latter may show various degrees of complexity. Circulatory and respiratory organs are absent. The excretory organs are protonephridia. The space between the ciliated epidermis and the internal organs is occupied by a semi-syncytial parenchyma. There are two longitudinal ganglionated ventral nerve cords and two cerebral ganglia, which are usually called the 'brain'. Paired or multiple cephalic sense organs in the form of eyes may be present. The great majority of species is hermaphroditic with each individual having separate male and female gonads, the latter usually represented by two ovaries with or without vitellaria, the former usually by numerous testes. (See Hyman, 1951a, vol. 2, pp. 64–218.)

For a proper understanding of the mode of origin of the germ cells in the Turbellaria it is necessary first to discuss the great regenerative capacity of the flatworms, which, in some species, is linked up with asexual reproduction. Several planarian species, for example *Dugesia* spp., can reconstitute a new individual from a tiny fragment taken from almost any level, whereas, in other species, regeneration is restricted to a particular portion of the body. Initially, a regeneration blastema is formed at the wound surface, while morphallactic processes only play a role in the ultimate adjustment of the 'new' to the 'old' parts of the flatworm.

Origin of the regeneration blastema

Stolte (1936) couched the controversial issue of the origin of the regeneration blastema in the following two opposing points of view: (1) the cells of the regeneration blastema are dedifferentiated parenchymal cells formed in the vicinity of the wound area, or (2) the blastema is formed by so-called 'formative cells' or 'neoblasts', which are undifferentiated embryonic reserve cells present in the parenchyma and are recruited whenever necessary from the entire worm or from large parts of it.

Wolff & Dubois (1948), Dubois (1949), Wolff & Lender (1962) and Lender (1962), using *Dugesia lugubris*, have provided extensive evidence for the recruitment of neoblasts from the entire planaria or from a large portion of it during the formation of the regeneration blastema. The evidence mainly derives from irradiation and transplantation experiments. The neoblasts are very susceptible to X-rays. Partial or complete X-irradiation leads to gradual disintegration of the worm. Such animals do not regenerate their lost cells. Repair and regeneration require the presence of a wound surface and the evocation of a wound reaction. By irradiation of the anterior portion of the animal prior to decapitation, or by implantation of a non-irradiated graft into a totally irradiated host prior to decapitation, they were able to show that the delay in the appearance of the regeneration blastema is longer with increasing distance between the wound surface and the non-irradiated part. They ascribed this to the longer distance to be covered by the neoblasts to reach the wound area. By using species with different pigmentation the migration of cells from the non-irradiated tissue towards the blastema could actually be demonstrated. They claimed that the entire blastema is formed by the migrating neoblasts, which give rise to all the various ecto-, meso- and endo-dermal derivatives. Flickinger (1964) could not demonstrate a transfer of cells from a ^{14}C-labelled graft to an unlabelled decapitated host and inferred a local origin of the regeneration blastema. Best *et al.* (1965), however, attributed Flickinger's negative results to the possible death of the neoblasts in the graft as a result of the high isotope level used.

Chandebois (1960, 1962, 1965, 1968, 1976) presented evidence for the essential role of the syncytial parenchyma in the formation of the regeneration blastema by emphasising its mitotic activity near the wound surface, and by demonstrating the absence of neoblasts after starvation, X-irradiation and multiple regeneration. Bautz (1978) states that six months' starvation does not appreciably affect the regenerative capacity of *Dendrocoelum lacteum*. Coward, Bennett & Hazlehurst (1974) could demonstrate the role of lysosomes in the dedifferentiation of cells involved in blastema formation. The purely trophic function which Chandebois attributes to the basophilic neoblasts in the regeneration process is not supported by ultrastructural analysis, however. The neoblasts show all the characteristics of undifferentiated cells with normal histogenic potentialities (see below, pp. 40 to 42).

Brøndstedt & Brøndstedt (1961*a,b*) called attention to the fact that the regional distribution of neoblasts along the longitudinal axis of the planarian body cannot be reconciled with the different rates of head regeneration at various cranio-caudal levels observed in *Dugesia* (*Euplanaria*) *torva* and in *Dendrocoelum lacteum*, which have a similar neoblast distribution. They concluded that the so-called 'time-graded regeneration field' cannot be explained by the locally available numbers of neoblasts. They therefore do not agree that the role of the planarian neoblasts and that of the coelenterate interstitial cells is analogous, as proposed by Wolff & Lender (see Brøndstedt, 1969). Stéphan-Dubois (1961) tried to explain the limited regenerative capacity of *Dendrocoelum lacteum* in comparison with *Dugesia* in terms of the different distribution of the neoblasts in the two species and the lower number of neoblasts in the former. However, her arguments do not fully invalidate Brøndstedt's criticism and, moreover, cannot be reconciled with the fact that certain marine polyclade Turbellaria, which show very little regenerative capacity, nevertheless contain neoblasts.

Although the origin of the neoblasts from embryonic stock is widely accepted (see 'Embryonic development and origin of the neoblasts', p. 44), Woodruff & Burnett (1965) state that neoblasts can also be formed *de novo* by dedifferentiation of specialised cells, for example intestinal gland cells. They state that this is a normal and continuous process in both normal and regenerating animals. There is an inverse relationship between the number of gland cells and the number of neoblasts. Moreover, mitotic divisions of existing neoblasts cannot account for their large numbers in regenerating worms.

Le Moigne *et al.* (1965) stated that, in *Dugesia*, the cells of the 48-h and 60-h regeneration blastema have the character of undifferentiated cells, while the neighbouring 'old' tissues of the flatworm show no signs of dedifferentiation. In our opinion, however, this does not exclude the possible involvement of the parenchyma in the initial stages of blastema formation. In *Dugesia dorotocephala*, Betchaku (1967) observed active migration of neoblasts towards the wound area as well as migration of gastrodermal cells in the opposite direction, the latter thus making space for the regeneration blastema. On the second day after decapitation in *Dugesia dorotocephala*, Morita (1967) found the majority of the neoblasts in the anterior regions of the 'old' tissues, where they multiplied actively, but not in the regeneration blastema itself, which consisted of already differentiating cells. He concluded that differentiation must already have started during migration. Best, Hand & Rosenvold (1968) observed mitoses in neoblasts of the regeneration blastema, although the rate of mitosis was not higher in the blastema than in the 'old' mesenchymal tissues. Coward, Hirsh & Taylor (1970), using thymidine kinase as an indirect measure of cell proliferation, found that, in *Dugesia*, the earliest and maximal proliferative response to wounding and regeneration was in the non-blastemal region of the worm, but that secondary waves of proliferation occurred in the blastema itself during the process of differentiation. In *Dugesia gonocephala* Banchetti & Gremigni (1973) observed neoblast migration towards the blastema, first in the vicinity of the wound and only later from more distant regions. Gremigni (1974) holds the opinion that the neoblasts activated by the wound stimulus are able to multiply and migrate and that they constitute the main source of cells for the blastema. A karyological investigation in a triploid–hexaploid biotype of *Dugesia lugubris* by Gremigni & Puccinelli (1977) points to the origin of the blastema from neoblasts or from dedifferentiated germ cells, or both. Krichinskaya & Efimova (1978) found no reduction in regenerative capacity in *Dugesia tigrina* after repeated regeneration (anterior and posterior regenerates removed up to eight times). The regenerates were normally formed by neoblasts, but dedifferentiation of intestinal cells was observed after repeated regeneration.

Summarising, it may be stated that neoblasts as undifferentiated cells apparently play an important role in the formation of the regeneration blastema and may be recruited from a large area or from the entire worm, but that the syncytial parenchyma may also contribute to the blastema by dedifferentiation, particularly in the initial phases of blastema formation and under conditions in which the normal neoblast population has been exhausted or destroyed (see Brien, 1964 and the critical review by Coward, 1969).

Neoblasts

Neoblasts are free parenchymal cells possessing amoeboid motility (Lender & Gabriel, 1960). They are generally considered as being undifferentiated (Lender & Gabriel, 1960; Klima, 1962; Woodruff & Burnett, 1965; Benazzi, 1966; Sauzin, 1966; Le Moigne, Sauzin & Lender, 1966; Morita, 1967; Gremigni, 1974). Hay & Coward (1975) claim, however, that the neoblasts seen in the light microscope in non-regenerating animals constitute a heterogeneous population of cells consisting of undifferentiated, so-called 'beta' cells (fig. 5.1) as well as glandular and muscle cell bodies. Neoblasts or 'beta' cells are small, ovoid cells (5–6 × 7–8 μm) with a large nucleus (4–5 × 5–6 μm) and a large nucleolus. Their cytoplasm is strongly basophilic due to an abundance of ribosomes.

The divergent notion of Chandebois (1960, 1962, 1965, 1968, 1976) that the so-called neoblasts only have a trophic function, divide amitotically, become either polyploid or aneuploid, and degenerate after transferring their RNA to the parenchymal syncytium of the blastema, is not supported by other authors. As already mentioned, many authors have observed normal mitotic divisions in neoblasts. Pederson (1959) demonstrated that the basophilia of the neoblasts is due to their high RNA content, not to an

Neoblasts 41

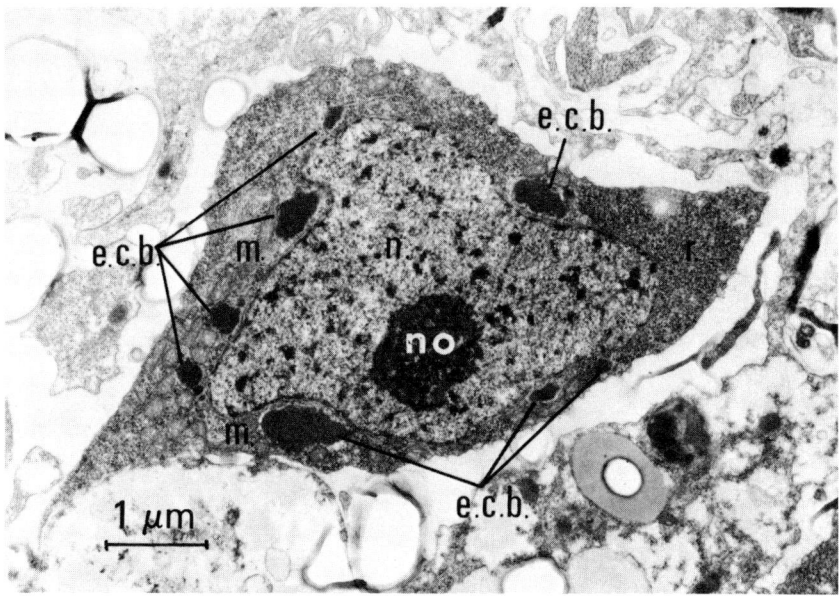

Fig. 5.1. Electron micrograph of an undifferentiatied 'beta' cell (neoblast) in the intact flatworm *Dugesia dorotocephala*, showing abundance of free ribosomes (r.) and electron-dense chromatoid bodies (e.c.b.) in the cytoplasm, commonly associated with mitochondria (m.), and a fine chromatin pattern in the nucleus (n.); no = nucleolus. (Courtesy E. D. Hay & S. J. Coward, 1975.)

increased amount of DNA or mucopoly-saccharides. Inhibition of RNA synthesis by actinomycin D blocks regeneration in *Polycelis nigra* (Gabriel & Le Moigne, 1971; Le Moigne & Gabriel, 1971). Neoblasts are also characterised by alkaline phosphatase activity (Lender & Gabriel, 1960). Betchaku (1967) mentions that neoblasts have a strong mutual affinity but little affinity with other cells.

Ultrastructurally, the nucleus, which is regular in shape, has a distinct nuclear membrane with few pores. The chromatin is found in clumps scattered in the karyoplasm and the nucleolus is mainly granular (Gremigni, 1974). The cytoplasm is filled with free ribosomes and contains only a few other types of organelle, the endoplasmic reticulum, mitochondria and Golgi apparatus. During activation as a consequence of wounding and subsequent blastema formation the neoblasts acquire more cytoplasm and contain numerous polysomes. Ergastoplasm, Golgi apparatus and lipid droplets increase in amount and mitochondria become more numerous and more highly structured as an expression of metabolic activity (Sauzin, 1966). Differentiation into nerve and muscle cells may start at 72 h after decapitation (Sauzin, 1967*a,b*; Pederson, 1972).

In our opinion, the most interesting observation made by Morita and co-workers is the presence of electron-dense lumps in the cytoplasm, which seem to originate from nuclear components and which are often found in the vicinity of the nuclear membrane, with strands extending to nuclear pores (Morita, 1967; Morita, Best & Noel, 1969). The lumps show a fibrillar–granular structure and are often associated with mitochondria. According to Morita and co-workers this material is characteristic of differentiating neoblasts and is not found in immature ones. However, Sauzin (1967a, 1968) has described electron-dense nuclear emissions in resting neoblasts of adult planarians. Their volume and number seem to increase at the beginning of regeneration but decrease again during cellular differentiation. Sauzin found such material in differentiating nerve cells, rhabdites, muscle cells and protonephridia. Le Moigne (1967a) mentioned the presence of nuclear emissions in embryonic neoblasts (fig. 5.2). Spiegelman & Dudley (1973) observed aggregates of granulo-fibrillar material in differentiating regeneration cells of older blastemas, but not in neoblasts of young blastemas. According to Hay & Coward (1975) undifferentiated 'beta' cells are characterised by lumps of electron-dense cytoplasmic material which they call 'chromatoid bodies' (see fig. 5.1). Although these phenomena may relate only to the initiation of cellular differentiation, the striking analogy with the observations by Noda & Kanai (1977) on interstitial cells in the coelenterates certainly calls for further investigation.

The scope of this book does not justify an extensive discussion of the spatial organisation of the regeneration blastema as a manifestation of the spatial relationships among the various organ systems of planarians. For this the reader is referred to the reviews by Lender (1962), Wolff & Lender (1962), Wolff, Lender & Ziller-Sengel (1964) and Chandebois (1976).

Asexual reproduction

Vandel (1921) described the phenomenon of scissiparity in *Polycelis cornuta* and in the *Planaria* species *alpina, vitta* and *subtentaculata*. A constriction appears at a given level in the worm, leading to independent development of the anterior and posterior regions. After separation each fragment regenerates the missing part. The level of constriction varies from the posterior one-eighth to the middle of the worm (in front of the pharynx). Decapitation accelerates scission but does not initiate it, because it is genetically determined. Sexual and asexual reproduction are, to some extent, mutually exclusive; the sexual state temporarily suppresses constriction. According to Vandel this points to the use of the same source of cellular material (neoblasts) for the two events. The switch from sexual to asexual reproduction is temperature dependent (see also the reviews by Stolte, 1936; Berrill,

Asexual reproduction 43

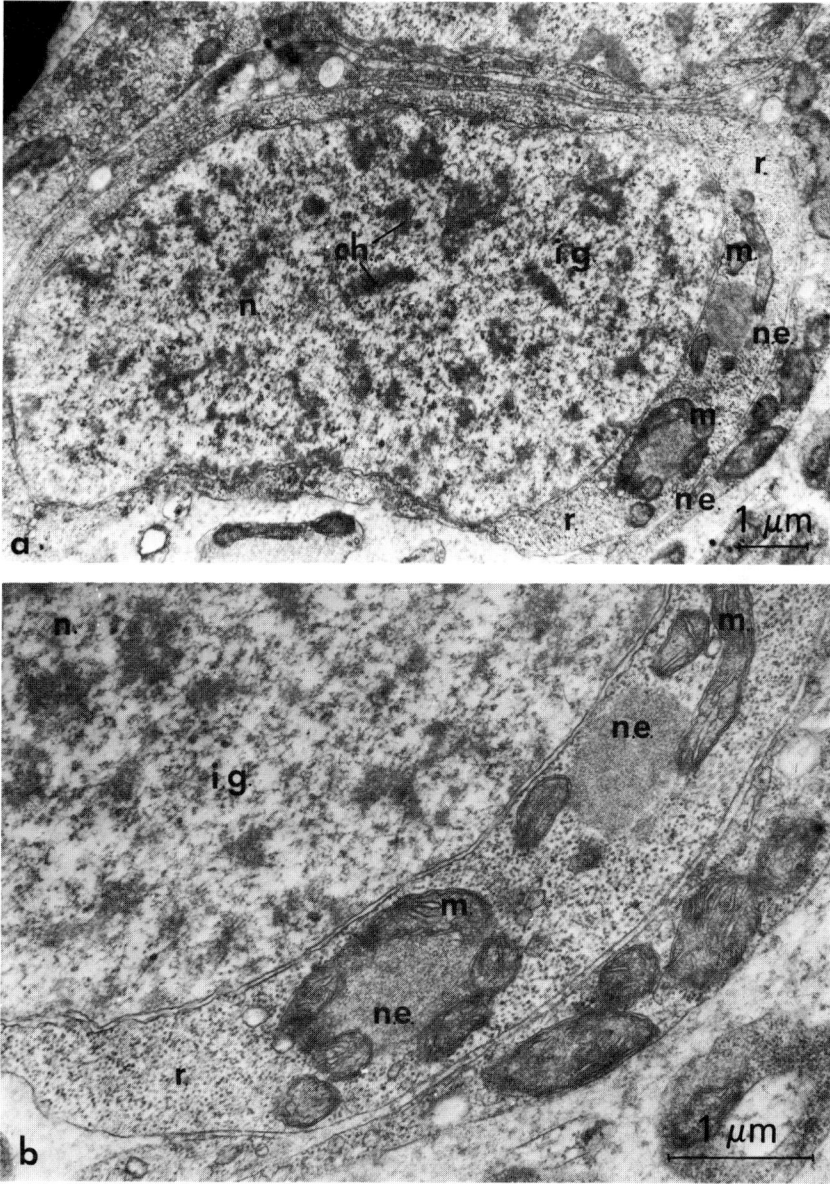

Fig. 5.2. (a) Electron micrograph of undifferentiated cell in parenchyma of ten-day embryo of *Polycelis*, showing numerous cytoplasmic ribosomes and large electron-dense nuclear emissions (electron-dense bodies), often surrounded by mitochondria. (b) Enlargement of part of (a), showing fibrillo-granular structure of electron-dense bodies. ch. = chromatin masses; i.g. = interchromatin granules; m. = mitochondria; n. = nucleus; n.e. = nuclear emissions (electron-dense bodies); r. = ribosomes. (Courtesy A. Le Moigne, 1969.)

1952c and Stéphan-Dubois, 1964). Kenk (1941), using different races of *Dugesia tigrina*, found that sexuality can be induced in the posterior two-thirds of an asexual animal when combined with the anterior one-third of a sexual animal. This could be due to the passage of hormones or to the migration of 'sexual' neoblasts. These observations were confirmed by Okugawa (1957) in reciprocal recombinates of sexual and asexual races of *Dugesia gonocephala*. He ascribed the phenomenon to a hormonal effect.

Embryonic development and origin of the neoblasts

The embryonic development of the various Turbellaria depends primarily on the egg type. In the Acoela, Rhabdocoela and Polycladida, where special yolk glands are missing, so-called 'simple' eggs are formed. They consist of one or a few ova surrounded by a thin shell membrane and a gelatinous layer. In the Acoela and Polycladida the eggs show a determinative, spiral type of cleavage, with micromere formation starting after the first or second cleavage, respectively. Gastrulation is achieved mainly by epiboly. A solid stereogastrula is formed. The micromeres of the second quartet, and possibly also those of the third quartet, form ectomesoderm as well as epidermis and pharynx, while the *4d* cell gives rise to endoderm ($4d^1$) and endomesoderm ($4d^2$). The parenchyma is formed from both ectoderm and endomesoderm. Neoblasts can be distinguished only rather late in development (see Stéphan-Dubois, 1964; Kato, 1968). It must be emphasised that the term 'determinate cleavage' only refers to the rigid subdivision of the egg cytoplasm and does not say anything about the actual state of determination of the segregating cells. The parasitic Rhabdocoela and Alloeocoela show a highly modified development with only minor traces of spiral cleavage.

The Tricladida, which possess a yolk gland or vitellarium, form so-called 'composite' eggs consisting of only a few ova surrounded by thousands of yolk cells, all enclosed in an egg shell secreted by the shell gland. No evidence of spiral cleavage is found in their strongly modified development. The blastomeres lose contact and the yolk cells fuse into a yolk syncytium (Seilern-Aspang, 1958). Some blastomeres turn into wandering amoebocytes, which migrate into the yolk syncytium, while the remaining blastomeres arrange themselves into a thin-walled vesicle around part of the yolk, in the wall of which an embryonic pharynx is formed. The latter is replaced by a second transitory pharynx, which in its turn is replaced by the definitive pharynx (Seilern-Aspang, 1956, 1958; Le Moigne, 1963; Melander, 1963). Le Moigne (1963) claims that, in *Polycelis nigra*, undifferentiated embryonic cells can be traced from the egg to the young worm by means of methyl-green pyronin staining for RNA. Le Moigne (1966a) distinguishes eight stages of embryonic development. At the stage when the definitive pharynx appears (stage 5) the embryonic cells have the ultrastructural characteristics of undifferentiated neoblasts and begin to multiply (Le Moigne, 1966b). From stage 5 onwards amputation of the anterior part of the embryo leads to so-called 'embryonic regeneration'. However, actual regeneration of the anterior structures does not occur before the differentiation of the nerve cords is completed (Le Moigne, 1963, 1969). Moderate irradiation at stage 4a does not affect normal development but suppresses embryonic regeneration (Le Moigne, 1966a). At stage 4b, when cellular differentiation begins, some X-ray resistance appears (Le Moigne, 1967b). After hatching the neoblasts take part in the growth of the young worm (Le Moigne, 1967a).

In the Tricladida, as in the other Turbellaria, the origin of the neoblasts cannot be localised to a particular cell lineage; undifferentiated embryonic cells seem to be present throughout development.

Origin of the germ cells

Vandel (1921) observed regeneration of genital organs from embryonic cells in gonad-free fragments of asexually reproducing individuals of *Polycelis cornuta* and the *Planaria* species *alpina, vitta* and *subtentaculata*. He considered the asexual state to be the undifferentiated, and the sexual one the differentiated state.

Stéphan-Dubois (1964) described normal gonad formation from asexual tissues in regenerates of pre-ovarian head fragments of triclade planarians with a high regenerative capacity. Moreover, after destruction of the germ cells in the anterior half of the worm by means of X-irradiation, and subsequent release of regenerative events by decapitation, normal ovaries regenerate from neoblasts immigrating from the non-irradiated posterior region. Stéphan-Dubois concluded that, in the triclade Turbellaria, the embryonic neoblasts are the source of both somatic and germ cells (see also Lender, 1962). In the polyclade Turbellaria, where the gonads appear late in development, the germ cells are formed *in situ* from parenchymal cells.

Fedecka-Bruner (1964) completely destroyed the dorsally situated testes by dorsal irradiation with superficially penetrating X-rays, and subsequently observed gonad formation from neoblasts migrating along the muscle fibres from the undamaged ventral into the irradiated dorsal region.

Benazzi (1966) presented genetic evidence for the formation of germ cells from neoblasts or from dedifferentiated somatic cells in anterior and posterior regenerating segments of polyploid biotypes of certain *Dugesia* and *Polycelis* species.

Melander (1963) and Gremigni (1974), studying the embryonic origin of the PCGs in the Paludicola and in *Polycelis nigra*, believe that they have found indications for a precocious segregation of the PCGs, as well as for early sex differentiation, in the presence or absence of large chromocentres in the chromosomes of blastomeres giving rise to oogonia and spermatogonia, respectively.

Banchetti & Gremigni (1973) and Gremigni & Puccinelli (1977) claim that, in *Dugesia*, young oogonia and spermatogonia in ovaries and testes situated close to the wound surface may interrupt their cytodifferentiation and may subsequently be transformed into neoblasts, which then migrate to the regeneration blastema. These observations have recently been confirmed on polyploid biotypes of *Dugesia lugubris* by Gremigni, Miceli & Puccinelli (1980) and Gremigni, Miceli & Picano (1980). The results of this work plead in favour of the possible retransformation of germ cells into neoblasts, as described previously for the coelenterates.

Sexual differentiation

Transplantation experiments have demonstrated that the posterior copulatory organ is formed under the influence of the testicular anlagen in the pharyngeal region (Wolff, Lender & Ziller-Sengel, 1964). Ghirardelli (1965) showed that gonad regeneration depends on initial brain formation; prevention of brain formation by amphetamine treatment interferes secondarily with gonad formation.

In the hermaphroditic triclade Turbellaria the ovaries and testes show a different regional distribution along the antero-posterior axis. Vannini (1963) has attempted to explain this different sexual regionality by an interaction of masculinising and feminising hereditary factors (acting upon the sexually bipotential germ cells) with an antero-posterior morphogenetic gradient set up by a cephalic organising centre. Ghirardelli (1965) showed that the morphogenetic substance involved is a brain factor.

Conclusion

It is evident from the above discussion that, in Turbellaria possessing a high regenerative capacity, germ cell formation occurs either from undifferentiated neoblasts or from parenchymal cells, and thus represents a typically *epigenetic* process which shows much resemblance to germ cell formation in the coelenterates. Unfortunately, it is not known whether the electron-dense material found in the cytoplasm of planarian neoblasts has anything to do with germ cell formation.

TREMATODA

General characterisation

Trematodes are dorso-ventrally flattened, acoelomate Bilateralia with a distinct cranio-caudal polarity and bilateral symmetry. As an adaptation to parasitic life the body is covered with a secreted cuticle which is probably of parenchymal rather than epidermal origin since, in the Monogenea, the ciliated epidermis of the miracidium larva is shed during metamorphosis and replaced by a cuticular layer. The trematodes are devoid of respiratory and circulatory systems and possess protonephridia. The intestinal system consists of an anterior mouth surrounded by a single sucker or flanked by two suckers, a pharynx and an intestinal canal consisting of two blindly ending tubes with or without lateral diverticula. The nervous system consists of paired cerebral ganglia from which longitudinal nerve cords extend caudally. The space between the cuticle and the internal organs is occupied by a parenchyma. The trematodes are hermaphrodites with separate male and female gonads, the latter with a vitellarium.

The Trematoda are subdivided into the Monogenea and the Digenea. The Monogenea are ectoparasites with two anterior and a varying number of posterior suckers. They have a larval stage which directly transforms into the adult. The Digenea are endoparasites with an anterior and an abdominal sucker. They have a complicated life cycle with several intermediate forms and a number of separate hosts (see Hyman, 1951a, vol. 2, pp. 219–310).

The egg of the Monogenea undergoes total, unequal cleavage, forming a mass of blastomeres, of which the smaller cells gradually enclose the larger ones by epiboly, thus forming a stereogastrula. The peripheral cells form a ciliated epidermis. Internally, an intestinal cavity is formed by disintegration of the central cells. The pharynx forms mid-ventrally from a ball of cells. The remaining cells furnish mesenchyme, muscle and gland cells, and the reproductive system. The initial development rather strongly resembles that of the rhabdocoelic Turbellaria. Attachment organs in the form of hooklets and suckers are formed at the posterior end of the embryo. The free-swimming larva attaches itself to the gills of, for example, frog tadpoles, and metamorphoses into the adult.

The recent literature concerns, almost exclusively, the Digenea, to which a number of important vertebrate and human parasites belong.

Life cycle and origin of the germ cells

The following life cycle of the Digenea is now generally accepted. Normal gametogenesis with complete meiosis takes place in the adult hermaphroditic worm, producing haploid eggs and sperm which are shed. The zygote develops into a free-living, ciliated miracidium larva, which infects a first intermediate host that is usually a mollusc. Inside this host the miracidium transforms into a mother sporocyst, in which a large number of germinal cells are formed. They either develop into daughter sporocysts or form one or more generations of rediae. The germinal cells of the daughter sporocysts and rediae develop into cercariae. The latter are free-living, trematode-like larvae, which infect a second intermediate host that may be a worm, a mollusc, an arthropod, a fish or an amphibian. Inside the second intermediate host the cercariae transform into resting metacercariae. When the latter is eaten by a higher vertebrate (the definitive host) the metacercariae are freed in its digestive tract and develop into hermaphroditic trematodes (Cort & Olivier, 1941, 1943; Cort, 1944; Woodhead, 1950, 1954, 1955, 1957; Pieper, 1953; Van der Woude, Cort & Ameel, 1953; Van der Woude, 1954; Ciordia, 1956; Schäller, 1960; Bednarz, 1962, 1973). Bednarz (1973) distinguishes the phylogenetically lower trematodes which have intermediate generations of rediae from the higher trematodes which form no rediae.

Opinions differ strongly on the nature of the intermediate sporocyst and redia generations. The majority of authors adhere to the so-called germ line hypothesis (fig. 5.3), which states that an uninterrupted germ line runs from the zygote, through the various intermediate forms, to the germ cells of the adult hermaphroditic trematode. According to this hypothesis the zygote divides unequally into a larger 'ectodermal' cell which will form all the somatic tissues, and a smaller cell which will form the germinal cells of the miracidium larva (Rees, 1940; Cort, 1944; Pieper, 1953; Van der Woude, 1954; Guilford, 1958; Dunn, 1959; Bednarz, 1962, 1973). During the transformation of the miracidium larva into a mother sporocyst the initially few germinal cells divide many times, giving rise to a large number of germinal

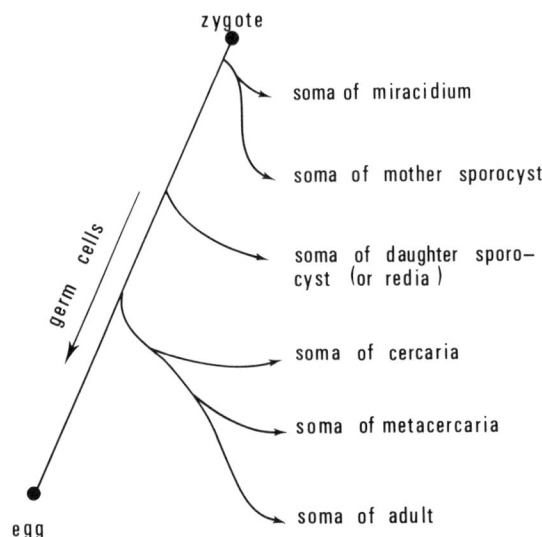

Fig. 5.3. Diagram of proposed germ cell lineage in the life cycle of the trematodes, showing segregation of germ cells and somatic cells in successive generations. (After Brooks, 1930.)

cells which replace the degenerating somatic tissues of the miracidium larva, turning the latter into a double-layered sporosac consisting of an outer epithelial layer and an inner germinative layer. The mature germinal cells of the sporocyst again divide unequally, giving rise to the somatic and germinal cells of either the daughter sporocysts or the rediae. In the daughter sporocyst both individual germinal cells and so-called 'germ masses' are found. The latter break up into smaller masses, which develop into cercariae. Unequal division of the germinal cells occurs again during the formation of the cercariae, which will finally develop into hermaphroditic adult trematodes via the resting metacercaria stage (Brooks, 1930; Cable, 1931, 1934; Rees, 1940; Cort & Olivier, 1943; Pieper, 1953; Cort, Ameel & Van der Woude, 1954; Van der Woude, 1954; Guilford, 1958; Dunn, 1959). Cercaria morphogenesis was described by Cheng & James (1960) for *Crepidostomum cornutum*. The multiplication of the germinal cells in the sporocysts and rediae is considered as an instance of polyembryony, that occurring in the mother sporocyst representing primary, and that in the daughter sporocysts and rediae secondary polyembryony.

Several variations of and deviations from this scheme have been described. Cable (1934), who stated that the full-grown germinal cells in the rediae closely resemble the oogonia of the adult, nevertheless considers the germ line to be temporarily interrupted in the cercaria, where a late segregation of germinal cells would occur. Van der Woude (1954) observed that some of the germinal cells of the cercaria form accessory, somatic structures, which would indicate that the germ line is not yet fully segregated. Rees (1940) and Bednarz (1962, 1973), though adhering to the germ line concept, regard germinal cell multiplication in the sporocyst and the redia differently from the authors mentioned earlier. Rees suggests two alternatives, i.e. internal fertilisation or parthenogenetic development of the germ cells. Bednarz defends diploid parthenogenetic development of the eggs in *Fasciola hepatica* and other Digenea. He found indications of meiosis and first polar body extrusion, but without reduction of the chromosome number, the germinal cells in sporocysts and rediae being diploid. The parthenogenetic egg undergoes unequal division as described above. Both authors deny the existence of polyembryony in the digenetic trematodes and consider their life cycle to consist of an alternation of a hermaphroditic adult generation with a number of parthenogenetic intermediate generations. However, Brooks (1930), Cable (1934), Pieper (1953), Van der Woude (1954), Guilford (1958) and Schäller (1960) deny the occurrence of maturation phenomena.

Schäller (1960) considers polyembryony as a form of asexual reproduction which leads to the alternation of sexual adult and asexual sporocyst and redia generations. He regards the propagatory cells as asexual and states that, outside sexuality, there is no question of a real germ line. His main arguments are the diploid character of the propagating cells and the absence of definite ovaries. However, in our opinion the absence of regenerative capacity and the occurrence of an unequal division of the mature germinal cells, giving rise to a larger somatic and a smaller germinal cell, as observed by many authors (see above), argues against such an interpretation.

Woodhead (1950, 1954, 1955, 1957) claims to have distinguished female germ cells on one side and male germ cells on the opposite side of the miracidium, sporocyst, redia and cercaria, and consequently advocates a hermaphroditic sexual character for all generations in the life cycle of the digeneic trematodes. He considers the second generation of sporocysts, which show 'germ ball' (or germ mass) formation, to be a precocious generation which extends only through the large germ ball stage and breaks up into a second generation of germ balls, which form cercarial embryos. Woodhead's interpretation, which has never been confirmed, is strongly criticised by Bednarz (1973).

Unfortunately, nothing is known about the ultrastructure of the germinal cells or the possible involvement of a germ plasm. Van der Woude (1954) speaks of an unequal distribution of 'germ plasm' among the daughter cells of the zygote, but means only a segregation of future somatic from future germinal cytoplasm. A better insight into the multiplication of the germinal cells may enable us to discriminate between parthenogenetic and polyembryonic development. (See James, Bowers & Richards, 1966, and James & Bowers, 1967, for *Cercaria bucephalopsis haimaena*.)

Conclusion

The facts mentioned above, though not fully conclusive, plead in favour of a *preformistic* mode of germ cell formation in the Trematoda, in contrast to the typically epigenetic mode characteristic of the Turbellaria and, as will be shown below, the Cestoda.

CESTODA

General characterisation

The cestodes are dorso-ventrally flattened endoparasites belonging to the acoelomate Bilateralia. They have a distinct cranio-caudal polarity and bilateral symmetry. The body is covered with a thick cuticular layer, the origin of which (epidermal or parenchymal) is not well understood. The cestodes lack an intestinal system and even a mouth. Food is taken up from the environment through the cuticular layer. The subcuticular muscle layer contains longitudinal, transverse and dorso-ventral fibres. The nervous system consists of a number of longitudinal nerve cords connected anteriorly by a commissure. The space between the cuticle and the internal organs is filled by parenchyma. The cestodes are hermaphrodites with separate male and female gonads. The male genital complex consists of numerous testicular vesicles scattered in the parenchyma and connected to a single vas deferens, often provided with a vesiculum seminalis, and with or without a penis, and ending in a single genital atrium. The female genital complex consists of double ovaries and a vitellarium, connected to a blind-ending uterus and a vagina which opens into the genital atrium.

The Cestoda are subdivided into the Cestodaria and the Cestoda (s.s.). The Cestodaria very much resemble the Trematoda. They consist of a cephalic region with a contractile proboscis and an elongated body with a single genital complex. The free-living, ciliated, spherical oncosphere larva has five pairs of hooks. The true Cestoda exhibit a variety of forms. The less evolved ones have a scolex with suckers and hooks and a body with a number of genital complexes, but no subdivision into segments. The more highly evolved forms show a well-developed scolex followed by a neck region containing a proliferative zone, where new segments (proglottids) are formed continuously. The entire segmented region is called 'strobila'. The proglottids, each of which is equipped with a separate genital complex, may break off at the caudal end of the worm. Fertilisation may occur by self-insemination in the genital atrium or by copulation of proglottids of the same or of different strobilae. (See Hyman, 1951a, vol. 2, pp. 311–416.)

Life cycles

Cestode life cycles show various degrees of complexity. The egg develops into a spherical oncosphere larva with three pairs of hooks. In more primitive forms the oncosphere infects a crustacean as the first intermediate host, which is eaten by a fish as the second intermediate host, and finally by another vertebrate as the definitive host. In the less primitive forms the oncosphere larva is no longer free-living but enclosed in an embryophore. There is only a single intermediate host, which may be a mollusc, an insect or a mammal. This is then eaten by another vertebrate, which becomes the definitive host.

The recent literature deals almost exclusively with the true cestodes, in particular with the more highly evolved forms which are parasites of the higher vertebrates.

Embryonic development

The so-called ootypes of *Acanthobothrium, Caulobothrium* and *Phyllobothrium* consist of an egg cell and a small number of vitelline cells, surrounded by an egg shell (Euzet & Mokhtar-Maamouri, 1975, 1976). The egg shell is secreted by the vitelline cells (Mackiewicz, 1968). The egg has no visible polarity or symmetry (Bazitov & Lapkalo, 1977). Nevertheless, cleavage follows a strictly determinate course leading, via a modified cleavage pattern, to a morula with three different types of blastomeres: large yolk-rich macromeres, medium-sized mesomeres, and small micromeres (Rybicka, 1964a,b, 1966a,b; Swiderski, 1967; Mokhtar-Maamouri & Swiderski, 1975; Euzet & Mokhtar-Maamouri, 1976; Bazitov & Lapkalo, 1977). In the so-called pre-oncosphere an outer and an inner syncytial envelope are formed inside the egg shell, the former by the residual vitelline cells and the latter by the macromeres. These envelopes contain much reserve food in the form of fat and glycogen. Some mesomeres give rise to an innermost, thin syncytial oncospheral membrane, while the other mesomeres multiply, giving rise to large so-called germinative cells and a number of small somatic cells forming the hooks and the associated musculature. The micromeres degenerate, a process which may start before the outer envelope is formed (Swiderski, 1968). In the late pre-oncosphere the majority of embryonic cells degenerate, leaving only a restricted number of viable cells. A pair of penetration glands is formed. The oncosphere larva finally consists of three pairs of hooks with muscle fibres attached to them, a pair of penetration glands and about two dozen embryonic cells, half of which develop into germinative cells, while the remainder form small somatic cells located near the bases of the hooks (Rybicka, 1964a,b, 1966a,b; Swiderski, 1967; Mokhtar-Maamouri & Swiderski, 1975; Euzet & Mokhtar-Maamouri, 1976). The pre-oncosphere is characterised by intense cell multiplication and high metabolic activity, whereas the oncosphere represents a typical resting stage with greatly reduced metabolic activity. The oncosphere is freed by the lytic action of the penetration glands and of the intestinal enzymes of the host. The further development of the oncosphere after infection of the host is described by Ogren (1962) as involving the internal organisation of the hexacanth oncosphere into a young cysticercoid, and the dedifferentiation of the epidermal gland cells into germinative cells.

Adult organisation and origin of the germinative cells and the germ cells

During embryonic development germinative cells are formed rather precociously, giving rise to the anlage of the proliferative zone of the adult worm. According to Bolla & Roberts (1971) the proliferative or germinative region of the adult worm is characterised by high DNA synthesis. The region begins about 200 µm behind the apex of the scolex and extends to the point where germinal primordia are formed. It contains undifferentiated germinative cells which have a large, lightly staining nucleus with a dense nucleolus and highly basophilic cytoplasm. Further away from the apex the

Adult organisation and origin of germinative and germ cells 53

number of germinative cells decreases; here they are restricted to the parenchyma–muscle zone.

In *Hymenolepis diminuta* Sulgostowska (1972) distinguishes three embryonic cell types, i.e. large 'germinative' cells found mainly in the neck region and in the developing reproductive system, medium-sized 'germinative–somatic' cells found in the neck region and in immature proglottids, and finally small 'somatic' cells, which accompany the excretory tubules and the nerve cords in the neck region and the strobila. In the neck region the reproductive system is formed from germinative and germinative–somatic cells, giving rise to gonads and gonadal ducts respectively. The primordia of the male and female ducts and the female gonads are found in the medullary region, those of the male gonads in the peripheral region (Sulgostowska, 1974). In the adult worm undifferentiated embryonic cells of all three types are still found next to differentiated cells (Sulgostowska, 1976).

Gustafsson (1976, 1977) describes the germinative cells in *Diphyllobothrium* as undifferentiated elements with a high nucleo-cytoplasmic ratio and basophilic cytoplasm due to an abundance of free ribosomes. Other organelles are scarce: small mitochondria with irregular cristae, granular

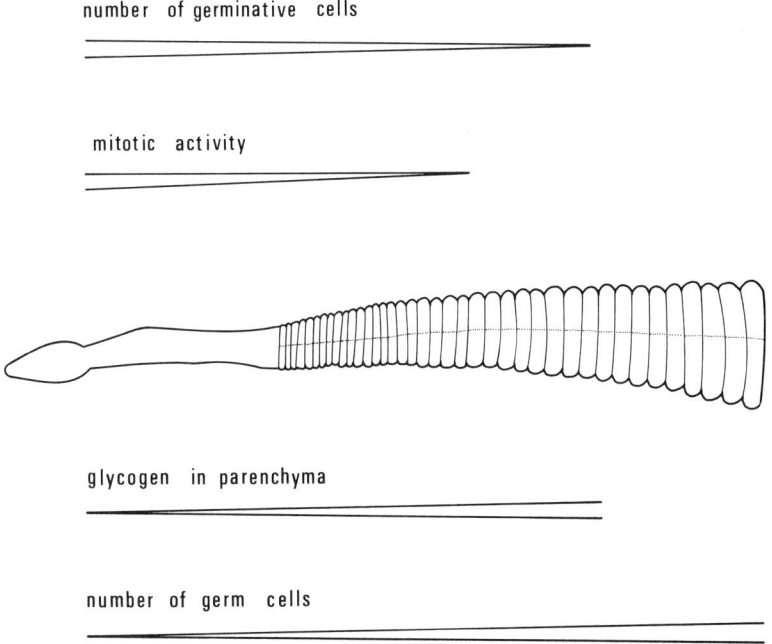

Fig. 5.4. Inverse relationship between number of germinative cells and mitotic activity on the one hand, and number of germ cells and glycogen content on the other, in the parenchyma of the neck and segment-forming region of *Diphyllobothrium dendriticum*. (After Gustafsson, 1977.)

endoplasmic reticulum (scarce or absent), and a poorly developed Golgi apparatus. The round nucleus has finely granular chromatin and one or two large nucleoli. The germinative cells are found scattered in the parenchyma and along the muscle fibres. Parenchyma cells are discrete elements of variable size with an irregular nucleus having sparse chromatin and an ill-defined nucleolus. They constitute only a minor portion of the cell population of the worm and are found mainly surrounding other structures. Their role seems to be the storage of glycogen for anaerobic metabolism. Gustafsson (1976) states that the germinative cells give rise both to parenchyma and muscle cells and to germ cells. Gustafsson (1977) has observed an interesting relationship in distribution between germinative cells and germ cells: in the proliferative zone a cranio-caudal decline in the number of germinative cells is accompanied by a rise in germ cell number (fig. 5.4 on p. 53). The cestodes, therefore, show a typical one-way transformation of undifferentiated germinative cells into differentiated cell types, both somatic cells and germ cells.

Conclusion

The undifferentiated germinative cells constitute the source of the germ cells, showing that germ cell formation in the cestodes is a typically *epigenetic* phenomenon. Unfortunately, nothing is known of the causal factors playing a role in germ cell development, nor of the possible involvement of 'germ plasm' in germ cell formation.

6

Nematoda

General characterisation

The Nematoda belong to the pseudocoelomate Bilateralia. They are cylindrical unsegmented worms, free-living or parasitic. The epidermis has a cuticle and is underlain by a muscle layer. There is no distinct head region. The nervous system consists of a circumenteric nerve ring and ventral and lateral longitudinal nerve cords. The digestive tract comprises an anterior stomodaeum with a terminal mouth, a triangular pharynx, a digestive tube, and a posterior proctodaeum with an anal opening. The nematodes possess protonephridia, but lack respiratory and circulatory systems. They may be dioecious or hermaphroditic (see Hyman, 1951*b*, vol. 3, pp. 197–454).

The life cycle can be divided into a progressive, a stationary and a regressive phase and usually is of a definite length. In *Anguillula aceti* the progressive phase lasts 18 days in the male and 20 days in the female (at 21 °C) and can be subdivided into a phase of intra-uterine development of five days and postembryonic phases of 13 and 15 days, respectively. The stationary phase lasts 28 and 26 days, respectively, and the regressive phase two and three days, respectively, so that the entire life-span of the male is 48 days and that of the female 49 days (Pai, 1928).

In many nematodes there is a cell constancy for the individual organs except the gonads (Pai, 1928; Tadano, 1968). The worms are not capable of regeneration but only of cytoplasmic repair during the progressive phase.

Embryonic development and origin of the germ cells

The embryonic development of *Ascaris equorum* was originally described by Boveri (1887*a,b*, 1888, 1891, 1893, 1899, 1904, 1909*a,b*, 1910*a*) and Zur Strassen (1896, 1898, 1903/6). The main facts were later confirmed by Pai (1928) for *Turbatrix*; by Pasteels (1948), Panijel & Pasteels (1951), Izumi (1953), Nigon (1965) and Chitwood (1974) for *Parascaris*; by Taylor (1960) for *Litomosoides* and *Dirofilaria*; by Chuang (1962) for *Rhabditis*; by Nigon (1965) for *Camallanus*; by McLaren (1973) and Hope (1974) for *Dipetalonema*, and by Malakhov & Cherdantzev (1976) for the free-living nematode *Pontonema vulgare*. The data reviewed below refer mainly to *Ascaris* species.

The egg is surrounded by a thick and resistant egg membrane, the outer layer of which is secreted by the uterus, the inner one by the egg cytoplasm (Tadano, 1968). The egg contains numerous organelles distributed uniformly in the cytoplasm (Taylor, 1960). Nevertheless, the egg has a rather rigid animal–vegetal polarity (Moritz, 1967a). Fertilisation occurs before the second maturation division. In all species fertilisation initiates irregular movements of the egg surface, particularly in the polar regions, involving displacement of a polar protuberance. These movements lead to cytoplasmic flow and to the displacement and axial orientation of the pronuclei (Nigon, Guerrier & Monin, 1960).

Cleavage is neither radial nor spiral but follows a characteristic rigid pattern. The first cleavage is transverse, separating the animal S_1 cell from the vegetal P_1 cell. At second cleavage S_1 divides meridionally into A and B blastomeres, while P_1 divides transversely into a median S_2 and a distal (vegetal) P_2 cell (fig. 6.1). The T-shaped embryo now changes into a rhomboid shape. S_2 forms an anterior and a posterior cell, while P_2 forms a dorsal S_3 and a ventral P_3 cell. Subsequently, A and B divide along a sagittal plane into right a and b and left α and β cells, respectively, marking the plane of bilateral symmetry. Alternate displacements occur during further divisions. P_3 divides into S_4 and P_4 (fig. 6.2). According to the majority of authors P_4 represents the first PGC, but Pai (1928) and Chitwood (1974) claim that P_4 divides once more before the first PGC, P_5, is formed, its companion S_5 being the precursor of the somatic cells of the gonad and the genital ducts.

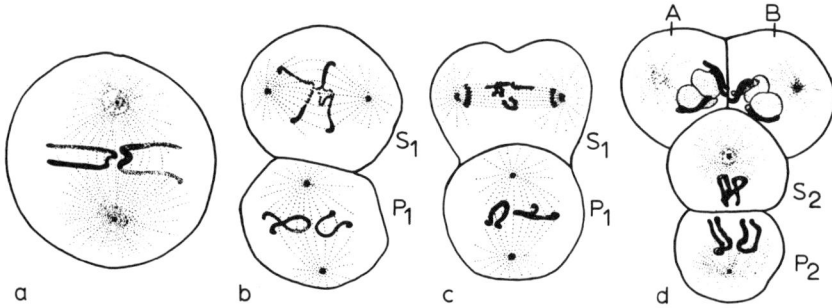

Fig. 6.1. Chromatin diminution during division of S_1 blastomere into somatic blastomeres A and B, and normal mitosis in first generative stem cell P_1, in *Ascaris equorum univalens*. (a) Metaphase of first cleavage; (b) metaphase of second cleavage, chromosomes broken up in middle region in S_1 blastomere; (c) anaphase of second cleavage, terminal regions of chromosomes left behind during cleavage of S_1 blastomere; (d) telophase of second cleavage, terminal regions of chromosomes not included in daughter nuclei A and B of S_1, but normal mitosis in P_1, forming S_2 and P_2. (After King & Beams, 1937/8.)

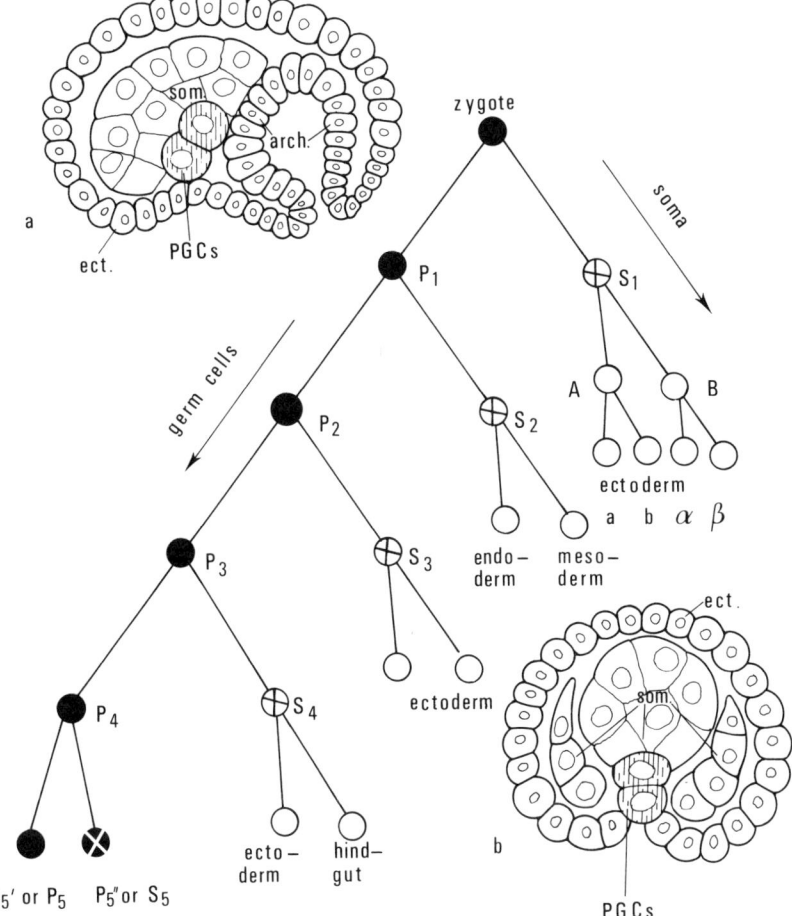

Fig. 6.2. Cell lineage in *Ascaris megalocephala*, with chromatin diminution in S_1 to S_4 (crossed open circles) and normal mitosis in P_1 to P_5, representing the generative cells (solid circles). Crossed solid circle indicates questionable chromatin diminution in S_5. (a) Sagittal, (b) frontal section through gastrula, PGCs hatched. arch.= archenteron; ect.= ectoderm; som.= somatic cells. (Redrawn from Tobler, 1976, after Boveri, 1910a.)

Kimble & Hirsh (1979) state that, in *Coenorhabditis elegans*, the somatic cells of the hermaphroditic and male gonads derive from two progenitor cells (Z_1 and Z_4) in the newly hatched larva. Their relationship with one of the division products of P_4 is unclear, however.

The lineage S_1 to S_4 exhibits chromatin diminution, marking the somatic cells of the embryo. It is not known whether S_5 also shows chromatin diminution and thus actually represents a somatic cell.

58 Nematoda

Not only is cleavage strictly determinate, but differentiation of the individual blastomeres is also cell-specific. The descendants of S^1 form the primary ectoderm and cephalic ganglia, those of S_2 form the endoderm, primary mesoderm and stomodaeum, those of S_3 the secondary ectoderm, and those of S_4 the tertiary ectoderm and hind-gut.

Gastrulation starts at the 24-cell stage with invagination of the primary endoderm and epiboly of the ectoderm. The initially round blastopore first becomes oval, then slit-shaped. Its anterior end is surrounded by stomodaeum cells, its posterior end by mesodermal and proctodaeum cells. Subsequently the stomodaeum anlage, the mesoderm, the PGC and the proctodaeum anlage invaginate. The mesoderm forms two cell bands on either side of the central endoderm. Stretching and swelling of the mesoderm bands leads to the formation of a worm-shaped embryo (Chitwood, 1974) and to closure of the blastopore in its middle region, leaving an anterior blastopore which becomes the mouth, and a posterior opening which becomes the anus (fig. 6.3) (Malakhov & Cherdantzev, 1976).

Schleip (1924) obtained fusion of two oocytes placed in parallel orientation, resulting in a single giant larva. Fragmentation of the uncleaved egg also leads to regulation and normal development. The uncleaved egg, therefore, still shows regulative capacity. Pai's (1928) observation that, in *Turbatrix*, the isolated P_1 blastomere can form a complete embryo, while the isolated S_1 blastomere degenerates, could not be confirmed by Seck (1938).

Centrifugation before first cleavage may prevent the rotation of the first

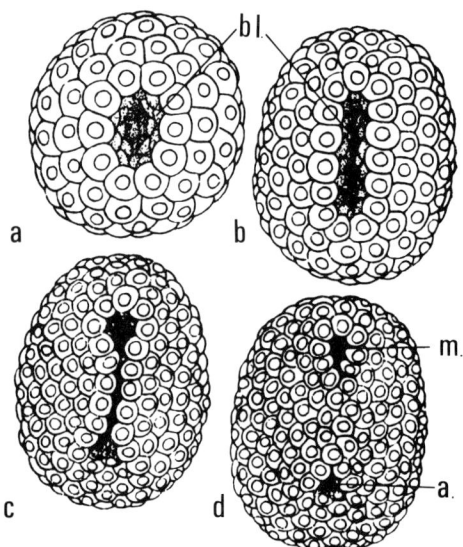

Fig. 6.3. Closure of blastopore (bl.) and formation of mouth (m.) and anus (a.) in the nematode *Pontonema vulgare*. (After Malakhov & Cherdantzev, 1976.)

cleavage spindle and may lead to complete inversion of the original polarity of the egg (Nigon, Guerrier & Monin, 1960; Guerrier, 1967). Depending upon whether the egg is divided unequally or equally by the first cleavage plane, a single embryo or twin embryos develop. A sharp discrepancy therefore exists between the regulative capacity of the uncleaved egg and the strictly determinate development of the individual blastomeres, suggesting the occurrence of determinative events during first cleavage; visible criteria for this are lacking, however.

Pasteels (1948) described an unequal distribution of RNA granules during first cleavage, leading to the formation of an animal and a vegetal cap-shaped cytoplasmic segregation, a process repeated during subsequent divisions. The vegetal cap of the presumptive PGC is denser and more closely associated with the egg cortex than that of the other vegetal blastomeres. According to Pasteels the presumptive PGC gradually acquires its 'germinal' state during successive cleavages, possibly as a form of epigenetic development. However, the observations can also be interpreted as a rigid but stepwise segregation of the germinal from the somatic cytoplasm.

Chromatin diminution

As already mentioned, chromatin diminution occurs in the lineage of the somatic S_1 to S_4 (or S_5) blastomeres, but not in the P_1 to P_4 (or P_5) lineage. The blastomeres P_1 to P_3 (or P_4) may be called pPCGs and P_4 (or P_5) the first PGC. Chromatin diminution was first described by Boveri and has been confirmed by many authors, for example King & Beams (1937/8), Pasteels (1948), Lin (1954), Zur Strassen (1959) and Moritz (1967a,b). The relevant literature has been discussed extensively by Nigon (1965) and Hope (1974).

At metaphase of the divisions in question the middle portions of the large chromosomes become fragmented to yield up to 30 small fragments, called karyosomes, while the end pieces of the chromosomes remain intact. At anaphase each karyosome divides and the division products are distributed to the two daughter nuclei, like normal chromosomes. When the new nuclear membrane forms, the large end pieces divide but do not become incorporated into the nuclei. They come to lie in the cytoplasm near the nuclear membrane, where they later disintegrate (see fig. 6.1). Goldschmidt & Lin (1947) showed that the discarded chromosome ends consist of heterochromatin.

Chromatin diminution is a rather variable phenomenon. Moritz (1967a) observed a weakening of chromatin condensation already in the S_1 cell, foreshadowing chromatin diminution in the next division. In some species chromatin diminution may be only partial during the division of S_1 into A and B, consisting of an elimination of parts of the heterochromatin without the middle chromosome section fragmenting into individual karyosomes. Chromatin diminution is then completed in the next division. In other species chromatin diminution in the S lineage does not occur before the third division.

In dispermic eggs Boveri (1910b) observed a tetrapolar first division, which led to an irregular distribution of two or three of the four large chromosomes; these then showed chromatin diminution during second cleavage. He concluded that it is the polar organisation of the egg, leading to a differential distribution of animal and vegetal cytoplasmic substances, that is responsible for chromatin diminution. Beams & King (1937) and King & Beams (1937/8) suppressed cytodiaeresis during the first division by prolonged strong centrifugation and observed chromatin diminution in all nuclei during the next division. They attributed chromatin diminution to an animal factor which becomes diluted during cell division. Von Ubisch (1943) traced the presumptive germ cell material back to the middle region of the vegetal half of the egg and postulated a differential distribution of a 'protective' substance, with a maximum in that region. Pasteels (1948) postulated that the regional accumulation of RNA during cleavage is responsible for the 'germinal' state of a particular cell line, thus preventing chromatin diminution. Nigon, Guerrier & Monin (1960) ascribed the polar differentiation of the egg, including germ cell segregation, to the structure of the cortex.

Spindle orientation seems to determine the fate of the blastomeres. Dispermic eggs with the spindle in an equatorial position show chromatin diminution in all blastomeres. The same occurs in eggs strongly damaged by ultraviolet (UV) irradiation, where the spindle is not parallel to the animal–vegetal axis. In so-called ball germs, with an animal–vegetal spindle orientation in both blastomeres of the two-cell stage, two P_2 cells are formed instead of one (see Moritz, 1967b).

Moritz (1967a) observed that RNA is periodically formed during the interphase of each cleavage and is transferred to the cytoplasm during mitosis. After chromatin diminution has occurred the somatic cells contain less RNA than the germ cells. He also found a deficit in basic proteins in diminution-competent chromosomes in comparison with germ cell chromosomes. He assumes that the change in chromosomal proteins is essential for chromosome fragmentation and somatic differentiation of blastomeres. Total UV irradiation leads to chromatin diminution in all blastomeres including the germ cells (Moritz, 1967b). Ultraviolet irradiation of the vegetal pole plasm of the P_1 blastomere results in a maximum rate of chromatin diminution. Moritz concludes that the behaviour of the germ cell nucleus in the P_1 blastomere is determined by particular conditions at the vegetal pole region. There is a UV susceptibility gradient along the animal–vegetal axis with a maximum in the presumptive PGC region. (Compare

Von Ubisch, 1943 and Pasteels, 1948, and see further the general review by Beams & Kessel, 1974.)

Tobler (1976) found that the amount of DNA in somatic cells is 73 per cent of that of the germ cells. There is no difference in base composition between eliminated and retained DNA, but the eliminated DNA is enriched in repetitious sequences. Moreover, germ cell and somatic cell DNA contain the same percentage of genes coding for 18S and 28S RNA, so that chromatin diminution does not represent an elimination of large amounts of ribosomal cistrons. On the other hand, the repetitious DNA sequences of germ cell DNA are qualitatively different from those of somatic cell DNA, so that germ cell DNA apparently contains repetitious sequences which are no longer present in the DNA of the somatic cells. Since Tobler could analyse only the repetitious sequences, nothing can be said about the non-repetitious sequences in germ cells and somatic cells.

Conclusion

It must be concluded that the Nematoda have a very pronounced *preformistic* type of development, including an extreme *preformistic* mode of germ cell formation characterised by an early and permanent segregation of the germ cells from the somatic cells. The latter lose some of their genetic information in the process of chromatin diminution.

7
Ectoprocta (Bryozoa)

General characterisation

Ectoprocta are colonial coelomates. They have a calcareous, chitinous or gelatinous exoskeleton which surrounds the zooecium. The latter consists of an ectomesodermal vesicle, the cystide, into which the so-called polypide can retract. The polypide is suspended in the single coelomic cavity, which extends into the trunk. The polypide is provided with a retractile, circular or horse-shoe shaped lophophore bearing ciliated tentacles around the mouth opening. The U-shaped digestive tract consists of the stomodaeum or pharynx, the esophagus, the stomach, the intestine, and the proctodaeum or rectum which opens on the oral side outside the lophophore, hence the name Ectoprocta.

The Ectoprocta lack nephridia and a circulatory system. There is a circumenteric ring of ganglia with ascending and descending nerve cords. The gonads are formed in the peritoneal wall. The retractor muscles as well as one or several genital strands (funiculi) connecting the base of the polypide with the cystidal wall traverse the coelomic cavity. The animals are usually hermaphroditic. (See Hyman, 1959, vol. 5, pp. 275–515.)

The Ectoprocta are subdivided into the Gymnolemata and the Phylactolemata.

Asexual reproduction

Colony formation occurs mainly during the summer by means of asexual budding while, at the approach of winter, asexual statoblast formation sets in. In the Gymnolemata budding usually involves the formation of a single cystidal bud near the distal end of the mother cystide, while in the Phylactolemata a more complex budding process on the ventral side of the mother polypide leads to the formation of a primary bud, a twin bud and an adventive bud, all of which are in different stages of development (fig. 7.1a).

Budding of a new cystide always involves the entire parietal wall of the cystide, consisting of an ectodermal layer with the underlying somatopleure and coelomic epithelium. The ectoderm forms a mass of embryonic cells

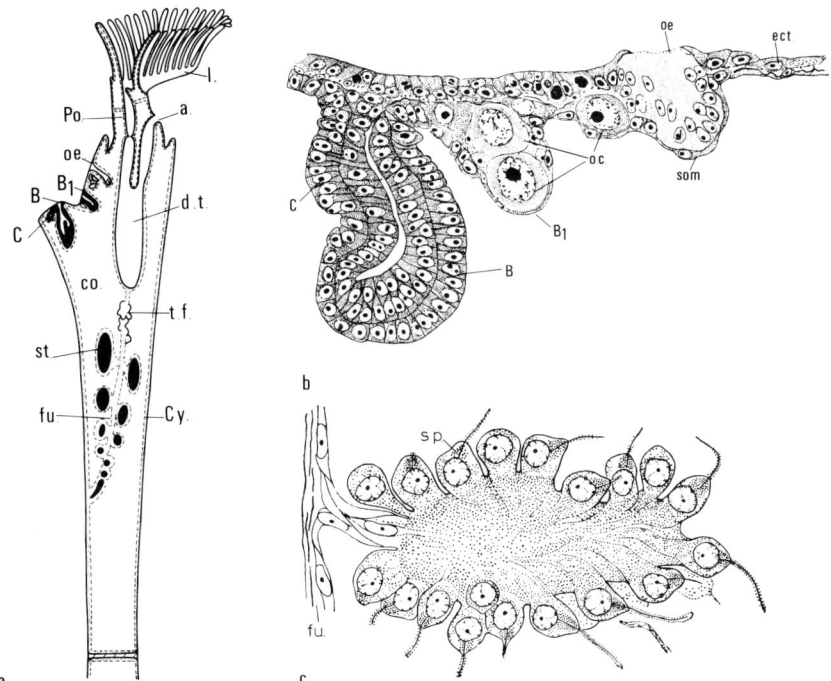

Fig. 7.1. (a) Diagrammatic representation of bud formation and gonad development in *Plumatella fungosa* (phylactoleme Bryozoa). a.= anus; b= primary bud; B_1= adventive bud; C= twin bud; co.= coelom; Cy.= cystide; d.t.= digestive tube; fu.= funiculus; l. = lophophore; oe. = ooecium; Po. = polypide; st. = statoblast; t.f. = testicular follicles. (b) Detail of primary (B), adventive (B_1) and twin (C) buds and ooecium (oe.) with location of oocytes (oc.). ect. = ectoderm; som. = somatopleural wall. (c) Detail of testicular follicle in upper portion of funiculus (fu.), with spermatids (sp.). (After Brien, 1966.)

which, together with the surrounding somatopleure, protrudes into the coelomic cavity. Inside the embryonic mass an atrial cavity is formed, in the floor of which a cellular ridge appears which develops into the digestive tract, while the tentacles of the lophophore grow out around the mouth. Behind the mouth ganglion formation occurs. An ectodermal strand surrounded by mesoderm, called the funiculus, traverses the coelomic cavity and connects the base of the retractile lophophore with the cystidal wall. All organs arise from the initial ectodermal embryonic mass, while the coelomic epithelium lines the outside of the digestive tract, the funiculus, and the inside of the hollow tentacles (Braem, 1897; Stolte, 1936; Brien & Huysmans, 1937; Brien, 1953, 1960, 1966).

At the approach of winter statoblasts or hibernacula are formed from the ectodermal cone of the funiculus in the form of a series of islets which differentiate into the deflated double-walled cystogenic vesicles, which enclose a mass of yolk-rich cells. The outer wall of the vesicle secretes a chitinous capsule around the yolky embryonic cells, protecting the latter from desiccation.

Sexual reproduction

Sexual reproduction is restricted to the formation of the so-called primary zooecide from a winter statoblast. In spring the embryonic cells of the statoblast begin to develop, forming a cystidal vesicle, in the wall of which a polypide bud develops. The box-like chitinous capsule of the statoblast breaks open and a primary zooecide is formed. This is the sexual form, which produces eggs and sperm. In Gymnolemata the ovary is usually situated more distally than the testis (Franzén, 1977), while in the Phylactolemata the testes develop from clusters of mesodermal cells along the ectodermal core of the funiculus and the ovaries from somatopleure cells in the blastogenic region between the twin and the adventive buds (Brien, 1960) (see fig. 7.1b,c). Silén (1966) described how sperm is discharged through pores at the tip of two dorso-median tentacles and temporarily adheres to the tentacle crown of another individual, till the latter discharges its eggs into the sea water. In other species the ova are transported to a brooding chamber or ooecium, where the eggs are fertilised by sperm penetrating through the hollow tentacles. Freely discharged eggs are usually small and develop indirectly, that is via planktonic, pelagic larvae, while large eggs situated inside the brooding chamber develop directly into the adult form (Silén, 1945). (See also the general review by Brien, 1964.)

Embryonic development and origin of the germ cells

The zygote, which shows total and equal cleavage, develops into a morula and next into a coeloblastula. The vegetal endoblasts invaginate through the blastopore and fill up the blastocoelic cavity. Subsequently the endoblasts cytolyse, leading to the appearance of the secondary cavity of the so-called pseudoblastula. After the 72-cell stage two mesoblastic cell masses are formed from the two telomesoblasts. The mesoblastic cells spread over the inner surface of the ectoderm, enclosing the coelomic cavity and thus forming an embryonic cystide, in the wall of which polypidal buds develop (fig. 7.2) (Brien, 1960).

In the larva of *Alcyonidium* Faulkner (1933) claimed to be able to distinguish germ cells from somatic cells by their larger size. This has never been confirmed, however, and according to Brien (1960) is difficult to reconcile with asexual zooecide formation during the growth phase of the colony and with the epigenetic nature of sexuality. Chrétien (1957) described the origin

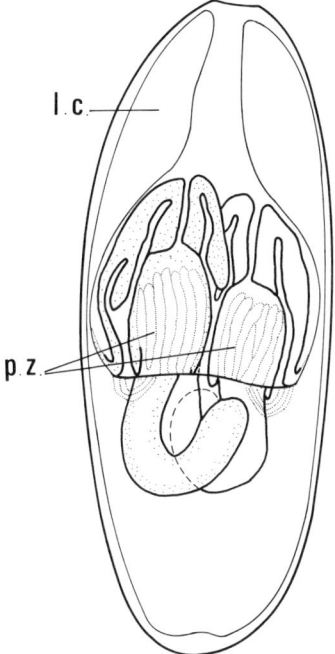

Fig. 7.2. Ciliated, free-swimming larva of *Plumatella fungosa*, with formation of two primary zooecia (p.z.) inside the larval cystide (l.c.). (After Brien, 1966.)

of the germ cells from ordinary mesenchymal cells. Franzén (1977) believes that he has recognised the germ cells at an early stage of zooecide budding in the form of what he calls 'neoblasts'. (See also the general review by Brien, 1966.)

Conclusion

In the Ectoprocta the available data plead in favour of a typically *epigenetic* mode of germ cell development during the phase of sexual reproduction, while germ cells seem to be totally absent during asexual reproduction.

8
Annelida

General characterisation

Annelida are metameric, coelomate Bilateralia belonging to the Protostomia. The epidermis may form chaetal sacs. There is a sub-epidermal muscular layer which consists of external circular and internal longitudinal sheets. Some muscle bundles traverse the coelomic cavities. The Annelida have anterior cerebral ganglia from which two ventral nerve cords descend. The latter have segmental ganglia which are connected transversely in a rope-ladder configuration. The Annelida have a centrally situated intestinal tube and dorsal as well as ventral blood vessels which form a closed circulatory system. The proto- or meta-nephridia and the gonads are arranged segmentally. There are left and right segmental coelomic cavities into which eggs and sperm are deposited. In lower forms these may be set free by the bursting of the coelomic wall, while higher forms have coelomoducts. Annelida are either hermaphroditic or dioecious.

The Annelida are subdivided into the Archiannelida, Polychaeta, Oligochaeta and Hirudinea. The small group of Archiannelida is characterised by a homonomic segmentation and an unsegmented nervous system, while chaetae are mostly absent. The marine Polychaeta have a homonomic or heteronomic segmentation. The skin bears numerous chaetae embedded in parapodia. The nervous system is segmented. The freshwater or terrestrial Oligochaeta lack parapodia and have only a few chaetae. There is a special gonadal region, the clitellum, in which male and female gonads are arranged in a particular configuration, allowing cross-fertilisation during copulation. The Hirudinea are dorso-ventrally flattened ectoparasites without parapodia and chaetae but with two suckers; they are provided with a clitellum. The coelomic cavities are strongly reduced. (See Ihle & Nierstrasz, 1928 and Borradaile & Potts, 1963.)

The four groups of Annelida show many similarities, so that the phylum will be treated as a whole, with subdivisions for the various groups.

Asexual reproduction and regenerative capacity

Although in the *Archiannelida* and *Polychaeta* sexual reproduction is the common type of propagation, asexual reproduction is also found in several species. It consists of simple fission or multiple fragmentation, followed by typical regeneration (Okada, 1968).

In the primitive polychaete *Filograna implexa*, Faulkner (1930) described separate phases of sexual and asexual reproduction, the latter occurring by transverse fission. The plane of fission may vary but the body usually tends to be divided into nearly equal parts. In the region of fission complete histological dedifferentiation occurs, followed by regeneration. From the anterior half new caudal segments grow out until a complete worm is formed. In the posterior half the first six or seven 'abdominal' segments are transformed into 'thoracic' segments and a new acron is formed. According to Faulkner (1930), neoblasts are involved both in phagocytosis and replacement of old tissues and in the formation of new ecto-, meso- and endo-dermal structures. He assumes that neoblasts are likewise responsible for regeneration of lost parts as well as for the normal posterior outgrowth of the larva. He therefore agrees with Iwanoff's (1928) statement about the essential role of neoblasts in regenerative events. Stolte (1936), reviewing the older literature, comes to a similar conclusion. Berrill (1952c), however, makes a distinction between anterior and posterior regeneration. In both cases the epithelial layer of the regenerate is formed by the 'old' epidermis, but the inner mesenchymal mass of the blastema has different origins. In anterior regeneration it is mainly the epidermis which is responsible for regeneration, with the formation of a new gut from the ectodermal stomodaem. In posterior regeneration the new gut is formed from the 'old' gut and posterior outgrowth as a whole is dependent upon the posterior extension of the intestine. Neoblasts tend to migrate posteriorly, not anteriorly. According to Berrill the cranio-caudal gradient in regenerative capacity observed in many species is correlated with the posterior outgrowth of the intestine as well as with the size of the regeneration blastema. Clark & Clark (1962), studying caudal regeneration in *Nephthys*, which proliferates new segments during its entire life-span, concluded that neoblasts are *not* involved in autotomy and regeneration in the polychaetes. Fibroblasts and coelomocyte-like cells are formed from nucleated fragments of dedifferentiated muscle cells. The mesenchymal blastema mainly consists of coelomocytes and fibroblasts, with contributions from the epidermis and the gut. Herlant-Meewis (1964) still postulates a reserve of so-called blastogenic cells in asexually reproducing polychaetes. Boilly (1968, 1969a,b,c) denies the existence of totipotent migratory neoblasts in the polychaetes. Intersegmental migration of regenerative cells does not occur even after repeated transection. All regenerative cells come from the segment contiguous to the wound. He concludes from his X-irradiation studies that the various epidermal, mesodermal and endodermal structures of the regenerate derive from the corresponding 'old' tissues. This was confirmed by Hill (1970). Boilly concluded further that the mesoderm plays an important role in growth, differentiation and segmentation of the regenerate, while the digestive tract exerts a morphogenetic role in caudal, but not in cranial, regeneration. Potswald (1969) claimed that, in *Spirorbis*, the so-called perivasal cells along the ventral blood vessel in the achaetous region are *not* neoblasts but highly differentiated secretory cells.

Clark & Evans (1961), in *Nereis diversicolor*, and Stagni (1961a), in *Spirorbis pagenstecheri*, demonstrated the important role of the supraesophageal ganglion in posterior regeneration during the first few days after amputation. Herlant-Meewis (1964) ascribed the cephalic influence to the presence of neurosecretory cells in the cephalic ganglia. Hill (1972), however, observed caudal regeneration in the absence of the brain in *Branchiomma nigromaculata* and *Chaetopterus variopedatus* and concluded that, in these species, the usual function of the brain is taken over by the large ventral ganglia.

Asexual reproduction in the *Oligochaeta* occurs either by the formation of a fission zone behind the fourth segment, or by pygidial bud formation as an accelerated form of scissiparity, which can lead to a virtually unlimited number of asexual individuals, as, for example, in *Aeolosoma hemphrichii* (Hämmerling, 1924) and *Chaetogaster diaphanus* (Meewis, 1934).

In the older oligochaete literature neoblasts as totipotent embryonic reserve cells were considered the chief source of blastogenic cells for regeneration and asexual reproduction (see Stolte, 1936 and Berrill, 1952c). Although Hämmerling (1924) still ascribed an important role to the neoblasts in the formation of the regeneration blastema, while chaetal sacs, ring musculature and brain would be formed by de- and re-differentiation of the 'old' epidermis, Meewis (1934, 1937) no longer believes in the unlimited role of totipotent embryonic reserve cells in regeneration and asexual reproduction in the oligochaetes. She considers regeneration to be predominantly 'germ layer' specific, replacement cells from each 'germ layer' being responsible for the formation of the corresponding structures. In later studies Herlant-Meewis (1946a,b, 1954, 1964) sincerely doubts the existence of totipotent neoblasts in oligochaete worms, considering regeneration after wounding, fission or budding to be a coordinated action of replacement cells from the various 'germ layers'. However, the idea of the involvement of neoblasts in regeneration finds support up to the present day: Stéphan-Dubois (1978a,b) concluded that, although anterior regeneration in *Tubifex tubifex* occurs without the involvement of neoblasts, posterior regeneration depends on neoblasts as the source of the regeneration blastema, except for gut formation, which takes place from the 'old' gut. She observed that, if amputation involves more than ten anterior segments, regeneration is abortive or absent, even though neoblasts migrate towards the wound area. Regional factors must, therefore, also be involved. In the oligochaetes neurosecretory cells of the cephalic ganglia play an essential role in pygidial regeneration, as in the polychaetes (Herlant-Meewis, 1964).

As far as we know the *Hirudinea* possess only restricted regenerative capacities, while asexual reproduction is unknown.

Summarising, it may be stated that, in polychaete regeneration, dedifferentiation of somatic cells near the wound surface certainly contributes to the formation of the regeneration blastema. Although the role of neoblasts is still debated, it is unlikely that a general reserve of totipotent cells exists even in species with asexual reproduction. The essential problem in both polychaete and oligochaete regeneration is the question of whether metaplasia is unlimited or restricted to derivatives of the corresponding 'germ layer'. At present the general opinion tends to favour restricted metaplasia in normal annelid regeneration, as well as in asexual reproduction (see the review by Okada, 1968).

Embryonic development

Embryogenesis in the *Polychaeta* has been extensively reviewed by Weygoldt (1963), Okada (1968), Cather (1971) and Anderson (1966a, 1973). Polychaetes show a typical indirect development with an intermediate free-living larval stage, the trochophore larva. This is characterised by a number of transitory larval structures, i.e. the apical tuft, the proto-, neuro- and telo-troch, and the ecto-mesenchyme. Permanent structures are the protonephridia and the digestive tract with stomodaeum, stomach, intestine and anus.

Cleavage is of the typical determinate, spiral type. The first two divisions are meridional and slightly unequal, the D blastomere being larger than the others. A markedly unequal dexiotropic third cleavage leads to the formation of the first quartet of micromeres. The fourth, fifth and sixth cleavages, which are likewise unequal, are alternatively laeotropic and dexiotropic.

Bezem & Raven (1975) were able to make a computer simulation of the early embryonic development of *Podarke obscura* by minor adaptations of their program developed for the mollusc *Limnaea stagnalis*, showing that, in eggs with spiral cleavage, typical blastula formation is the result of a rigid cleavage pattern of the animal-vegetally and dorso-ventrally polarised egg. We shall return to this study in our discussion of molluscan development (chapter 10, p. 88).

In the polychaetes ooplasmic segregation seems to start after breakdown of the germinal vesicle. Eighty per cent of the clear cytoplasm of the egg is later found in the CD blastomere and is subsequently passed on almost exclusively to the larger 2d and 4d blastomeres, 2d representing the primary ectoteloblast and 4d the primary mesoteloblast.

70 Annelida

Some polychaete eggs show a transient polar lobe which fuses with the CD, and later with the D, blastomere. This occurs, for example, in *Sabellaria* and *Sternaspis*. In the polar lobe of *Sternaspis* pleiomorphic mitochondria are found as the main organelles, together with minor quantities of dense granules, glycogen and lipids (Villa, 1976).

Anderson (1966a, 1973) emphasises that cell lineage is of only relative significance, since similarly designated blastomeres may have different fates in different groups and even species. The only thing spiral cleavage does is to subdivide the egg cytoplasm in a rigid pattern with a minimum number of cell divisions, leading to the formation of a hollow coeloblastula, as in *Polygordius*, or a solid stereoblastula, as in *Nereis* and *Arenicola*. It is the subdivision of the blastula according to organ anlagen which is important for the comparison of different groups and species (fig. 8.1). In the polychaete embryo the main organ anlagen are already segregated at the 64-cell stage.

Depending upon the amount of yolk, gastrulation may be chiefly by invagination or mainly by epiboly. The endoderm formed from blastomeres 3A, 3B, 3C and 4D sinks in first, followed by the mesoblasts and the stomodaeum anlage. The originally roundish blastopore first becomes elongated and then slit-shaped. Its antero-ventral part becomes the mouth, the posterior part the anus. A thickened cell mass around the anus will form the pygidium or terminal segment. Eight ectoteloblasts and two mesoteloblasts are situated near the pygidium and will give rise to the adult organs. In some species a few primary segments with special chaetal sacs are formed in the

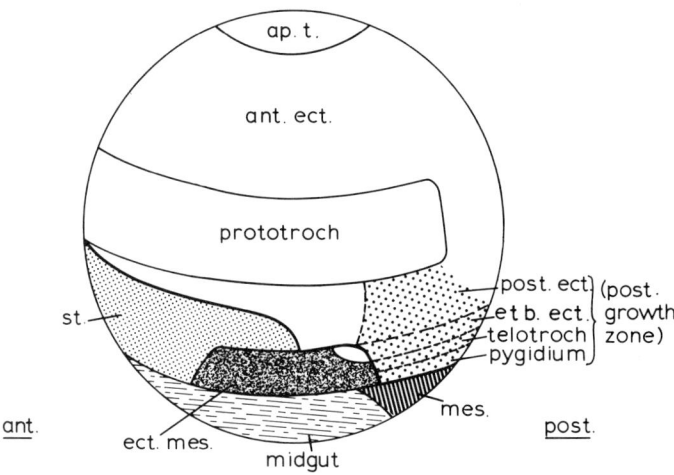

Fig. 8.1. Fate map of blastula of *Podarke* (Polychaeta), drawn in left lateral view. <u>ant.</u> = anterior; ant.ect. = anterior ectoderm; ap.t. = apical tuft; ect.mes. = ectomesoderm; etb.ect. = ectoteloblast ectoderm; mes. = mesoderm; **post.** = posterior; post.ect. = posterior ectoderm; st. = stomodaeum. (After Anderson, 1966a.)

larva, after which secondary segments are formed by the teloblasts during metamorphosis (heteronomy). In other species homonomic segmentation occurs during the outgrowth of the worm.

Guerrier (1968, 1970b, 1971) discriminates sharply between the determinate cleavage pattern and the actual state of determination of the blastomeres. From centrifugation and compression experiments with *Sabellaria* eggs before and after first cleavage he concluded that bilateral symmetry does not depend on a preformed dorso-ventral polarity in the uncleaved egg, but on the position of the first cleavage spindle. When under experimental conditions the polar lobe becomes equally divided over the first two blastomeres, twin embryos develop. When the first cleavage plane is equatorial instead of meridional, morphogenesis is severely disturbed, but when the second cleavage plane is equatorial, development is normal. Centrifugation resulting in an abnormal distribution of endoplasmic material does not cause abnormal development. According to Guerrier, the polychaete egg, like that of the spiralian molluscs (see chapter 10, pp. 95–7), manifests an essentially epigenetic development notwithstanding the determinate cleavage pattern.

Normal development in the *Oligochaeta* and *Hirudinea* has been reviewed extensively by Weygoldt (1963), Anderson (1966b), Okada (1968), Cather (1971) and Anderson (1973). There is no larval stage; direct development into a juvenile animal with many body segments occurs.

Embryogenesis takes place inside a protective cocoon, which is secreted by the clitellum during copulation of the hermaphroditic animals. Eggs of aquatic species are usually large and yolky, while those of terrestrial species are often small and contain little yolk. As in the polychaetes, cleavage is total and of the spiral type, with formation of a 2d micromere as the primary ectodermal somatoblast and a 4d micromere as the secondary, mesodermal somatoblast. Development may be shortened, however, with the initial cleavage pattern already completed after third quartet formation. In such cases the 3d micromere and not the 4d blastomere may become the second somatoblast. Oligochaete eggs contain animal and vegetal polar plasms, which are accumulations of yolk-free or yolk-poor cytoplasm. The main portion of the animal polar plasm ends up in the 2d micromere and that of the vegetal plasm in the 4d cell or its equivalent. The polar plasms of *Tubifex* are rich in mitochondria and rough-surfaced membrane vesicles (Weber, 1958, 1960). In the Hirudinea development is sometimes accelerated even more, so that 2D may become the second somatoblast.

Small eggs depend largely for their development on the albumin secreted by the clitellum. In such eggs the A, B and C blastomeres have become albuminotrophic cells, in contrast to large eggs where the A, B, C and 4D blastomeres form the midgut. When comparing the fate maps of different groups and species (fig. 8.2) it becomes evident that the basic pattern is the

72 Annelida

same for all annelids; the oligochaetes and hirudineans, with their direct development, only lack the anlagen of the larval organs, such as the apical tuft, prototoch and ectomesoderm.

The ecto- and meso-teloblasts may start their teloblastic activity before gastrulation. Of the eight ectoteloblasts the medial ones form neuroblasts, and the lateral ones ectoderm, chaetal sacs and cells of the circular muscle layer. The two mesoteloblasts give rise to two mesodermal bands, each cell of which forming a separate somite anlage. The cerebral ganglia develop from the extreme anterior ends of the ectoblastic anlagen.

In small eggs the albuminotrophic cells invaginate, while in large eggs the mid-gut cells are overgrown by the ectoderm. Subsequently, the mesoblasts and stomodaeum anlage invaginate, the latter forming the pharynx. In albuminotrophic eggs a precociously functional pharynx may be formed, which is replaced by a definite pharynx formed by the outer cells of the pharynx anlage.

Penners (1936, 1938) could demonstrate by means of UV-microbeam treatment that, in *Tubifex rivulorum*, elimination of the ectoteloblasts does not influence the development of the mesoderm. A partial regulation of the defect from ordinary ectoderm cells occurs. Destruction of the mesoteloblasts, however, leads to strongly aberrant development. No regulation of the mesoderm occurs. Although several ectodermal structures such as cephalic ganglia, ventral nerve cords, circular muscles, chaetal sacs and stomodaeum develop, segmentation of the worm and morphogenesis of the endoderm are impaired, demonstrating the organising role of the mesoderm in annelid development. Devriès (1968, 1969, 1970, 1971, 1973, 1974*a,b,c*), working with *Eisenia* (where the A, B and C blastomeres do not take part in embryogenesis and only the D quadrant forms the blastula), could confirm Penners' result but observed new, though delayed, mesoderm formation

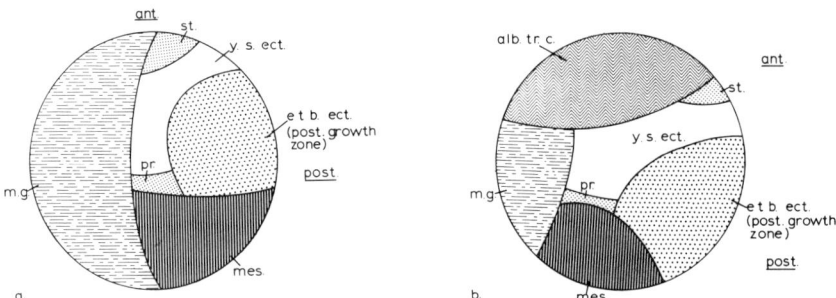

Fig. 8.2. Fate maps of blastulae of (a) *Tubifex* and (b) *Eisenia* (Oligochaeta), drawn in left lateral view. alb.tr.c. = albuminotrophic cells; ant. = anterior; etb.ect. = ectoteloblast ectoderm; mes. = mesoderm; m.g. = midgut; post. = posterior; pr. = proctodaeum; st. = stomodaeum; y.s.ect. = yolk sac ectoderm. (After Anderson, 1966*b*, 1973.)

after destruction of the mesoteloblasts. He could also demonstrate a mitogenic influence exerted by the mesoderm on the ectoteloblasts and on the endoderm.

Origin of the germ cells

In the *Polychaeta* the origin of the PGCs is a rather controversial issue. Malaquin (1925, 1934) and Iwanoff (1928) advocated a rather early segregation of the PGCs. In the polychaete *Limnodrilus* Iwanoff distinguished two pairs of so-called PGCs between the splanchnopleure and the endoderm at the level of the eighth and the ninth segments of the mesodermal bands, whence they migrated to the gonadal dissepiments 9/10 and 10/11 by amoeboid movement. In *Salmacina dysteri* Malaquin observed the splitting-off of a pair of small cells from the two so-called M cells which derive from the 4d cell, and claimed that these are the PGCs. They become attached to the blastoporal lip a short distance from the future anus, and in the young trochophore larva are found at the junction of the mid- and hind-gut. They migrate through the blastocoel to the pygidial region, where they come into contact with the two mesodermal bands. They begin to multiply in the meta-trochophore larva.

In species showing asexual reproduction, for example *Filograna implexa*, and in species where the gonadal segments can regenerate from more anterior sterile segments, for example *Spirorbis*, the germ cells are said to develop from cords of neoblasts running along the longitudinal muscle bundles or the ventral blood vessel. These would form oogonia in the first and second 'abdominal' segments and spermatogonia in succeeding segments (Faulkner, 1930; Stolte, 1936; Vannini, 1947; Reggiani, 1956, 1957; Stagni, 1959, 1961a,b). Dorsett (1961) suggested that, in *Polydora ciliata*, the PGCs derive from the ventral epithelium and migrate to the gonadal anlagen, where they mature. In many other polychaete species the PGCs even appear as late as metamorphosis (Vannini & Stagni, 1959). (See also Anderson, 1966a and Stéphan-Dubois, 1964.)

Stolte (1938) claimed that, in ageing animals, PGCs are transformed into phagocytosing amoebocytes, which break down the intestine and musculature. One cannot be sure, however, that amoebocytes did not first infiltrate the gonads, erroneously suggesting such a transformation.

In the *Oligochaeta*, Iwanoff (1928) described two pairs of PGCs in the early *Tubifex* larva at the level of the eighth and ninth segments of the mesodermal bands. Penners & Stäblein (1930), Penners (1934), Herlant-Meewis (1946a) have the PGCs in *Tubifex*, *Limnodrilus* and *Peloscolex* descend from the 4d cell during the first division of the mesoteloblasts, and claim that they migrate to their definitive location in dissepiments 9/10 and 10/11. Meyer (1931) believed, however, that the migratory cells described by previous authors are not PGCs but that these are formed *in situ* in the genital segments. Devriès (1971) believes that the small cells which in *Eisenia* split off from the mesoteloblasts perish during their first division. Ultraviolet microbeam and micro-needle deletion experiments by Devriès (1971) demonstrated that the PGCs must be formed from mesodermal cells prior to the formation of the mesoteloblasts, but after the segregation of the endoblasts. According to Relexans (1970a,b) the results of temperature-shock experiments performed during early development, which may disturb postembryonic gonad formation, plead in favour of an early segregation of the germ cells. Apart from a preliminary study by Chapron & Relexans (1971) in *Eisenia*, little is known of the ultrastructure of the PGCs.

In *Tubifex*, *Psammocystus* and *Limnodrilus* Iwanoff (1928) observed that, in regenerates where neoblasts are present in most segments, PGCs are, in fact, absent in the gonadal anlagen before sexual maturity, when they are newly formed from the peritoneum. Herlant-Meewis (1946b) transected *Lumbricillus* worms at the level of the eighth segment, the anterior part being devoid of germ cells. Posterior regeneration occurred in about 10 per cent of the cases, the germ cells in the gonadal segments being reconstituted from ordinary parietopleural cells. This was confirmed by Gates (1943, 1951) in *Perionyx*. These results demonstrate that *de novo* formation of PGCs from somatic cells may occur when the normal germ cells are experimentally removed. In addition, no trace of germ cells is found in Naidiformes showing scissiparity, so that true asexual reproduction seems to occur in some oligochaete annelids. (See also Stolte, 1936; Stéphan-Dubois, 1964 and Anderson, 1966b, 1973.)

In the *Hirudinea* the PGCs are recognisible only in the genital dissepiments. It is assumed that, in analogy with the oligochaetes, the presumptive PGCs may segregate from the somatic cells at an early developmental stage, splitting off from the M cells before mesoteloblast formation, then passing a quiescent phase among the mesodermal cells of the splanchnopleure layer, before becoming recognisable as PGCs. Relevant evidence for this assumption does not exist, however. (See Stéphan-Dubois, 1964 and Okada, 1968.)

Sexual differentiation

In most *polychaetes* the first and second 'abdominal' segments carry ovaries and the subsequent segments testes. After removal of the head and subsequent regeneration of the cephalic ganglia in the polychaete *Spirorbis*, Stagni (1961a) observed an extension of the ovarian region over four instead of the normal two 'abdominal' segments. This may be interpreted in terms of an interaction of male and female hereditary sex factors with a cephalo-caudal morphogenetic gradient acting on bipotential germ cells (see Vannini, 1963). Recently, Porchet, Dhainaut & Porchet-Hennere (1979) reported the appearance of two peptides in the coelomic fluid of sexually differentiating animals of *Perinereis cultrifera*, which would be responsible for sexual differentiation.

Somewhat controversial observations have been made in the oligochaetes. In *Eisenia foetida* Relexans (1970a, 1973, 1975) observed reversal of the male sex under conditions of starvation after hatching. Transplantation of undifferentiated gonads showed that both sexes are reversible. He postulated a dominant masculinising influence in the ventral region of segments 10 and 11 and a dominant feminising influence in that of segment 13, both influences acting on the bipotential germ cells. The ventral nerve cord, moreover, exerts a trophic influence on the expression of male potentiality. Lattaud (1973, 1974) observed that, in *Eisenia*, ovaries develop normally in a culture medium without hormones, suggesting autodifferentiation, but testes show sex reversal in the absence of hormones. When male and female gonads are cultured together with nervous system, masculinisation of the ovary occurs under the influence of a testicular androgenic hormone, which in its turn is formed under the influence of the neurosecretory activity of the nervous system. However, in the same species André & Davant (1977) found that the male and female reproductive organs differentiate from determined territories of the parietal mesoderm, independent of each other and of axial organs such as the digestive tube, splanchnopleure and nerve cord. Extra-gonadal germ cells are liable to be formed from the somatopleural coelomic layer in the ventral region of the genital segments, but differentiate only in the male direction, the female germ cells being incapable of complete self-differentiation. Zinc chloride ($ZnCl_2$) treatment may cause sex inversion both in testes and ovaries and may lead to the formation of extragonadal supernumerary germ cells in more anterior as well as more posterior segments (André & Davant, 1979).

Conclusion

Although the origin of the PGCs in the annelids remains a controversial issue, it may be stated that they are certainly of mesodermal origin. They may segregate from the mesodermal cells at different stages of embryonic development or may be formed epigenetically from undifferentiated mesodermal cells. Although in some forms presumptive PGCs seem to segregate during early embryonic development, germ cell formation looks to be mainly *epigenetic*. This epigenetic character is particularly evident in the cases of gonadal regeneration and asexual reproduction.

9

Echinodermata

General characterisation

The Echinodermata are enterocoelous coelomates with pentaradiate symmetry. The larval stage shows bilateral symmetry, however. There is no head or brain. The body is differentiated into an oral and an aboral surface. The echinoderms, which are marine animals, have a calcareous endoskeleton consisting of separate plates which bear external appendages. The nervous system consists of a ganglionic ring around the esophagus with five radiate ganglionated strands and a more weakly developed aboral ring. There is a digestive system with an orally situated mouth, a digestive tract with digestive glands, and an anus which may be situated centrally or eccentrically at the aboral side, or may even be displaced to the oral side. A spacious coelom is present around the internal organs in the form of a typical endocoel, bilateral in origin. The two coelomic cavities are separated by dorsal as well as ventral mesenteries. There are no special excretory and respiratory organs, but there is a well-developed water-vascular system of coelomic origin with numerous external protrusions in the form of podia, which are arranged in five grooves or bands, the so-called ambulacra. The podia are connected with a radial canal in each radius, which communicates with a ring canal around the esophagus. The water-vascular system communicates with the exterior by a separate canal through a single pore or a cluster of pores, the hydropore, situated in the so-called madreporal interradius. Most echinoderms are dioecious, but some are hermaphroditic.

The Echinodermata are subdivided into two subphyla, the Pelmatozoa containing the Crinoidea and a number of extinct classes, and the Eleutherozoa, which comprise the Holothuroidea, Echinoidea, Asteroidea and Ophiuroidea, as well as one extinct group.

The Crinoidea are either sessile, stalked animals or stalkless, free-moving animals provided with cirri. The oral surface, which faces upward, has a central or nearly central mouth and an eccentric anus. The oral surface is surrounded by pentamerous, simple or branched flexible arms with ambulacra. The branched arms end in so-called pinnules. The coelom is filled with mesenchyme. The crinoids are dioecious.

The Holothuroidea are soft, orally–aborally elongated animals with an anterior mouth, surrounded by a set of tentacles connected with the water-vascular system, and a posterior anus. The animals are radially symmetrical but show secondary bilateral symmetry due to the development of a flat ventral creeping surface provided with locomotory podia, and a curved dorsal surface with only warts and papillae. Large, internal, tree-like respiratory organs, the so-called water lungs, extend from the cloaca. The endoskeleton is much reduced. The animals are dioecious or hermaphroditic.

The Echinoidea are spherical or cordiform animals with a well-developed pentagonal endoskeleton. The oral surface, which faces downward, has a central or anteriorly displaced mouth, surrounded by a membranous peristome with external gills. The anus, which is surrounded by a membranous periproct, is situated aborally or is displaced posteriorly and in some species even all the way to the oral side. There are five ambulacra with double rows of podia extending from the oral to the aboral side. The mouth is provided with a complex chewing apparatus, the so-called 'lantern of Aristotle'. The external appendages comprise spines, pedicellariae, sphaeridia, podia and gills. The echinoids are strictly dioecious.

The Asteroidea are orally–aborally flattened, pentagonal, star-like animals. The ambulacral grooves are limited to the downward-facing oral surface and extend to the tips of the arms. There is a flexible endoskeleton. The short digestive tract consists of a centrally situated mouth, an esophagus with ten esophageal pouches, and a stomach with ten glandular appendages which extend into the arms. The animals are dioecious.

The Ophiuroidea are orally–aborally flattened animals with a central disc and five slender, flexible arms with ambulacral grooves. The digestive system begins with a large pentagonal aperture, the so-called preoral cavity, which is provided with teeth and leads to the true mouth. There is a short esophagus and a sacciform stomach, but no intestine and digestive glands and no anus. The animals are dioecious or hermaphroditic. (See Hyman, 1955, vol. 4, pp. 1–746.)

The gonadal complex in the Asteroidea, Ophiuroidea and Crinoidea consists of an aboral ring situated between the dermis and the coelomic wall and representing the primary genital cord, and two genital canals in each interradius, extending towards the oral side. A genital gland lies at the base of each genital canal. A duct leads from each gonad to a corresponding pore on the oral surface. The Crinoidea discharge their sexual products through rupture of pinnules. The Echinoidea have no communicating genital cord but only separate genital glands in the interradii. The Holothuroidea possess a single genital gland in the form of a large bunch of caeca protruding into the coelomic cavity; it is situated in the anterior part of the animal near the esophagus, the gonoduct ending in a so-called pharyngeal bulb.

Regenerative capacity

All echinoderms possess a more or less extensive regenerative capacity. The crinoids regenerate arms which have broken off by trauma or autotomy. The echinoids show regeneration of external appendages only. In the other three classes the regenerative capacity is much more pronounced and even leads to a form of asexual reproduction. In the ophiuroids and asteroids fission of the central disc occurs along preformed lines. The fission products, which usually have one or more intact arms, regenerate the missing parts. In some asteroids arms may be cast off at a more or less preformed level some distance from the central disc. The isolated arm fragment first regenerates a disc and subsequently four new arms. The holothuroids may eviscerate almost their entire digestive tract together with the respiratory organs and gonadal tubules as a defence mechanism. In addition, asexual reproduction occurs by transverse constriction into anterior and posterior halves, each of which then regenerates the missing parts. (See Hyman, 1955, vol. 4, pp. 1–746.)

Huet (1966) demonstrated that, in the asteroid *Asterina gibbosa*, regeneration involves only the tissues bordering the wound surface. In normal regeneration the new tissues are formed by cells originating from the corresponding 'old' tissues by dedifferentiation. Regeneration can be prevented by total or partial X-irradiation. An intact oral nerve ring and intact radial nerve connections with the wound area are prerequisites for blastema formation. The stimulating influence of the nervous system, leading to a high rate of mitosis in the wound tissue, is required during the entire regeneration process (Huet, 1967, 1972, 1975).

It is not known whether the same type of regeneration occurs in the other groups, particularly during asexual reproduction, where the fission products may not contain all organ systems and may, for instance, lack gonads.

Embryonic development

The various classes of the echinoderms are so similar in development, particularly during early stages, that a common treatment is justified. In this book we will restrict ourselves to the main facts.

The oocytes are attached to the ovarian wall, initially with a broad base, later with a narrow stem. The pole opposite the stem represents the future animal pole, the stem side the future vegetal pole. After detachment the oocyte becomes surrounded by a jelly layer provided with a micropyle near the animal pole.

Eggs and sperm are shed into the sea water, where fertilisation occurs. After the entry of the sperm through the micropyle a fertilisation membrane is formed by the extrusion of the cortical granules into the perivitelline space. The eggs are monospermic due to a fast block to polyspermy, which arises prior to the formation of the fertilisation membrane and resides in the new surface layer, the hyaline layer. The eggs produce 'fertilizin', which reacts with the 'antifertilizin' of the sperm. It agglutinates homologous sperm and precludes fertilisation by sperm of other species.

The first two cleavages are equal and meridional. The third cleavage is equatorial and gives rise to four animal and four vegetal blastomeres of nearly equal size. The fourth cleavage in the animal quartet is meridional, while the vegetal quartet splits off four micromeres near the vegetal pole by a markedly unequal equatorial division. The first three cleavages are fully synchronous, but a vegetal–animal gradient in cell division becomes manifest during the fourth division, the micromeres acting as 'pacemaker' in mitotic activity (Parisi et al., 1978). The fifth cleavage of the animal octet, the so-called mesomeres, is equatorial, giving rise to two tiers of eight cells, called the an_1 and an_2 tiers of blastomeres. The fifth cleavage of the four vegetal macromeres is meridional, giving rise to a third tier of eight blastomeres. At the same time four smaller micromeres are split off from the first micromere quartet (32-cell stage). At sixth cleavage the micromeres divide again, as well as the eight macromeres. The latter divide equatorially, giving rise to the so-called veg_1 and veg_2 tiers of blastomeres. Consequently, a 64-cell stage with five tiers of cells is formed, called an_1, an_2, veg_1, veg_2, and micromeres respectively (fig. 9.1). During the cleavage process a blastocoelic cavity filled with a colloidal substance is formed. Cilia develop on the surface of the radially symmetrical blastula, including an apical tuft of long, immobile cilia. In the sea-urchins hatching occurs at the blastula stage, when the fertilisation membrane dissolves.

The vegetal surface of the blastula flattens. Strong pulsatory activity of the micromeres around the vegetal pole leads to the detachment of the micromeres from the blastular wall and their subsequent accumulation in the blastocoelic cavity, thus forming the so-called primary mesenchyme (Gustafsson, 1963). They move along the inner wall of the blastula by pseudopodial movement and concentrate at two nearly opposite sites under the directing influence of the ectoderm. The primary mesenchyme will form the future endoskeleton. Invagination of the archenteron occurs on the flattened vegetal side of the blastula. Subsequently, cells are budded off from the tip of the advancing archenteron, forming the secondary mesenchyme. These cells send out long pseudopodia to the ectoderm at the animal pole, which contract and pull the archenteron further inwards. A flattening of the animal–ventral side of the gastrula, indicating the oral field, is the first indication of bilateral symmetry. (See further Giudice, 1973; Hörstadius, 1973 and Davenport, 1979.) Holland (1976) described the ultrastructural features of the crinoid gastrula. When the invagination of the archenteron is completed the larva flattens along the dorso-ventral axis and becomes

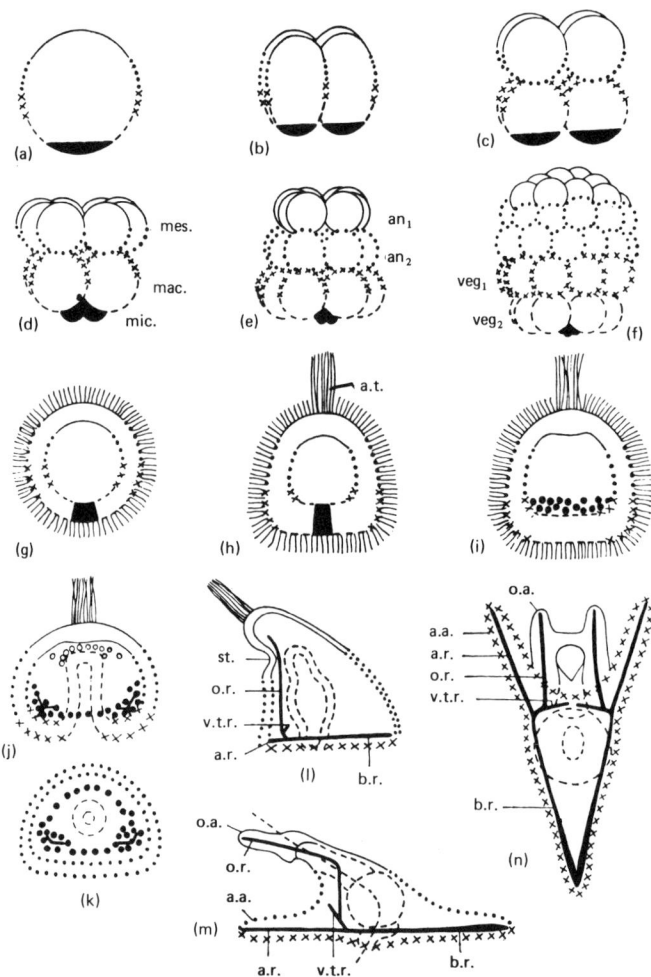

Fig. 9.1. Diagram of normal development of the sea-urchin *Paracentrotus*. (a) Uncleaved egg; (b) 4-cell stage; (c) 8-cell stage; (d) 16-cell stage with mesomeres (mes.), macromeres (mac.) and micromeres (mic., in black); (e) 32-cell stage with an_1 in continuous lines and an_2 in dotted lines; (f) 64-cell stage with veg_1 in crosses and veg_2 in broken lines; (g) early blastula stage; (h) older blastula stage with apical tuft (a.t.); (i) blastula with primary mesenchyme; (j) gastrula with primary and secondary mesenchyme and triradiate spicule formation; (k) transverse section through same gastrula stage; (l) so-called prism stage with invaginating stomodaeum (st.); (m) pluteus larva seen from the left, with position of original egg axis indicated by broken line; (n) same pluteus larva seen from anal side. a.a. = anal arm; a.r. = anal rod; b.r. = body rod; o.a. = oral arm; o.r. = oral rod; v.t.r. = ventral transverse rod. (After Hörstadius, 1935.)

roughly triangular in sagittal section; this is the so-called prism stage. The cells of the animal pole then thicken and the formation of the skeleton begins. Skeleton formation proceeds differently in the various groups, but will not be discussed here.

The distal end of the archenteron constricts and is pinched off, forming the first coelomic anlage. Microfilaments in the coelomic pouch cells and filopodia of the mesenchymal cells play a leading role in this segregation process. Cytochalasin B treatment, affecting the microfilament system, leads to regression of the pouch (Crawford & Chia, 1978). Concomitant with coelomic pouch formation a stomodaeal invagination develops on the ventral side, close to the original animal pole. The tip of the intestinal portion of the archenteron bends towards the stomodaeal invagination, guided by pseudopodia of the secondary mesenchyme. Thus, the clear bilateral symmetry of the echinoderm larva is established. The primary coelom anlage divides into two vesicles, which are pulled to the left and right of the intestine by their own pseudopodia. Each of the two vesicles extends posteriorly, forming an anterior and a posterior portion. The two posterior vesicles are the so-called somatocoels. From the left anterior vesicle a canal extends towards the dorsal side, where it communicates with the exterior through the future hydropore. The posterior portion of the left anterior vesicle forms the hydrocoel, which first becomes horseshoe-shaped and later ring-shaped and sends out five extensions, the future radial canals. The anterior portion of the left vesicle, called the axocoel, is connected with the hydrocoel by the so-called stone canal connecting the water-vascular system with the hydropore. The right anterior vesicle also divides into anterior and posterior portions but does not develop further and may even disappear completely (fig. 9.2). The left and right somatocoels constitute the definitive body cavities and are separated by dorsal and ventral mesenteries. The various groups of echinoderms show special features as additions to or alterations of this general scheme. (For further details see Hyman, 1955; Giudice, 1973; Hörstadius, 1973 and Davenport, 1979.)

Although cleavage follows a definite pattern, the echinoderm egg is among the most regulative or epigenetic animal egg types. A dramatic demonstration of this epigenetic character was recently provided by Dan-Sohkawa & Satoh (1978), who obtained complete dwarf embryos from isolated blastomeres of 2-, 4- and 8-cell stages and even from some blastomeres of the 16-cell stage of *Asterina gibbosa*, showing that essential differences between animal and vegetal blastomeres are not yet established at the 8-cell stage. The dwarf embryos developed synchronously with the controls and the gastrulae contained numbers of cells strictly proportional to the sizes of the isolated blastomeres.

Hörstadius (1973) gave an extensive review of the experimental data on

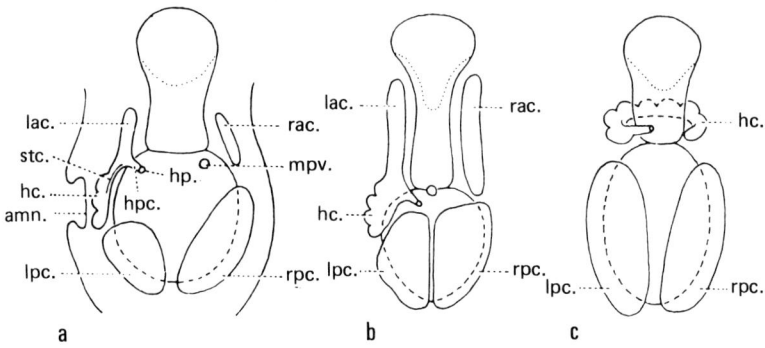

Fig. 9.2. (a) Diagram of development of coelomic cavities in older echinoid pluteus larva; (b) same stage of development in asteroid bipinnaria larva; (c) same stage of development in holothurian auricularia larva. amn. = amnion; hc. = hydrocoel; hp. = hydropore; hpc. = hydropore canal; lac. = left anterior coelom; lpc. = left posterior coelom; mpv. = madreporic vesicle; rac. = right anterior coelom; rpc. = right posterior coelom; stc. = stone canal. (After Hörstadius, 1973.)

which the so-called double gradient hypothesis is based, which was initially proposed by Runnström (1928) and later strongly supported by Hörstadius (1935). The hypothesis postulates the presence in the sea urchin egg of two gradients, an animal and a vegetal one, extending decrementally towards the opposite poles. Hörstadius (1927, 1928) demonstrated that the micromere-forming capacity is not restricted to the vegetal region of the embryo but extends decrementally to the middle of the subequatorial region. In his 1928 paper he also showed that animal and vegetal embryo halves form different larval types, the isolated animal halves showing an enlarged apical tuft and the vegetal halves an enlarged archenteron. The hypothesis received extensive corroboration from experiments involving recombination of different tiers of blastomeres of the 32- or 64-cell stages (Hörstadius, 1935), as well as from the results of treatment with animalising and vegetalising agents such as thiocyanate and lithium ions, respectively (Hörstadius, 1936a,b). Hutchins & Brandhorst (1979) have shown that, in *Strongylocentrotus*, commitment to vegetalised development as a result of Li^+ treatment occurs during the first five hours after fertilisation. However, they found no early changes in protein synthesis. Biochemical analysis of the sea urchin egg has led to the isolation of endogenous animalising and vegetalising agents, which have antagonistic effects on animal and vegetal half embryos (Josefsson & Hörstadius, 1969; Hörstadius & Josefsson, 1972; Hörstadius, 1973).

The primary axis of the echinoderm egg reflects the way in which the oocyte is attached to the ovarian wall, and is apparently firmly fixed in the architecture of the fertilised egg, since stratification of the egg contents by centrifugation at any angle to the primary axis does not affect the development of the egg, which always conforms to the original axis. The animal–vegetal polarity must, therefore, already be firmly established before fertilisation (Hörstadius, 1939).

The formation of four complete dwarf embryos from the isolated blastomeres of the 4-cell stage demonstrates that the dorso-ventral polarity cannot be firmly determined at that stage. This is supported by the observation that bilateral symmetrisation can be suppressed by a number of agents such as Li^+, absence of sulphate, changes in the Na^+, K^+, Ca^{2+} and Mg^{2+} concentrations of the external medium, and detergents, leading in all cases to a radially symmetrical embryo. The effect of these agents is restricted to the first nine hours of development, during which the embryo reaches the blastula stage (see Giudice, 1973; Hörstadius, 1973 and Davenport, 1979). Lallier (1978) showed that surface material located at the vegetal pole, which is responsible for the differentiation of the endomesodermal structures, is accessible to detergents only during the first few minutes after fertilisation. In our opinion an interesting parallel exists between the radialisation of the sea urchin embryo and that of the molluscan embryo (see chapter 10, p. 96).

The recombination experiments involving animal–vegetal tiers of blastomeres of the sea urchin embryo by Hörstadius and others have demonstrated that the micromeres, when combined with animal blastomeres representing presumptive ectoderm, can induce endodermal structures in the latter. This shows, on the one hand, that the mesodermal micromeres constitute the first specialisation of the echinoderm embryo and, on the other hand, that the mesoderm/ectoderm interaction is the first epigenetic interaction in development leading to the formation of the endoderm.

Origin of the germ cells

According to Lender & Delavault (1964), all echinoderms have an initial gonadal anlage in the form of a genital bud at one end of the hydroporal canal in the madreporal interradius. This initial anlage always disappears, however, and a second gonadal anlage is formed near the axial organ at the other end of the hydroporal canal. PGCs seem to be recognisable for the first time in the coelomic wall of the somatocoel, where the initial genital bud is formed. They are free-moving cells which migrate through the genital canals to the definitive gonadal anlagen.

Bruslé (1968a,b), Bruslé & Delavault (1968) and Delavault & Bruslé (1968), studying *Asterina gibbosa*, which functions either as a male or as a hermaphrodite (Delavault, 1961), concluded that there is no question of a 'germinative epithelium' consisting of somatic cells and undifferentiated generative cells (PGCs). Oogonia and spermatogonia arise directly as two separate stocks. A sharp distinction must, however, be made between the radio-sensitive germinal cells and the radio-resistant somatic cells (Bruslé, 1970).

After surgical extirpation of an entire interradius in *Asterina gibbosa*, regeneration begins with the formation of the aboral genital ring sinus, followed by the development of the genital canals and the formation of the definitive gonadal anlagen. PGCs migrate from the genital sinus towards the gonadal anlagen (Lender & Delavault, 1964; Lender, 1965). Huet (1965) destroyed the germinal tissue by X-irradiation of the entire animal, after which no gonad was reformed. He concluded that a distinct germ line exists in *Asterina gibbosa*. Later, it was found that regeneration of the gonad does take place when the aboral ring sinus is not irradiated. The absence of gonad formation after local irradiation of the aboral ring demonstrates that the latter constitutes the reserve of PGCs from which cells migrate to the regenerating gonad(s) (Huet, 1972, 1974) (fig. 9.3).

Kille (1942) stated that, in *Holothuria parvula*, reproducing asexually by means of transverse constriction, the caudal half, which lacks a gonad,

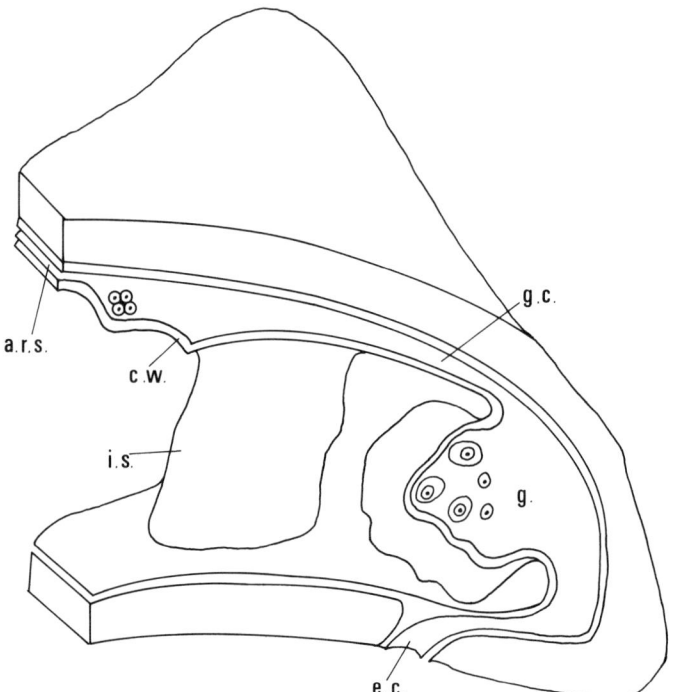

Fig. 9.3. Diagram of genital structures in interradius of body disc. of *Asterina gibbosa*. a.r.s. = aboral ring sinus; c.w. = coelomic wall; e.c. = ejaculatory canal; g. = gonad with germ cells; g.c. = genital canal; i.s. = interradial septum. (After Huet, 1974.)

remains sterile after regeneration. This, at first sight, seems to be in agreement with the idea of a distinct germ line. However, in our opinion, asexual reproduction without the reconstitution of germ cells in both fission products is a very unlikely situation, so that this observation certainly calls for confirmation. The same problem holds for animals which, as a normal defence mechanism, eviscerate their entire digestive tract including the gonad; such animals cannot be expected to remain sterile.

Conclusion

In the echinoderms the PGCs originate late in development. They arise in the wall of the somatocoelic cavities. There is thus a long initial phase of development during which no germ cells can be distinguished. Nothing is known of their actual mode of origin. Although some evidence pleads in favour of a distinct germ cell population once the germ cells are formed, the occurrence of asexual reproduction in various groups seems to be at variance with such a conclusion. Asexual reproduction certainly needs to be studied more closely with respect to germinal tissue in the different fission products. Our tentative conclusion is that the echinoderms show a mode of germ cell formation that is *intermediate* between the typically epigenetic and the typically preformistic modes.

10

Mollusca

General characterisation

Molluscs are coelomate Bilateralia without evidence of metamerism, with an initial bilateral symmetry which is often secondarily altered into an asymmetrical configuration. The soft body is often protected by a calcareous shell of one or more pieces, secreted by the mantle or pallium, which represents a fold of the body wall. The ventral body wall is adapted for locomotion in the form of a foot. The nervous system generally consists of a circum-esophageal ganglionic complex of three to five pairs of ganglia. The pharynx is commonly provided with a radula. The animals mostly have an open circulatory system with heart and gills, and possess metanephridia.

The phylum Mollusca is subdivided into seven classes: (1) The marine Aplacophora are bilaterally symmetrical, vermiform molluscs without head, mantle, foot, shell or nephridia. Their cuticle is provided with calcareous spicules. There is a straight digestive tract with a radula. The paired gonoducts open into the terminal part of the intestine or debouch independently. (2) The marine Polyplacophora are bilaterally symmetrical, dorsoventrally flattened molluscs with a head without eyes and tentacles, a mantle, a foot and external gills. The shell consists of eight successive pieces. There is a pharynx with radula, a pair of metanephridia, and a pair of separate gonoducts. (3) The deep-sea Monoplacophora are bilaterally symmetrical molluscs with head, foot, mantle and radula and with a bilateral univalve shell. They have serially repeated external gills, nephridia, muscles and nerve branches. (4) The marine, freshwater and terrestrial Gastropoda are torted or detorted, asymmetrical molluscs with a well-developed head with eyes and tentacles, with a generally spirally wound univalve shell, and with the ventral part of the foot flattened as a creeping sole. The Gastropoda are subdivided into the dioecious Prosobranchia, the hermaphroditic, marine Opistobranchia (with often reduced or wanting shell) and the hermaphroditic, freshwater and terrestrial Pulmonata (with or without shell and with the mantle transformed into a pulmonary sac). (5) The dioecious, marine Scaphopoda are torted, symmetrical molluscs without eyes, tentacles or gills, enclosed in a tusk-like shell open at both ends, with a tubular mantle and an elongated foot. (6) The dioecious or hermaphroditic Pelecypoda or bivalves are untorted, bilaterally symmetrical molluscs with a bilobed mantle, enclosed in a shell with two lateral valves hinged together mid-dorsally. The animals are without head, pharynx, jaws and radula and without tentacles, but possess paired gills, nephridia and gonads. (7) The dioecious, marine Cephalopoda are externally bilaterally symmetrical molluscs with a spiral, chambered shell, with a reduced shell embedded in the dorsal mantle, or without a shell. The animals are highly cephalised, with a large brain and large, highly organised eyes and other sense organs, with jaws and radula, and with the foot altered into a set of arms encircling the mouth and, except for the Nautiloidea, beset with suckers. They have a closed circulatory system with a well-developed heart. (See Hyman, 1967, vol. 6, pp. 1–792.)

Molluscs show only sexual reproduction; asexual reproduction and possible additional forms of propagation such as parthenogenetic development or polyembryony are completely unknown (see Beeman, 1977, and Berry, 1977). Regenerative capacity is restricted to the reconstruction of appendages such as tentacles, gills, mantle and foot. These features set the molluscs apart from many other invertebrate groups. Although some of the more primitive molluscs are not very highly organised and others, like the subterranean bivalves, are strongly adapted to a particular habitat, the cephalopods are among the most highly evolved invertebrates, with a level of organisation comparable to that of the vertebrates.

Early molluscan development falls into two categories, the holoblastic, spiralian type characteristic of all molluscs except the cephalopods, and the meroblastic, discoidal or superficial type characteristic for the cephalopods (Fioroni, 1966, 1967, 1971, 1974, 1977). Except for the latter, which must be treated separately, the development of the other groups of molluscs shows so many similar features that these groups can be discussed together. The descriptive and experimental literature on molluscan development is so extensive that we must restrict ourselves to the major facts.

The literature on the development of the spiralian molluscs shows two rather contradictory aspects: on the one hand, the careful description of normal development, with accurate establishment of cell lineages as the expression of a strictly determinate type of cleavage; on the other hand extensive experimental analysis demonstrating the existence of rather pronounced regulative or epigenetic features of development. This remarkable paradox finds its explanation in the fact that the Spiralia have of old been considered as the prototype of mosaic development. When the concept of strict mosaic development began to give way in other animal groups, the molluscs served more or less the last 'test case' for preformistic development.

Embryonic development of the spiralian molluscs

Normal embryogenesis of the molluscs is characterised by strictly determinate spiral cleavage. After two meridional cleavages, leading to a four-cell stage, at least four quartets of micromeres are split off on the animal side of the four macromeres, alternatively in dexiotropic and laeotropic directions. Molluscs with yolk-poor eggs usually show equal division during the first two cleavages, so that the subsequent micromere formation leads to a radially symmetrical configuration in the blastula. Bilateral symmetry appears relatively late in development in the archaeogastropods (primitive Gastropoda) and the chitons (Polyplacophora) (64- to 72-cell stage) (Wada, 1968), but much earlier in the pulmonates, where it manifests itself at the 24- to 28-cell stage in the formation of an enlarged 4d micromere. In other molluscs the first cleavage is unequal leading to the smaller AB and a larger CD blastomere. The latter again divides unequally into a smaller C and a larger D blastomere. Here, bilateral symmetry is established very early. It is evident that, in all these cases, bilateral symmetry is achieved by the particular configuration of the D quadrant (for references see below).

The very rigid cleavage pattern in molluscan development has led to extensive studies of normal development and cell lineage (see Dautert, 1929 in *Paludina*; Crofts, 1937 in *Haliotus*; Peltrera, 1940 in *Aplysia*; Clement, 1952 in *Ilyanassa*; Raven, 1946 and Verdonk, 1965 in *Limnaea*; Rattenbury & Berg, 1954 in *Mytilus*; Ghose, 1962 in *Achatina*; D'Asaro, 1966 in *Thais* and 1969 in *Bursa* and *Distorsio*; Guerrier, 1970a,d in *Limax* and 1970c in *Pholas* and *Spicula*; Van Dongen & Geilenkirchen, 1974 in *Dentalium*; Moor, 1977 in *Bradybaena* and Van den Biggelaar & Guerrier, 1979 in *Patella*).

The rigid cleavage pattern of the molluscan egg has led Raven & Bezem (1971, 1972, 1973), Raven, Bezem & Baretta-Bekker (1973), and Bezem & Raven (1975) to develop a computer simulation programme for the cleavage pattern in *Limnaea stagnalis*, which also incorporates mutual cell flattening and differentiation of larval structures (see also the reference to a similar programme for the annelid *Podarke*, chapter 8, p. 69).

Unfortunately, no fate maps have been prepared for the various groups of molluscs. This is the more regrettable because embryonic development in the various groups shows different adaptations as well as different degrees of suppression of larval structures. Important differences seem to exist in cell lineage among the various groups. In the great majority of molluscs the D quadrant takes the leading role in the organisation of the embryo by establishing bilateral symmetry and by determining the development of several adult organ systems, so much so that Cather (1971) compares the D quadrant of the molluscan egg to the primary organiser in the vertebrates. However, in the prosobranch *Bithynia* this function seems to be equally divided between the C and D quadrants (Verdonk & Cather, 1973; Cather, Verdonk & Dohmen, 1976). This evident discrepancy does not seem to affect the supposed overall configuration of the fate map, so that the latter may actually be much more suitable for comparison than the cell lineage (cf. Anderson's (1973) comparison of annelid and arthropod development).

All molluscs, with the exception of the cephalopods, undergo metamorphosis, but the extent to which larval structures are replaced by adult structures varies considerably. Free-swimming trochophore, or rather preveliger, larvae (Fioroni, 1971, 1977) are found among the Aplacophora, Polyplacophora, Scaphopoda, Pelecypoda and Archaeogastropoda (Fioroni, 1971). In many molluscs with more yolky eggs intracapsular development leads directly to a so-called veliger larva. Intracapsular nutrition of the embryo is common among the Gastropoda, leading, on the one hand, to special adaptations like micropinocytosis and albumen sac formation and, on the other hand, to a markedly shortened larval development, so that metamorphosis is virtually restricted to the reduction of the transitory nutritive organs (Fioroni, 1971).

Ooplasmic segregation has been studied by Raven (1946, 1964, 1966, 1967, 1970, 1974, 1975, 1976) in the pulmonate *Limnaea stagnalis*, which shows equal cleavage. Shortly after oviposition a special vegetal pole plasm moves towards the animal pole, where it fuses with the animal pole plasm that appears after the completion of the maturation divisions. The resulting plasm passes into the successive quartets of micromeres and is finally distributed in a simple animal–vegetal gradient in the blastula. Raven (1967, 1974) also observed the presence of subcortical patches of cytoplasm, which are imprinted upon the oocyte during oogenesis by the nuclei of adjacent follicle cells. During development the patches move towards the vegetal pole region, forming the so-called 'ectosomes' in the 4A–4D macromeres. Inside the latter they subsequently move towards the distal extremity of the cell (see also Dohmen & Van der Mast, 1978). As far as can be seen, neither the animal and vegetal pole plasms nor the subcortical patches show a distinct orientation foreshadowing dorso-ventral polarity (see the review by Raven, 1970).

In *Limnaea*, cleavage is arrested for a couple of hours at the 24-cell stage. This stage is, moreover, characterised by the disappearance of the blastocoelic cavity, causing a direct contact of two of the macromeres, the vegetal cross-furrow macromeres, with the animal micromeres. Soon one of the macromeres, the so-called D macromere, shifts to a central position and establishes intimate contact with the micromeres of the first quartet (Van den Biggelaar, 1976) (fig. 10.1). From that moment on dorso-ventral polarity is evident; it is further accentuated by the internal splitting-off of the 4d micromere. Dorso-ventrality is subsequently expressed in a divergent division of the basal cell of one of the arms (the dorsal arm) of the so-called

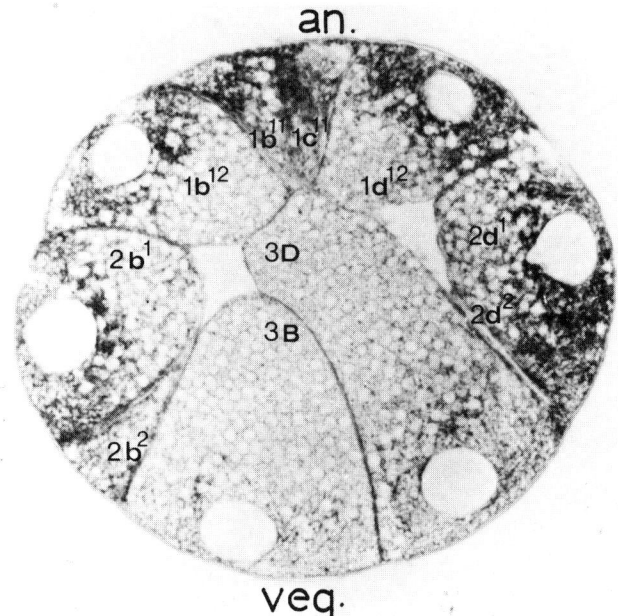

Fig. 10.1. Sagittal section through 32-cell stage of *Patella vulgata*, with 3D blastomere in central position bordering micromeres of first quartet. an. = animal pole; veg. = vegetal pole. (Courtesy J. A. M. Van den Biggelaar, 1977.)

molluscan cross, formed by the first quartet and cells of the second quartet of micromeres (Verdonk, 1965) (fig. 10.2). The 4d cell divides into the two mesoblasts M, which subsequently give rise to the two bands of mesodermal cells from which the internal organs develop.

Depending on the amount of yolk, gastrulation is by invagination or by epiboly. Further development is characterised by the formation of two types of cells in each of the germ layers: the usually large and early-differentiating larval cells, and the still undifferentiated cellular anlagen of the adult organs. Raven (1963b) states that ooplasmic segregation is primarily responsible for the early differentiation of the larval structures. The larval organs of the pre-veliger larva are presented by (1) the apical tuft of cilia, (2) the ciliated prototroch, which divides the larva into pre-trochal and post-trochal parts, respectively formed by the descendants of the first quartet of micromeres and by those of the second and third quartets, and (3) the albumen sac in the case of intracapsular nutrition. The prototroch later gives rise to the ciliated velum of the veliger larva. The adult organ anlagen are represented by the symmetrically situated pretrochal head anlagen (in the form of two cephalic plates from which the cerebral ganglia, eyes and antennae develop), the dorsal shell gland, the ventral foot, the stomodaeum, and the internal mesodermal organs (Raven, 1964, 1970, 1975).

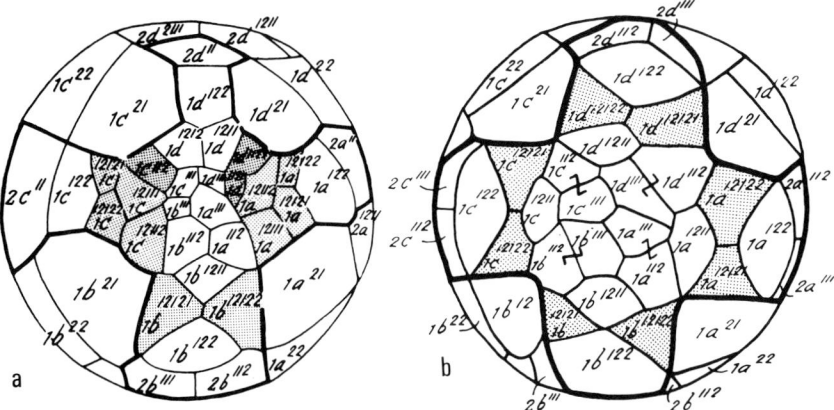

Fig. 10.2.(a) Cell lineage of late 'molluscan cross' stage of *Limnaea stagnalis* (24 h after first cleavage), with basal and inner median cells divided in lateral arms of the cross; dorsal arm clearly distinguishable from the other three. (b) Similar stage of Li$^+$-treated embryo (30 h after first cleavage), with inner median cells of all four arms divided; identical situation in all arms. Stippled areas = cells contributing to cephalic plates. (After Verdonk, 1968a.)

Polar lobe formation and its possible role in development

In a number of molluscs the first cleavage is essentially equal but characterised by the formation of a polar lobe on the vegetal side of the egg, which subsequently fuses with one of the two blastomeres. The same process occurs during the second division, leading to a four-cell stage with one enlarged blastomere, the D blastomere. Thus, polar lobe formation determines the dorso-ventrality of the embryo. Polar lobe formation is not restricted to a particular group of molluscs but occurs in the prosobranchs *Bithynia*, *Nassarius*, *Ilyanassa*, *Ocinebra* and *Urosalpinx*, the bivalves *Pecten*, *Mytilus*, *Modiolaria* and *Ostrea*, and the scaphopod *Dentalium* (Raven, 1966; Conrad, 1973). The polar lobe may be of the same size as each of the first two blastomeres, leading to the so-called trifoil stage, like in *Ilyanassa* (Clement, 1952, 1956, 1960, 1962, 1963, 1967, 1968, 1976), or it may comprise only a very small fraction of the original egg, for example only one per cent in *Bithynia* (Cather & Verdonk, 1974).

The external surface of the polar lobe region of the egg is clearly distinct from the rest of the egg surface, a fact which is most evident in SEM pictures. The polar lobe of *Ilyanassa* lacks microvilli, that of *Nassarius* has many microvilli, that of *Buccinum* has surface ridges, while that of *Crepidula* shows twisted folds (Crowell, 1964; Dohmen & Lok, 1975; Dohmen & Van der Mey, 1977). Conrad & Williams (1974a,b) observed in the polar lobe of *Ilyanassa* a circular subcortical band of microfilaments, which disappears after cytochalasin B treatment, and microtubules in the non-cortical cytoplasm, which dissolve upon colchicine treatment. Polar lobe formation apparently depends on predetermined properties of the egg cortex in the polar lobe region, since the polar lobe in *Dentalium* cannot be displaced by centrifugation either in 'normal' or in 'forced' egg position, in which latter case the contents of the polar lobe are drastically changed (Verdonk, 1968b; Verdonk, Geilenkirchen & Timmermans, 1971).

Berg & Kato (1959) found that the polar lobe of *Ilyanassa* contains non-nucleic acid material soluble in perchloric acid and with a UV absorption spectrum characteristic for purine and pyrimidine compounds. Collier (1960a,b) found less RNA in the polar lobe and in the CD blastomere than in the AB blastomere, but more total, acid-soluble and phospholipid phosphorus, which is probably localised in the yolk component. Davidson (1976) suggests that the polar lobe may contain maternal mRNA. Dohmen & Verdonk (1979), in addition, suggest that polar lobe morphogens may control development by influencing differential transcription in the embryo. Crowell (1964) found many yolk platelets in the polar lobe of *Ilyanassa*, as well as a restricted amount of E.R. and vesicles with a double membrane, which may contain RNA. Conrad (1973) stated that, in *Ilyanassa*, the polar lobe only contains normal cell organelles such as yolk granules, mitochondria, ribosomes, glycogen and multimembranous vesicles, though possibly in higher quantity. On the other hand, Dohmen & Verdonk (1974) and Cather, Verdonk & Dohmen (1976) found a cap-shaped mass of small vesicles with an electron-dense content in the tiny polar lobe of *Bithynia*, which they called the 'vegetal body' (fig. 10.3), while Dohmen & Lok (1975) and Dohmen & Verdonk (1979) found that the polar lobes of *Crepidula* and *Buccinum*, respectively, apart from small yolk platelets, mitochondria and multivesicular bodies contain aggregates of vesicles filled with electron-dense material intermingled with granular bodies (fig. 10.4). In *Buccinum* the latter resemble the polar granules of insects, in *Crepidula* the germinal granules of anuran amphibians. Unfortunately, nothing is known of their fate, particularly in connection with germ cell formation.

Although Raven (1964) and Verdonk (1968c) assumed a predetermined relationship of the polar lobe with the CD blastomere during first cleavage and with the D blastomere during second cleavage, Schleip (1925) had suggested, from the behaviour of dispermic *Dentalium* eggs, that the fusion of the first polar lobe with one of the two blastomeres is an arbitrary event. Guerrier (1970a) concluded from his experiments involving compression of *Limax maximus* eggs parallel to the primary axis that dorso-ventral polarity has not yet been determined in the uncleaved egg. He later confirmed this for eggs of *Pholas* and *Spicula* (Guerrier, 1970c). Guerrier et al. (1978) treated uncleaved eggs of *Dentalium* with cytochalasin B and found that this led to an equal rather than an unequal first division and to subsequent

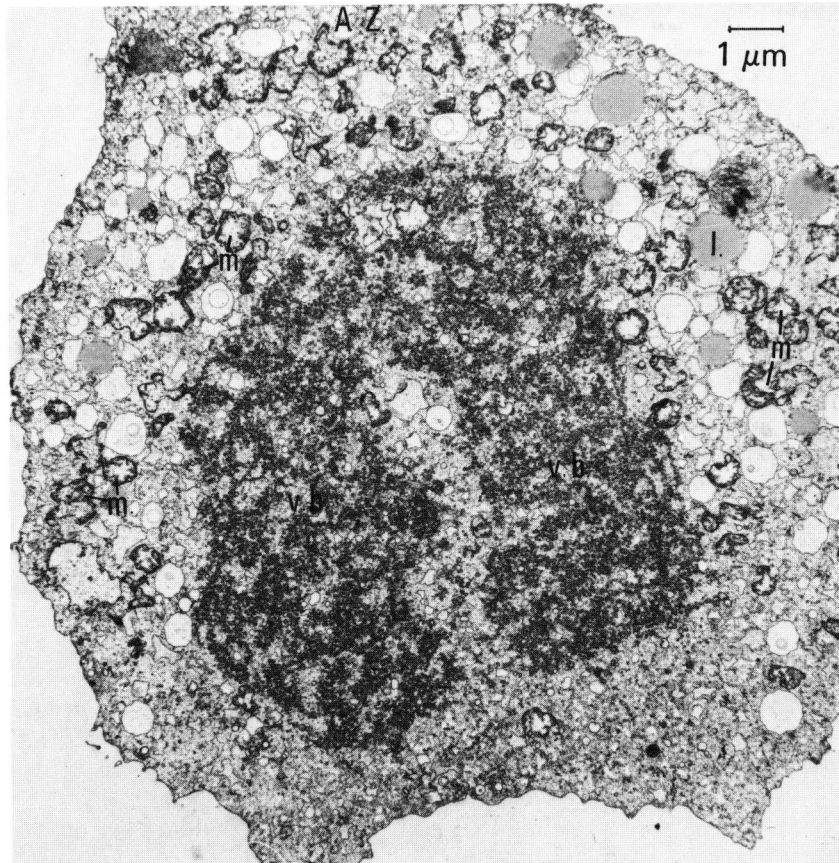

Fig. 10.3. Electron micrograph of first polar lobe of egg of *Bithynia tentaculata*, showing cap-shaped vegetal body (v.b.). l. = lipid droplet; m. = mitochondria; A.Z. = attachment zone of polar lobe to lower side of egg. (Courtesy J. N. Cather, N. H. Verdonk & M. R. Dohmen, 1976.)

duplication of the lobe-dependent structures. They concluded that the establishment of dorso-ventral polarity is due to the arbitrary fusion of the polar lobe with one of the first two blastomeres. (See Guerrier, 1970d, 1971, and Van den Biggelaar, 1976.)

Many authors have studied the effect of removal of the polar lobe. Clement (1952) was the first to remove the polar lobe of *Ilyanassa*. He observed the disappearance of the special features of the D quadrant, including 4d cell formation. This means a radialisation of the embryo. After removal of the polar lobe in *Dentalium* several adult structures were absent, viz. the foot, shell, eyes, etc. (see Raven, 1964; Van Dongen & Geilenkirchen, 1975; Van Dongen, 1976). *Ilyanassa* forms polar lobes during the first,

94 *Mollusca*

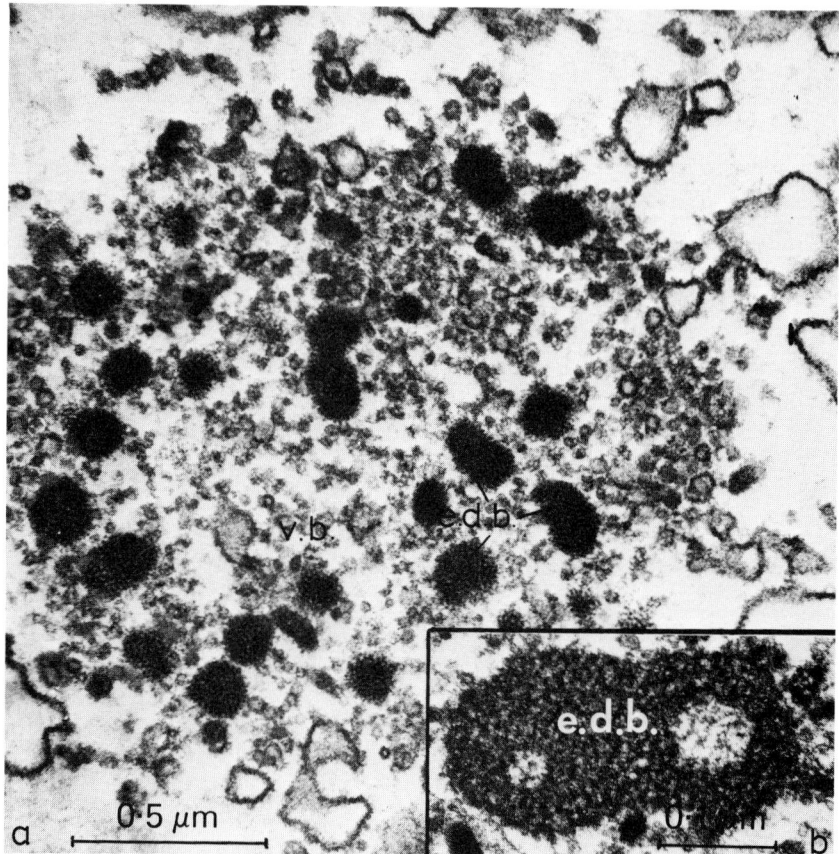

Fig. 10.4. (a) Detail of complex aggregate found in polar lobe of *Crepidula formicata*, with electron-dense bodies (e.d.b.) in vegetal body (v.b.). (b) Inset: higher magnification of electron-dense body, resembling polar granule of insects. (Courtesy M. R. Dohmen & D. Lok, 1975.)

second and third cleavages. The material of the polar lobe ends up chiefly in the 4d micromere. The isolated blastomeres A B and A, B or C form only a ciliated vesicle with muscle tissue and a ciliated enteron, whereas C D or D blastomeres give rise to more or less complete larvae with foot, shell and eyes (Clement, 1956). Removal of D or 1D leads to the absence of the shell, foot, eyes, heart and intestine; after removal of 2D, shell formation occurs occasionally; after removal of 3D, velum, shell, eyes and foot develop, but no heart and intestine, while removal of 4D does not affect development at all (Clement, 1960, 1962). Individual micromeres of the first three quartets and the 4d micromere play separate distinctive roles in development (Clement, 1963, 1967, 1976). These observations have recently been confirmed for *Dentalium* by Cather & Verdonk (1979). After separation of *Ilyanassa* eggs into animal and vegetal halves by means of strong centrifugal forces, only nucleated vegetal halves formed lobe-dependent structures (Clement, 1968). Clement assumes that, during cleavage, factors successively pass from the polar lobe into the various quartets of micromeres through cytoplasmic segregation, those ending up in the second quartet leading to weak velum formation, those in the third quartet to velum, eye, foot and shell formation, and those in the 4d micromere to heart and intestine formation. According to Raven (1964) larval apical tuft formation requires factors residing in the first polar lobe, while the development of the post-trochal region requires factors residing in the first and second polar lobes. These results have in general been confirmed by Rattenbury & Berg (1954) in *Mytilus*, by Cather (1967) and Atkinson (1971) in *Ilyanassa*, by Geilenkirchen, Verdonk & Timmermans (1970) in *Dentalium* and by Cather & Verdonk (1974) in *Bithynia*. Collier (1961) found that lobeless embryos of *Ilyanassa* are capable of only reduced protein synthesis (see Collier, 1965). In 1975 he observed a delay in RNA accumulation and a decrease in the rate of DNA synthesis in lobeless *Ilyanassa* embryos.

Further experimental analysis

Studies on the effect of centrifugation have shown that stratification of the uncleaved egg of *Limnaea stagnalis* does not interfere with normal development, due to the re-establishment of a more-or-less normal distribution of the cellular inclusions (Raven, 1948). Raven concluded that the directing forces of this redistribution must be localised in the egg cortex, the only structure which is not visibly affected by the centrifugal force (Raven, 1963a; see also Elbers, 1959). The animal–vegetal polarity of the egg also seems to be fixed in the egg cortex, since only very strong centrifugal forces of up to 100 000 g can change the primary polarity (Pease, 1940; Peltrera, 1940). Depending upon the time of treatment, centrifugation and compression of *Limax*, *Pholas* and *Sabellaria* eggs may affect animal–vegetal polarity. Treatment is effective before but no longer after first cleavage (Guerrier, 1968). From these experiments Guerrier concluded that the primary axis is definitively determined only after completion of the second maturation division. In eggs of *Limnaea stagnalis* centrifuged immediately after oviposition an animal pole plasm appears, but this no longer occurs when the egg is centrifuged two hours later (Raven & Van der Wal, 1964). Raven concluded that ooplasmic segregation is guided by some mosaic pattern in the egg cortex. Chemical agents like lithium ions (Li^+) strongly affect development. The effect is stage dependent. Lithium ions may interfere with animal and vegetal pole plasm formation and, according to Elbers (1959, 1969), acts primarily on the egg surface. Under the influence of Li^+ typical head malformations may occur, leading to cyclopia and acephalic development. Stronger effects lead to exogastrulation. Raven (1964) concluded that Li^+ acts upon a cortical gradient field having a maximum at the animal pole. According to him, the egg cortex is therefore characterised by both continuous and discontinuous variations in structure, which constitute 'blueprint' information, in contrast to the 'executive' information present in the nucleus and cytoplasm.

The most typical effect of Li^+ is the prevention of the bilateral symmetrisation or dorsalisation of the egg, the resulting embryo (which usually does not gastrulate) remaining radially symmetrical (Verdonk, 1965, 1968a) (see fig. 10.2, p. 91). Raven ascribes this to interference with normal cytoplasmic segregation, Verdonk to a change in the cleavage pattern, so that a distinct dorsal arm of the molluscan cross is not formed. Van den Biggelaar (1976) has recently put forward the hypothesis that dorsalisation of the *Limnaea* egg is purely a question of the attainment of an equilibrium position by the macromeres, one of them, then called the D macromere, achieving the central position (see fig. 10.1, p. 90). A similar situation develops in *Patella vulgata* at the 32-cell stage, when one of the four macromeres acquires the central position (Van den Biggelaar, 1977). Removal of one of the two vegetal cross-furrow macromeres in *Limnaea* does not interfere with normal development, since the remaining vegetal cross-furrow macromere automatically takes up the central position and gives off a 4d micromere. Van den Biggelaar assumes that the interaction between the central macromere and the micromeres of the first quartet induces the former to become the D blastomere. The observation by Dohmen & Van de Mast (1978) of the presence of annular gap junctions between the D macromere and the micromeres at the 24-cell stage in *Limnaea* may indeed be an indication of such an inductive interaction. The 'ectosomes' disperse in the D macromere, but not in the A, B and C macromeres (Raven, 1970). Guerrier & Van den Biggelaar (1979) conclude from experiments in which the orientation of the first cleavage spindle was affected, that dorsoventrality does not depend on the cytoplasmic composition of the blastomeres but on the inclusion of the vegetal cortex in the CD and D blastomeres successively.

Raven (1952) found that, in Li^+-treated *Limnaea* eggs, the shell gland develops only when and where the archenteron makes contact with the ectoderm. Hess (1956a,b) observed a delay in shell gland formation in *Bithynia* after experimentally delayed invagination. Cather (1967) found that in *Ilyanassa*, isolated ectoderm never formed a shell, but a recombination of ectoderm and one arbitrary macromere formed a small shell gland. These experiments demonstrate conclusively that the shell gland develops epigenetically under the inductive action of the endodermal archenteron.

Another argument which pleads against a strictly mosaic development of the molluscan egg is that instances of regulation have been reported. Hess (1957) observed regulation phenomena in 1/2 and 2/4 embryos of *Limnaea*. Morrill & Gottesman (1960) observed the development of normal larvae from one of the first two blastomeres in *Limnaea* in 2.5 per cent of cases. Although the percentage is low, the experiment nevertheless demonstrates the possibility of bilateral regulation. Hess (1962) reported almost normal development of 1/2 and 2/4 blastomeres in *Bithynia*. Morrill (1963) observed the formation of an enlarged polar body comprising 20 to 40 per cent of the egg volume after centrifugation. The remaining portion of the egg formed normal larvae in 37 per cent of cases, demonstrating the considerable regulative capacity of the uncleaved egg. In *Limnaea palustris*, Morrill, Blair & Larsen (1973) observed complete regulation in 1/2, 3/4 and 2/4 embryos, as well as upon removal of one to three micromeres of the first quartet, one micromere of the second quartet, or one macromere. Only 1/2 eggs cleaved as normal eggs, the other partial embryos continued the original cleavage pattern. According to Morrill, the regulation of partial embryos depends on a successful rearrangement of the remaining cells with respect to the D macromere before the 24-cell stage, indicating the occurrence of inductive interactions at that stage. These observations markedly restrict the validity of Raven's concept of 'blueprint' information localised in the egg cortex.

Although cleavage is strictly determinate, it must be concluded that molluscan development is, to a considerable extent, epigenetic in character. According to Guerrier (1970d, 1971) three factors are involved in the organisation of morphogenetic processes in the molluscan egg: (1) the initial structure of the egg, which is under direct control of the maternal genome, (2) the egg's animal–vegetal polarity, which orients the processes leading to the definitive stabilisation of the cortical structures, and (3) the orientation of the first two cleavage divisions (controlled by the former two factors), which leads to the establishment of dorso-ventral polarity. Normal development, in which the successive processes are rigorously chained up, gives the impression of a highly preformistic process, but in reality it is largely epigenetic in character. Guerrier & Van den Biggelaar (1979) come to the conclusion that the cell membrane is not a mosaic structure, but acts as a dynamic relay system for epigenetic signals.

Embryonic development of the Cephalopoda

Sacarrão (1962), Mangold-Wirz & Fioroni (1970) and Fioroni (1974) emphasised the separate position of the Cephalopoda in the phylum of the Mollusca. Mangold-Wirz & Fioroni even proposed to consider the cephalopods as a separate subphylum rather than a class (see also Boletzky, 1978b). The embryonic development of the Nautiloidea being completely unknown, only the development of the Octopoda can be discussed here. (See the reviews by Arnold, 1971; Cather, 1971; Arnold & Williams-Arnold, 1977; Haven, 1977 and Wells & Wells, 1977.)

The octopods are characterised by direct development into a miniature adult. The unfertilised egg has a thin layer of cytoplasm surrounding the entire yolk mass. During the first three hours after fertilisation a blastodisc is formed by the concentration of this cytoplasm around the animal pole of the egg. The first and second polar bodies are formed after fertilisation. Cleavage is superficial or meroblastic and is restricted to the blastodisc. The cleavage furrows run from the centre outwards (Arnold, 1961). According to Arnold (1968b), the third cleavage is slightly asynchronous, the 'anterior' (probably meaning dorsal) furrow appearing before the 'posterior' (probably meaning ventral) one. The animal pole of the egg corresponds to the future posterior end of the embryo, the vegetal pole to the oral yolk sac protruding from the future mouth. During cleavage the central blastomeres become separated from the yolk first. The outer blastomeres, which have no definite peripheral boundary, are called blastocones; they give off cells to the central mass and supposedly also supply nuclei to the thin layer of cytoplasm surrounding the yolk and forming the syncytial yolk epithelium. De Leo (1972) has tried to establish a cell lineage for the central cell mass up to the 64-cell stage. The initially single-layered blastoderm becomes double-layered by the internal splitting-off of cells from the peripheral blastocones, thus forming a horseshoe-shaped (*Loligo*) or a ring-shaped (*Octopus*) intermediate layer, which subsequently extends towards the centre of the blastoderm. This layer probably segregates ultimately into an endodermal and mesodermal layer (Wells & Wells, 1977). The yolk syncytium likewise extends underneath the blastoderm (Arnold, 1961). According to Arnold (1971) the blastoderm extends peripherally by marginal divisions of the blastocones. The double-layered blastoderm expands towards the vegetal pole. This expansion stops at different levels in different species, that is above, at or below the equator of the egg. The remaining portion of the egg becomes covered only by extra-embryonic ectoderm and syncytial yolk epithelium and forms the yolk sac. In its wall, blood cells and contractile elements develop (see fig. 10.5a–d).

Organogenesis occurs entirely in the multi-layered blastoderm, consisting of an outer ectodermal and an intermediate endo-mesodermal layer lying on top of the syncytial yolk epithelium (Arnold, 1971; Wells & Wells, 1977). The anlagen of the arms appear near the distal boundary of the blastoderm, while shell gland and mantle anlagen arise at the apical end of the embryo, near the original animal pole. The mantle initially is a ring-like structure, which grows out distally and ultimately covers the entire embryo except the head region. The eye anlagen, which go through a complex morphogenesis that cannot be discussed here, appear as thickenings on either side of the

Embryonic development of the Cephalopoda 99

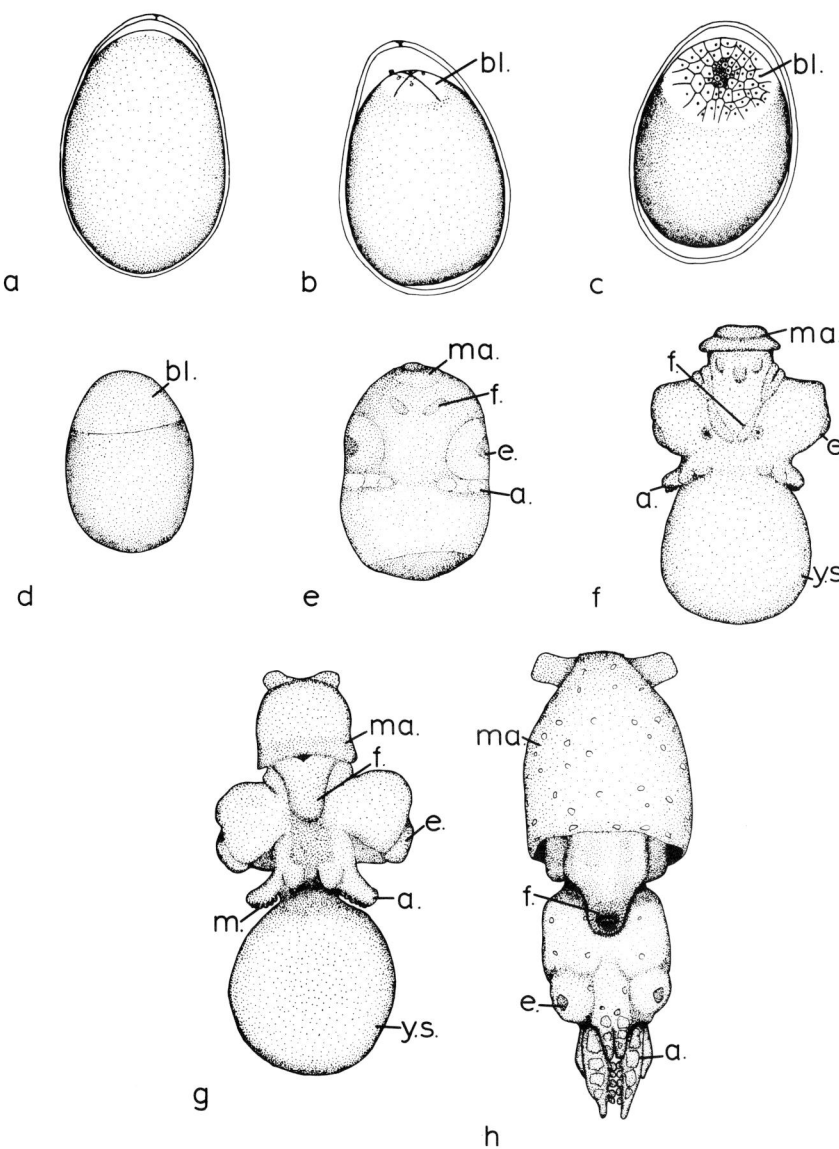

Fig. 10.5. Stages 1 (a), 5 (b), 9 (c), 13 (d), 17 (e), 21 (f), 25 (g) and 30 (h) of Normal Table of *Loligo pealii* (Arnold, 1965a), showing development of blastodisc (bl.), and later of eyes (e.), arms (a.), mantle (ma.), funnel (f.) and external yolk sac (y.s.). (Reproduced from original drawings, courtesy J. M. Arnold, 1965a.)

embryo. The funnel anlage, which is formed by fusion of anterior and posterior folds, is situated above (apically to) the eye anlagen. The gills form median to the posterior funnel (see fig. 10.5e–h). A branchial heart develops at the base of each of the gills. The stomodaeum is formed between the eye anlagen on the dorsal side. The midgut arises independently as a bipartite tube surrounding the internal portion of the yolk sac. The anus is formed late on the ventral surface, where the midgut makes contact with the ectoderm (Boletzky, 1978a). The brain is formed by fusion of a number of ganglia situated around the esophagus. Two otocyst anlagen sink in medial to the eye anlagen. When the main organ anlagen have appeared a constriction develops, separating the oral yolk sac from the organogenetic region of the embryo which surrounds the internal portion of the yolk sac (see fig. 10.5g,h). The function of the branchial hearts is taken over by the anlage of the systemic heart. A single coelomic cavity is formed, containing the anlagen of the kidney and the gonads. Arnold (1965a) and Lemaire (1970) have prepared normal tables of *Loligo pealii* and *Sepia officinalis*, respectively, each consisting of 30 stages.

According to Arnold (1963, 1965b), the outer layer of the blastoderm has no developmental potentialities, since organogenesis depends on influences from the yolk syncytium. Arnold (1968a) observed that constriction or UV irradiation of portions of the egg surface beyond the boundary of the expanding blastoderm leads to specific local defects in the future embryo. In 1963 he had concluded that the yolk syncytium contains a morphogenetic map of pre-programmed information, inducing the various organ systems at particular sites in the overlying blastoderm. As Raven had in his work on *Limnaea* (1964), Arnold (1968a) assumed that the yolk syncytium acquires its morphogenetic pattern from a prepattern in the egg cortex. Arnold & Williams-Arnold (1974) applied cytochalasin B at various stages of development and concluded that the drug interferes with the streaming pattern of the expanding blastoderm, preventing contact between the blastoderm and specific regions of the yolk syncytium, thus causing absence or defective development of particular organs (Arnold & Williams-Arnold, 1976). In our opinion, such reasoning is rather dubious; undoubtedly, the egg cortex will be incorporated into the outer, embryonic and extra-embryonic ectodermal layer when the blastoderm expands by peripheral cell division, and not into the underlying syncytial yolk epithelium.

Marthy (1972) obtained dwarf embryos from the blastodermal cap of the *Loligo* egg at stage III (Naef, 1928), which is comparable with about stage 13 (Arnold, 1965a). Marthy (1973) showed that the syncytial layer does not induce the eye anlagen in the overlying blastoderm, as claimed by Arnold (1963). Later, Marthy (1975, 1976) obtained complete dwarf embryos after ligation at 40 per cent of the egg axis in *Loligo* and *Octopus*, demonstrating that Arnold's interpretation of cephalopod organogenesis cannot be correct. Recently, Marthy (1977, 1978) demonstrated that ablation of mesentoderm cells causes the absence of a whole complex of organ rudiments at corresponding sites; consequently, it may, in fact, be the mesentoblast that plays the leading role in organogenesis, acting as the inductor for particular organ rudiments in the ectodermal layer. This suggests that the mesoderm serves as the primary organiser in cephalopod development, much as the dorsal blastoporal lip in the amphibians. The yolk syncytium then serves only to maintain the contact between the blastoderm and the yolk as the food source, and at the same time serves as a substrate for the expanding blastoderm (Marthy, 1976). Although the experimental evidence is still rather scanty, Marthy's postulate seems much more plausible than Arnold's interpretation.

Origin of the germ cells and gonad formation in the molluscs

In addition to a number of more-or-less consistent observations, the literature on the origin of the germ cells in the *spiralian molluscs* contains two markedly contradictory findings. Woods (1931, 1932) described an early segregation of germ cells in the lamellibranch *Sphaerium stricticum*, where a pair of PGCs would be set aside in the third division of the paired M cells, which are the direct descendants of the 4d micromere. (See the review by Tardy, 1970, concerning the classical idea of germ line origin from the 4d micromere.) The cells in question would divide once more before entering an inactive phase, which would last until the beginning of gonad formation. He even stated in 1932 that the oocyte is characterised by a concentration of granular mitochondria which would act as a germ cell determinant. This feature was also seen in the PGCs but disappeared during gonad formation. These observations have never been confirmed, however, so that it seems questionable whether the interpretation of his histological material was correct. The other finding is that of Tardy (1967), who claimed that, in the Aeolidiidae (nudibranch gastropods), gonads may regenerate from the wall of the hermaphroditic canal and the seminal vesicle after extirpation of the functional gonads. He assumed that the germ cells of the regenerated gonad originated from dedifferentiated or still undifferentiated cells of the gonadal ducts, but did not even exclude the possibility of colonisation of the gonads by migrating totipotent cells. Lubet, Herlin-Houtteville & Mathieu (1976), studying the annual cycle in the pelecypode molluscs, and Hogg & Wijdenes (1979), analysing gonadal regeneration in the pulmonate *Deroceras*, confirmed Tardy's observations. However, Hogg & Wijdenes state that regeneration occurs only as long as the hermaphroditic duct is in an immature state and consists of pluripotent gonadal stem cells. Notwithstanding the above findings, it is striking that, in all recent reviews, the origin of the germ cells in the molluscs is stated to be virtually unknown (Beeman, 1977; Berry, 1977; Webber, 1977).

Ghose (1963) localised the gonadal anlage in the pulmonate *Achatina fulica* as a portion of the mesodermal mass on the right side of the hindgut, the remainder giving rise to the pericardium, heart and kidney. Luchtel (1972*a,b*) stated that, in the pulmonates *Arion* and *Deroceras*, the embryonic gonads are of ectodermal origin and contain germinal and non-germinal cells. At hatching, the germinal cells differentiate into oogonia and spermatogonia, the non-germinal cells into follicle cells and sertoli cells. Luchtel found that the PGCs contain large granular cytoplasmic bodies of unknown nature. Guyard (1970) found that, in the pulmonate *Helix aspersa*, the gonads are formed in between the peritoneal complex and the intestine in the first week after hatching, when torsion and shell formation are completed. Similar observations were made by Guyomarc'h-Cousin (1976) in the prosobranch *Littorina saxatilis*. Brisson & Regondaud (1971) described as PGCs two to four voluminous cells found in the advanced trochophore stage of several pulmonate gastropods. These cells, which are situated on the ventral side, adjacent to the reno-pericardial mesoderm, have abundant cytoplasm and a large nucleus with a well-developed nucleolus and dispersed chromatin. A similar observation was made by Brisson (1973) in the pulmonate *Acroloxus*; the cells in question have a highly lobulated nucleus, finely dispersed chromatin, and cytoplasm with numerous mitochondria, a restricted number of dictyosomes, and poorly developed endoplasmic reticulum. Brisson & Besse (1975) identified the embryonic gonad anlage of *Limnaea stagnalis* in an advanced trochophore larva, claiming that it is a proliferation of ectodermal cells. Brisson & Regondaud (1977) found a similar situation in the pulmonate *Acroloxus*, but described the PGCs as mesodermal cells. Grifford (1977) claimed that the gonad of the prosobranch *Viviparus*, which consists of germinal and non-germinal cells, arises from multiplying pericardial cells. There is an observation by Brisson & Regondaud (1971) which may throw some light on some of the contradictory statements mentioned above. These authors describe gonad formation in the pulmonates as occurring from a special group of mesodermal cells giving rise to the hermaphroditic gonadal anlage (the ovotestis) as well as to part of the hermaphroditic duct, while the remainder of the genital tract seems to be of ectodermal origin. This may explain, on the one hand, the ectodermal origin of the gonadal complex postulated by Luchtel and by Brisson & Besse and, on the other hand, the possible regeneration of a gonad from the hermaphroditic duct, as described by Tardy and others.

Luchtel (1972*a,b*) explained sex determination in the pulmonates as being due to the distribution of sexually bipotential germ cells among two physiologically isolated compartments, a male medullar and a female cortical one. Grifford (1978) has described gonadal differentiation in the bisexual prosobranch gastropod *Viviparus viviparus*.

Our knowledge of germ cell origin in the *Cephalopoda* is almost as restricted as for the other molluscs. In *Octopus vulgaris* the gonad anlage forms part of the coelomic complex, which consists of the paired gonadal rudiments, the kidney rudiments, and the pericardial anlage with the pericardial gland rudiments (Marthy, 1968). Subsequently, the two gonadal anlagen fuse into a single rudiment (Wells & Wells, 1977; Marthy, 1978). In *Sepia* only a single (left) gonadal anlage is described (Richard & Lemaire, 1975; Arnold & Williams-Arnold, 1977). The gonadal anlage of *Sepia* contains, in addition to connective tissue cells, two other cell types, the smaller 'true' germ cells and large cells which disappear at hatching and may be involved in gonad differentiation (Lemaire & Richard, 1970; Lemaire, 1972*b*; Richard & Lemaire, 1975).

All cephalopods are dioecious. The sex of the embryo is precociously determined, since explants of undifferentiated gonads form either ovaries or testes, each in 50 per cent of cases (Richard & Lemaire, 1975). Primary sex differentiation, which usually occurs after hatching, is apparently not under hormonal control, but sexual maturation is controlled by a hormone formed in the optic gland (Lemaire, 1972*a*; Richard & Lemaire, 1975). Male and female animals are morphologically distinguishable by secondary sex characteristics. *Argonauta* is characterised by extreme sexual dimorphism, having a normal-sized female and a minute male (Wells & Wells, 1977). (See further the reviews by Fretter & Graham, 1964; Beeman, 1977; Berry, 1977; Webber, 1977.)

Conclusion

In the molluscs the PGCs become recognisable very late in development, at an advanced trochophore or pre-veliger stage. They probably arise from mesodermal cells in close association with the pericardial anlage. Unfortunately, no connection can yet be made between the germ plasm-like granular constituents of the polar lobe of *Crepidula* (Dohmen & Lok, 1975) and *Buccinum* (Dohmen & Verdonk, 1979) and the not yet fully characterised granular cytoplasmic bodies found in the PGCs of *Arion* and *Deroceras* (Luchtel, 1972*a*,*b*); a true germ plasm has not been described so far. In our opinion, germ cell formation in the molluscs is best classified tentatively as *intermediate* between the typically epigenetic and preformistic modes.

11

Onychophora

General characterisation

The Onychophora cannot be grouped formally under the Arthropoda since they have unsegmented limbs. They are therefore placed in a separate phylum. However, they have so much in common with the Myriapoda and the Hexapoda, with which they form a natural assemblage, that we prefer to consider them as forerunners of the Arthropoda, more or less bridging the gap between annelids and arthropods.

The Onychophora are land animals having a skin with a thin chitinous cuticle, which is moulted from time to time. The most anterior region is called the head; it carries preoral antennae and mouth tentacles, and comprises three segments. The trunk consists of homonomous segments, each with a pair of unsegmented limbs with terminal claws. Underneath the skin with its subepidermal connective tissue lies a muscular layer which consists of circular, diagonal and longitudinal smooth muscle fibres, and is comparable to the muscular sac of the annelids. The nervous system consists of a pair of supra-esophageal cerebral ganglia, connected to two widely separated ventral nerve cords with many commissures, which converge and unite again near the anal opening. The head has a pair of simple, annelid-like eyes. The digestive tract has an antero-ventral mouth provided with chitinous jaws, and consists further of an esophagus with a pair of large slime glands, a midgut and a hindgut. The segments contain distinct remnants of coelomic cavities. The respiratory organs are segmental tracheae. The animals are dioecious, with a single pair of ovaries or testes in the caudal part of the trunk. The gonoducts debouch in the preanal segment. (See Anderson, 1973, pp. 93–126.)

Embryonic development

The Onychophora are found only in the southern hemisphere. Most of the Australian species are ovo-viviparous, with yolky eggs; only a few species are oviparous. The eggs show intralecithal cleavage with blastoderm formation. The Australian forms are considered to be primitive. The South African species are viviparous, with maternal supply of nutrition to the yolk-poor eggs. The egg breaks up into a number of yolk spheres and a single nucleated blastomere. The latter divides repeatedly and forms a disc of blastomeres on one side of the egg, against the egg membrane. The embryonic anlage becomes saddle-shaped at the 64-cell stage and finally covers the entire egg surface (see Manton, 1949 for *Peripatopsis*) (fig. 11.1.). The South American species are also viviparous, with placental nutrition. The eggs contain little yolk and show total cleavage. The morula attaches itself to the wall of the oviduct with its future dorsal side, forming a ventral blastoderm and a dorsal placental stalk and plate region arising from the dorsal extra-embryonic ectoderm (see Weygoldt, 1963 and Anderson, 1966b, 1973).

Although the development of the various species differs quite markedly, their embryonic fate maps show a constant, phylogenetically conservative pattern (fig. 11.2). The anterior midgut anlage becomes internalised during cleavage and gastrulation, followed by the invagination of the mesoderm and posterior midgut. The proliferating posterior midgut forms an epithelium around the yolk mass. The anterior stomodaeum and the posterior proctodaeum invaginate separately. In embryos which have ventrally exposed yolk, the ventral extra-embryonic ectoderm extends towards the ventral midline, enclosing the yolk mass. The invaginating mesoderm splits into two anlagen, each of which forms a mesodermal band which extends forward underneath the embryonic ectoderm, forming the anlagen of the various segments. The formation of the segments proceeds in strict antero-posterior sequence, first the three head segments with the antennae and mouth tentacles, then two modified trunk segments, that is the jaw segment and the segment bearing the papillae of the slime glands, and finally the trunk segments. Each somite segregates into a dorso-lateral, a medio-ventral, and an appendicular portion. With the consumption of the yolk

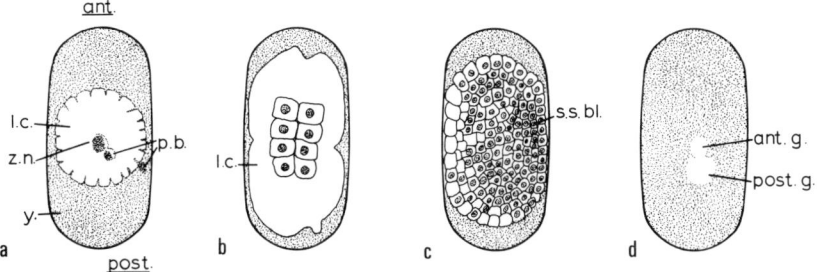

Fig. 11.1. Surface view of early embryonic stages of the onychophore *Peripatopsis* sp. (a) Uncleaved egg with two polar bodies (p.b.) on top of, and zygote nucleus (z.n.) inside lateral mass of cytoplasm (l.c.); <u>ant</u>. = anterior; <u>post</u>. = posterior; y. = yolk. (b) Eight-cell stage. (c) Saddle-shaped blastoderm stage (s.s.bl.). (d) Blastoderm surrounding entire egg; germinal anlage with anterior and posterior thickenings (ant.g. and post.g.). (After Manton, 1949.)

106   Onychophora

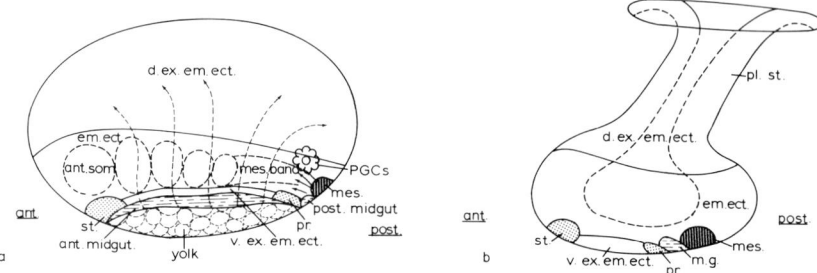

Fig. 11.2. (a) Fate map at stage of gastrulation and early segment formation in *Peripatoides orientalis*, seen in left lateral view, with arrows indicating direction of invagination of mesoderm (full arrows) and midgut cells (dashed arrows). (b) Fate map of blastoderm of a viviparous placental onychophore with placental stalk and plate, seen in lateral view. ant. = anterior; d.ex.em.ect. = dorsal extra-embryonic ectoderm; em.ect. = embryonic ectoderm; mes. = mesoderm; m.g. = midgut; pl.st. = placental stalk; post. = posterior; pr. = proctodaeum; som. = somites; st. = stomodaeum; v.ex.em.ect. = ventral extraembryonic ectoderm. (a and b after Anderson, 1966b and 1973 respectively.)

both the ventral and the dorsal extra-embryonic ectoderm shrivel up, the embryo closing first ventrally and later dorsally.

Although the Onychophora do not show any trace of spiral cleavage, Anderson (1973) postulates a phylogenetic relationship between the ancestors of the Onychophora and those of the clitellate Annelida on the basis of their anatomical structure and of the general configuration of their embryonic fate maps (compare fig. 11.2 with fig. 8.2 on p. 72).

Origin of the germ cells

Very little is known about the origin of the germ cells in the Onychophora. In *Peripatopsis*, Manton (1949) found a relatively early segregation of the germ cells during gastrulation. The gonadal rudiment develops from some mesodermal cells situated posterior to the invaginated mesodermal mass. Anderson (1966b) stated that the origin of the PGCs is essentially unknown. They are first identifiable at a fairly late stage, when they are closely associated with the presumptive mesoderm and the posterior midgut. However, Anderson (1973) later mentions that the PGCs segregate during gastrulation and migrate to the seventh and subsequent pairs of somites, finally settling in the splanchnic wall of the dorso-lateral somite compartments.

Conclusion

Although little is known of the actual origin of the germ cells, it seems rather likely that they are of mesodermal origin. They become recognisable only at a relatively late stage of development. (See further the related Myriapoda and Hexapoda in chapter 12 on the Arthropoda, and the conclusion on p. 155.)

12

Arthropoda

Introduction and general characterisation

Tiegs & Manton (1958) and Manton (1964, 1970, 1972) presented extensive anatomical and embryological evidence in favour of a polyphyletic origin of the arthropods. Anderson (1973) pointed out that the various groups of the arthropods cannot be compared adequately on the basis of egg structure and embryonic development alone, since both aspects show great variation among, as well as within, the various groups, for example as to yolk content and cleavage pattern. He strongly emphasises that comparison should be made on the basis of the fate maps of the various groups (see the figs 11.2, 12.1, 12.3, 12.5, 12.8, 12.12 and 12.15). Comparing the fate maps it is evident that, on the one hand, the Myriapoda and Hexapoda are undoubtedly related and, in addition, form a natural assemblage with the Onychophora. The latter again show a distinct relationship with the clitellate Annelida. Anderson therefore assumes that the ancestors of the Onychophora had a spiral cleavage pattern. On the other hand, the Crustacea and the Chelicerata are neither related to each other nor to the Onychophora–Myriapoda–Hexapoda assemblage. Both Manton and Anderson hold the opinion that the latter two groups should be considered as separate phyla. Leaving aside the question of the taxonomic classification of the various groups, it seems, nevertheless, logical to discuss the Crustacea and the Chelicerata separately. The Myriapoda and the Hexapoda will therefore be treated first, followed by the Crustacea and the Chelicerata. The special characteristics of each group will be given at the beginning of each subchapter. It must be realised that the individual groups have received very unequal attention, so that our knowledge of some groups is very scanty, whereas the literature on the higher insects, for instance, is immense (see also Ivanova-Kasas, 1979).

All arthropods are bilaterally symmetrical coelomates with a heteronomous, external metamery. The internal metamery has largely been lost; there are only traces of segmental coelomic cavities. There is an external, chitinous skeleton secreted by the epidermis. Each metamere is provided with at least one pair of appendages, which are composed of segments, hence the name Arthropoda. A varying number of rostral metameres forms the head with the preoral acron. The head contains a supra-esophageal ganglion of varying complexity which is connected circumenterically with the ventral rope-ladder-like ganglion system. The muscles, which insert on the exo-skeleton, are metamerically arranged. There is an open vascular system with a dorsal heart. A system of tracheae forms the respiratory organs, hence the other name, Tracheata, for the Arthropoda. The excretory organs may be proto- or meta-nephridia. All arthropods are dioecious.

MYRIAPODA

General characterisation

The Myriapoda comprise a number of Tracheata with homonomous trunk segmentation and with segmental tracheal invaginations as the respiratory organs. Each segment bears one or two pairs of segmented limbs. All myriapods have large post-oral antennal appendages and only transient pre-antennal and pre-mandibular rudimentary appendages. The mandibles flank the mouth, while the maxillae are situated just behind it. The post-maxillar segment may be incorporated into the head.

The Myriapoda comprise the Diplopoda, in which each body segment bears two pairs of limbs, the Symphyla, with two pairs of maxillae and relatively few trunk segments, the Pauropoda, with one pair of maxillae, a pair of specialised antennae and a relatively small number of trunk segments, and the dorso-ventrally flattened Chilopoda, with two pairs of maxillae and an extra pair of jaws. The three latter groups bear a single pair of limbs on each body segment (Manton, 1964, 1970, 1972; Anderson, 1973, pp. 127–73).

Embryonic development

The yolky eggs of the chilopods are ovoid or spherical, with a centrally situated nucleus. The latter divides repeatedly. The resulting intralecithal energids migrate to the surface where they form a uniform blastoderm, while some energids remain in the yolk and develop into vitellophages. The smaller eggs of diplopods, pauropods and symphylans show total cleavage and form pyramidal blastomeres arranged around a central blastocoel. In the various groups polygonal yolky cells become segregated from the yolk-free peripheral cells at different stages of development, as follows: in the small eggs of *Pauropus* at the two-cell stage, in the diplopods at the 16-cell stage and in the symphylans at about the 100-cell stage. Slightly later the cell boundaries between the blastomeres disappear and a unitary yolk mass is formed with peripheral and more central energids. The peripheral energids give rise to cuboidal blastomeres, which subsequently form a uniform blastoderm. (See Dohle, 1964, and Bodine, 1970, for the diplopods *Glomeris* and *Narceus*; Tiegs, 1940, for the symphylan *Hanseniella*; and Tiegs, 1947, for the pauropod *Pauropus*.)

The fate map of the Chilopoda is very similar to that of the Onychophora, while the Diplopoda, Symphyla and Pauropoda show variants of this scheme, in particular a diffuse immigration of the mesoderm along the entire length of the embryonic anlage (fig. 12.1). In the Symphyla and Pauropoda the midgut is formed by vitellophages, whereas in the Chilopoda and Diplopoda a separate postero-ventral midgut anlage exists. There are separate stomodaeal and proctodaeal invaginations. The stomodaeum forms the foregut, the proctodaeum the hindgut, posterior midgut and Malpighian tubules. In the Chilopoda the paired mesodermal bands proliferate mainly in a forward direction. There is, however, no strict antero-posterior segregation of segments. The embryo consists of cephalic lobes, labrum, pre-antennal, antennal, mandibular, maxillary and post-maxillary segments, a number of trunk segments, and a growth zone on either side of the proctodaeum giving rise to the caudal trunk segments and the telson. In the chilopods with yolk-rich eggs there is dorsal and ventral extra-embryonic ectoderm, while in the other groups and in the chilopods with less yolky eggs there is only dorsal extra-embryonic ectoderm. The chilopod germ anlage folds in two by dorso-ventral flexure (fig. 12.2). Diplopods, symphylans and pauropods hatch with only a few trunk segments and a large posterior

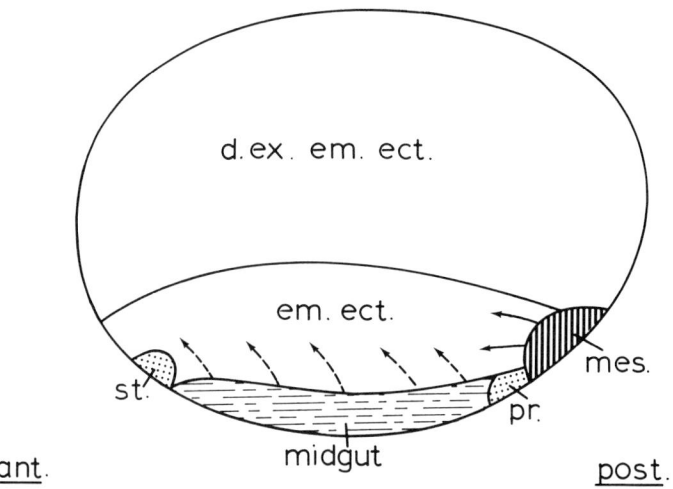

Fig. 12.1. Fate map of chilopod blastoderm, seen in left lateral view, with arrows indicating direction of invagination of mesoderm (full arrows) and midgut cells (dashed arrows). **ant.** = anterior; d.ex.em.ect. = dorsal extraembryonic ectoderm; em.ect. = embryonic ectoderm; mes. = mesoderm; **post.** = posterior; pr. = proctodaeum; st. = stomodaeum. (After Anderson, 1973.)

Origin of the germ cells 111

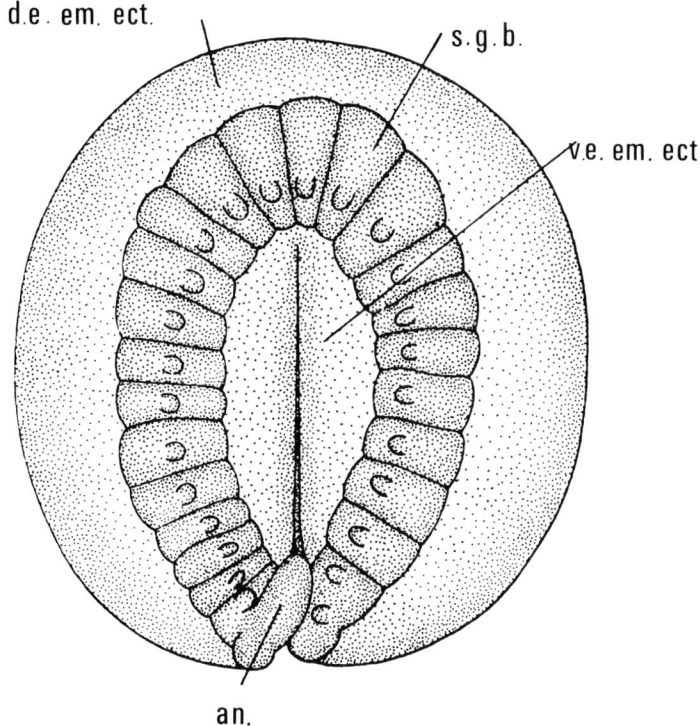

Fig. 12.2. Segmented germ band (s.g.b.) of chilopod *Scolopendra* sp. after flexure of embryo, showing dorsal and ventral extraembryonic ectoderm (d. and v.e. em.ect.), seen in left lateral view. an. = antenna. (Redrawn from Anderson, 1973, after Heymons, 1901.)

growth zone, whereas the chilopods show so-called epimorphic embryogenesis prior to hatching. In all myriapods the somitic mesoderm segregates into dorso-lateral, medio-ventral and appendicular lobes. In the chilopods the last two pairs of medio-ventral lobes form coelomoducts and represent the genital segments, while in the other groups the gonads lie at various levels in the trunk.

Origin of the germ cells

Very little is known about the origin of the germ cells in the Myriapoda. In the Diplopoda the gonads are formed in the medio-ventral somitic lobes of the eighth and ninth pairs of somites, where the PGCs appear (Dohle, 1964). In the Chilopoda the PGCs form in the medio-ventral somitic lobes of the last two pairs of somites, representing the genital segments

(Anderson, 1973). In the Pauropoda a single medio-ventral genital duct, which connects with the ectodermal gonoduct, is formed in the mesoderm of the second to sixth trunk segments. The PGCs form at the posterior end of the somites of the genital segments.

HEXAPODA (Insecta)

General characterisation

The hexapods are heteronomously segmented animals with clearly delimited head, thorax and abdomen, each consisting of a number of segments. The head bears a pair of antennae and has composite facet eyes and sometimes also single ocelli. The mouth parts consist of a median upper lip and three pairs of modified appendages, viz. the mandibles, the maxillae and the fused labrum or lower lip. The thorax comprises three segments, the pro-, meso- and meta-thorax, each of which bears a pair of segmented limbs, while the meso- and meta-thorax may also bear a pair of wings. The wings of the mesothorax may be transformed into wing covers. The abdomen consists of about ten segments with vestiges of appendages on the ninth and tenth segments, which have copulatory and egg-depositing functions. The hexapods have a complete chitinous exoskeleton on which the various muscles insert. The intestinal tract consists of a chitinous pharynx, esophagus and stomach, a secretory and resorbing midgut, and a chitinous hindgut with Malpighian tubules. There is an open vascular system and respiration occurs through a system of tracheae. All insects are dioecious, with segmental testes or a single pair of testes in the male, and a varying number of ovarioles in the female.

Post-embryonic development is characterised by successive larval stages, culminating in the imago. Each larval stage is preceded by a moult of the exoskeleton. All insects go through a complex process of metamorphosis. This occurs either in the form of a number of successive steps or in the form of a drastic reorganisation which takes place during a specific stage, the pupa.

The Hexapoda comprise the wingless Apterygota and the usually wing-bearing Pterygota. The latter are subdivided into the Hemi- and Holometabola, each comprising a large number of orders.

APTERYGOTA

The apterygote insects comprise the entognathous Collembola, Protura and Diplura and the ectognathous Thysanura (see Anderson, 1973, pp. 175–208).

Embryonic development

The embryology of the Collembola is best known, that of the Thysanura and Diplura less well known, and that of the Protura completely unknown (Jura, 1972).

Among the Apterygota all transitions from total cleavage to strictly superficial cleavage are encountered. The minute, spherical egg of the collembolan *Tetrodontophora bielanensis* is centrolecithal, with a centrally located nucleus, but without distinct periplasm. The first three cleavages are total, after which segregation of a peripheral zone with nuclei and a central yolk region occurs. A so-called periblastula is formed, in which superficial cleavage is initiated. Extrusion of yolk from the peripheral blastomeres or splitting-off of yolk-rich blastomeres towards the central yolk leads to the formation of a typical peripheral blastoderm and inner yolk mass (Jura, 1965). The blastomeres that contribute to the yolk become vitellophages and, according to Jura (1966), later form the midgut. Subsequently, the blastoderm differentiates into a region of large cells, which will form the so-called dorsal organ, and a region of small cells, which will form the germ anlage and the extra-embryonic membranes. The dorsal organ plays an important role in blastokinesis, probably as an organ for fluid secretion, and may also be involved in yolk utilisation (Jura, 1967b). The germ anlage is formed as a ventral thickening of the blastoderm. It extends along the greater part of the egg circumference, so that its anterior and posterior ends nearly touch the dorsal organ. The extra-embryonic serosa, situated between the germ band and dorsal organ, expands over the anterior and posterior tips of the embryo but not over the ventral surface. The fore- and hind-guts are formed as invaginations at both ends of the germ band, and both contribute to the formation of the midgut (Jura, 1966). Segmentation of the germ band starts early with the cephalic lobes and the gnathal and first thoracic segments, but the formation of the six abdominal segments is retarded. The mesoderm, which shows diffuse ingression, forms two lateral cords along the ventro-lateral margins of the germ band. The cords subsequently divide into 13 pairs of somites, which differentiate into dorso-lateral, medio-ventral and appendicular lobes. Blastokinesis starts with the maxillary and first abdominal segments.

The spherical, centrolecithal dipluran egg shows intralecithal cleavage; all energids migrate to the periphery during the sixth cleavage cycle, after which a blastoderm is formed. The blastoderm thickens at the vegetal pole, forming an irregular double layer. Cells accumulate at the periphery and form a ring-shaped thickening, while the centre becomes single-layered. Endodermal cells split off from the outer edge of this thickening, while mesoderm cells segregate diffusely from its inner edge (Jura, 1972). Subsequently, the blastoderm thickens at the vegetal pole, forming the germ

114   *Arthropoda/Hexapoda/Apterygota*

anlage and the extra-embryonic serosa, the former covering three-quarters of the egg circumference. Blastokinesis starts when the body segments and appendages are clearly visible. The dorsal organ forms as a condensation of the serosa but degenerates after completion of blastokinesis.

The elongated thysanuran egg has a central nucleus and a thin layer of periplasm. Cleavage is intralecithal with loss of synchrony during the fourth cleavage cycle (Woodland, 1957). With the exception of some yolk nuclei, the cleavage energids migrate towards the egg periphery where they form a continuous blastoderm. Numerous mitoses at the posterior end of the egg lead to the formation of a minute, disc-shaped germ anlage which is initially single-layered (fig. 12.3). Mesoderm formation may occur by random inward migration of germ band cells or by local immigration from the centre of the germ band. Segmentation of the germ band begins with the formation of the cephalic lobes, followed either by the gnathal and first thoracic segments, for example in *Lepisma*, or the gnathal, thoracic and first two abdominal segments, for example in *Thermobia*. Ultimately, eleven abdominal segments are formed. The stomodaeum and proctodaeum invaginate,

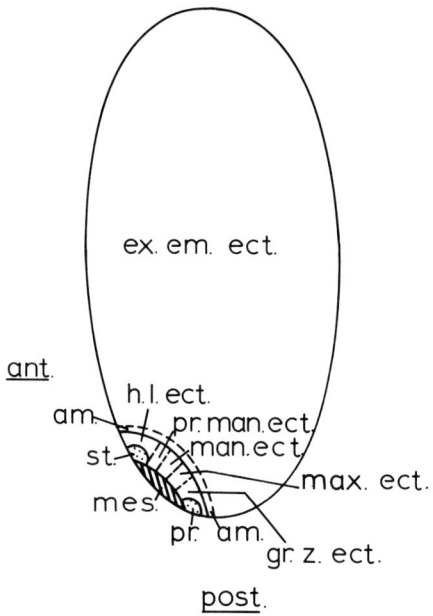

Fig. 12.3. Fate map of apterygote thysanuran blastoderm, seen in left lateral view. am. = amnion; ant. = anterior; ex.em.ect. = extraembryonic ectoderm; gr.z.ect. = growth zone ectoderm; h.l.ect. = head lobe ectoderm; mes. = mesoderm; man.ect. = mandibular ectoderm; max.ect. = maxillary ectoderm; post. = posterior; pr. = proctodaeum; pr.man.ect. = premandibular ectoderm; st. = stomodaeum. (After Anderson, 1973.)

while the midgut is formed partially by vitellophages and partially from the tips of the stomodaeal and proctodaeal invaginations. In *Machilis* the 'proamnion' zone, with small nuclei, directly surrounds the germ band, while the 'proserosa' zone, with large nuclei, is situated more peripherally. In *Lepisma*, *Thermobia* and *Ctenolepisma* the extra-embryonic part of the blastoderm forms large, flat serosa cells, while the small-celled amnion is formed from the margin of the germ band when the latter sinks into the yolk. A small dorsal organ is formed in the extra-embryonic ectoderm. During blastokinesis the amnion proliferates and gradually replaces the serosa.

Origin of the germ cells

Little is known about the origin of the germ cells in the apterygote insects. While in the collembolan *Isotoma cinerea*, Philiptschenko (1912a,b) first identified the PGCs at the 16- to 32-cell stage, in *Tetrodontophora bielanensis*, Jura (1967a) found the PGCs originating from two to five superficial cells at the 64-cell blastoderm stage. After segregation they sink into the yolk, where they multiply and form first one and later two groups of cells, which migrate to the abdominal part of the germ band, where they attach to the mesodermal cords. In the collembolan *Anurida maritima* total cleavage during preblastular development is directly responsible for the intravitelline segregation of the PGCs. In this species a cortical plasm of so-called oosomal nature, initially situated at the vegetal pole of the uncleaved egg, is transported along the first cleavage furrows to the centre of the embryo, where the PGCs segregate at the 128-cell stage (Garaudy-Tamarelle, 1969, 1970).

In the thysanuran *Lepisma*, at the time of blastokinesis, the PGCs form a knob-like protuberance at the extreme posterior end of the germ band, from where they migrate to the abdominal segments two to six in the female or four to six in the male (Heymons, 1897). This late appearance of the PGCs in the thysanurans was confirmed by Woodland (1957) and Sharov (1966) for *Lepisma*, *Thermobia* and *Ctenolepisma*, where the PGCs are found in association with the abdominal coelomic sacs. The segmental gonadal anlagen later fuse on either side into a single gonad (see also Anderson, 1973).

Klag (1977) described the morphological changes which the PGCs of *Thermobia domestica* undergo during embryonic development. Vesicles with a light interior and others with an electron-opaque content, called 'dense bodies', bleb off from the nucleus. The nucleolus disperses and reassembles by the end of embryonic development, when sex differentiation of the PGCs occurs. Throughout embryonic and post-embryonic development nucleolus-like bodies and so-called nuage material, which are considered to

PTERYGOTA

act as germ cell determinants, are found in the cytoplasm of the PGCs (fig. 12.4).

The pterygote insects are subdivided into Hemimetabola, with a gradual and step-wise metamorphosis, and the Holometabola, which pass through a phase of more-or-less complete reorganisation, the pupal stage, during which the larval structures are replaced by adult structures.

The Hemimetabola, which are less highly evolved than the Holometabola, comprise a large number of orders which are classified into three main groups: (1) the primitive Palaeoptera, to which belong the Ephemeroptera and Odonata; (2) the likewise primitive Polyneoptera, to which belong the Dictyoptera, Isoptera, Plecoptera, Cheleutoptera, Orthoptera, Dermaptera and Embioptera; and (3) the less primitive Paraneoptera, which comprise the Psocoptera, Mallophaga, Anopleura, Thysanoptera, Homoptera and Heteroptera.

The highly evolved Holometabola comprise the Coleoptera, Megaloptera, Neuroptera, Mecoptera, Lepidoptera, Diptera, Siphonaptera and Hymenoptera (see Anderson, 1973, pp. 209–62).

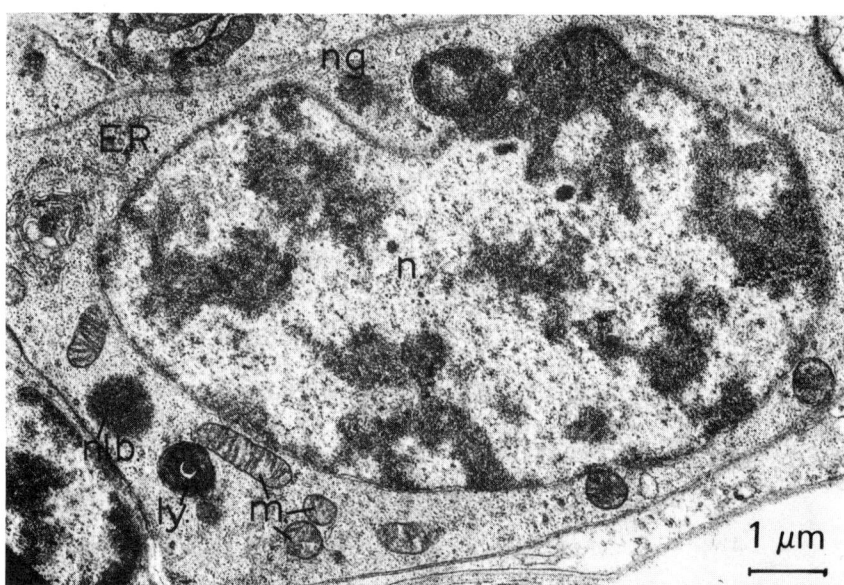

Fig. 12.4. Electron micrograph of a PGC in a ten-day embryo of *Thermobia domestica*, its cytoplasm containing nuage material (ng.) and a nucleolus-like body (nlb.), as well as mitochondria (m.), a lysosome (ly.), endoplasmic reticulum (E.R.) and ribo- and poly-somes. n. = nucleus. (Courtesy J. Klag, 1977.)

Embryonic development of the Hemimetabola: descriptive data

In the Hemimetabola, development from the rather large, yolk-rich egg is rather slow and epimorphic, all segments being fully developed at hatching. The embryonic anlage is small in comparison with the size of the egg. The egg is enclosed in an inner vitelline membrane, secreted by the oocyte with possible contributions from the follicle cells, and a thick outer chorion, secreted by the follicle cells only. The chorion is perforated by one or more micropyles (Counce, 1959, 1961). The chorion determines the shape of the egg, which shows both an antero-posterior and a dorso-ventral polarity.

Although parthenogenesis is quite widespread among the insects (see below, p. 144), normal zygote formation usually occurs. The freshly laid egg is in the metaphase of the first maturation division. After sperm entrance through the micropyle the two maturation divisions are completed in an antero-dorsal location. Four maturation nuclei are formed, of which three will form polar bodies. The female pronucleus, with its surrounding cytoplasm, and the innermost male pronucleus migrate to the interior of the egg for syngamy, acquiring a nearly central position. Although the egg is usually polyspermic, only one sperm nucleus fuses with the female pronucleus, the remaining sperm nuclei being resorbed during cleavage (see Counce, 1961). In the fertilised, yolk-rich eggs the yolk-free cytoplasm generally occupies only a small island around the zygote nucleus. A thin periplasm has been described in the more primitive Odonata, Orthoptera and Dermaptera, while the eggs of the more highly evolved orders possess a reticular plasm which permeates the entire egg and merges with a thin periplasm.

Cleavage is intralecithal and proceeds synchronously. The cleavage energids, consisting of a nucleus and a small surrounding island of cytoplasm, spread in all directions through the yolk mass (fig. 12.5a). They subsequently move towards the egg surface. They may arrange temporarily into a spheroid configuration in the endoplasm before entering the periplasm. The time when the cleavage energids enter the periplasm varies considerably among various groups and species, from the 64- to the 1024-nuclei stage (sixth to tenth cleavage). It depends, among other things, upon the size of the egg. In the majority of orders a number of energids remain in the yolk, forming the 'primary' vitellophages, but in the Dictyoptera, Plecoptera and gryllotalpid Orthoptera all cleavage energids migrate to the egg surface, while during later proliferation some division products migrate back into the yolk, thus forming the 'secondary' vitellophages. The superficial energids either form a uniform cellular blastoderm, or directly constitute a differentiating blastoderm with a local concentration of blastoderm cells representing the germ anlage (fig. 12.5b,c). In the Paraneoptera a uniform syncytial blastoderm is formed, which then transforms into a uniform cellular blastoderm. Subsequently, differentiation begins with the formation of

118   *Arthropoda/Hexapoda/Pterygota*

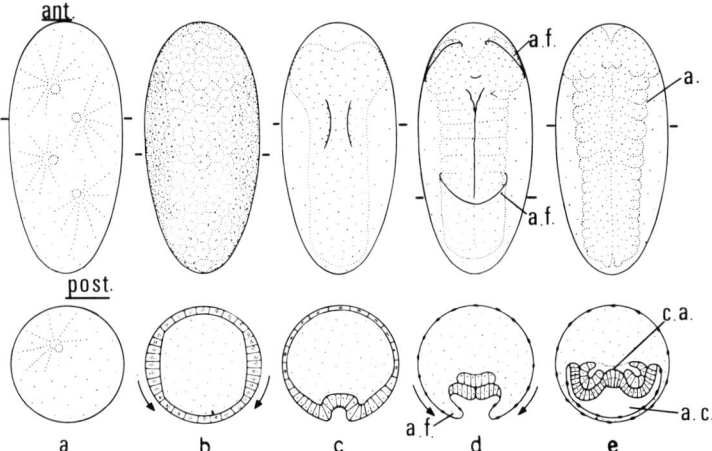

Fig. 12.5. Diagrammatic representation of early insect embryogenesis. Top row: ventral view with anterior (ant.) and posterior (post.) egg poles indicated. Bottom row: diagrammatic cross-sections at levels indicated in top row. (a) Multiplication and distribution of nuclei through ooplasm; (b) cellularisation of superficial layer of ooplasm, with arrows indicating ventral shift of lateral primordia for formation of definitive germ anlage; (c) invagination of presumptive mesoderm along midline of germ anlage; (d) germ band after gastrulation, with segment borders and formation of amniotic folds (a.f.); (e) advanced germ band stage with appendage buds (a.), transient coelom anlage (c.a.) and amniotic cavity (a.c.). (After Sander, 1976*a*.)

an embryonic primordium of densely packed cells and an attenuated extra-embryonic ectodermal area of varying size. The aggregation of the embryonic primordium usually begins bilaterally; its location is postero-ventral in most cases, but may be posterior or mid-ventral.

The germ anlage of the Hemimetabola belongs either to the 'short' or to the 'semi-long' germ type, according to Krause's (1939) classification (see also p. 121). The embryonic primordium often shows precocious head-lobe formation in front of the post-antennal region (fig. 12.5c). In the Cheleutoptera, Orthoptera and the majority of Dictyoptera, which have a short, heart-shaped embryonic primordium, the post-antennal region represents the segment-forming growth zone. In the Ephemeroptera, Odonata, some Dictyoptera, Embioptera and Dermaptera the long post-antennal region directly forms the gnathal and thoracic segments, while the abdominal segments are formed indirectly from the posterior segment-forming growth zone. During further development the embryonic primordium forms an elongated germ band. The germ band subsequently begins to segregate into segments. The onset of the elongation of the embryonic anlage coincides with the formation of a mid-ventral gastrulation groove, the site of proliferation of an inner layer, the cells of which spread laterally and forward beneath the embryonic anlage (fig. 12.c,d). In gastrulation both inward movements of presumptive mesoderm and overgrowth by the ectoderm seem to play a role (see Sander, 1976b). At the two ends of the gastral groove the stomodaeal and proctodaeal anlagen invaginate. Except for the anterior and posterior midgut anlagen, situated close to the stomodaeal and proctodaeal invaginations, the inner layer represents mesoderm. In embryos with a short post-antennal region, segmentation usually occurs in a strict anteroposterior sequence while, in embryos with a long post-antennal region, segmentation begins with the third thoracic segment and progresses forwards as well as backwards (fig. 12.5d,e).

Extension of the superficially situated germ band over the posterior pole and along the dorsal side towards the anterior pole occurs in the Dictyoptera, gryllotalpid Orthoptera and Cheleutoptera. In the Isoptera and Embioptera an additional submersion of the abdominal segments into the yolk takes place, while complete blastokinesis occurs in the Ephemeroptera, Odonata, most Orthoptera, and all Paraneoptera. During anatrepsis the onset of growth is accompanied by a flexure of the posterior end of the germ band, which subsequently extends through the yolk towards the anterior pole. Initially, the anterior end of the germ band remains at the surface, but it eventually also submerges. In many Heteroptera the germ band secondarily rotates 180° on its longitudinal axis.

During the growth and segmentation of the germ band the appendages of the head and thorax form, while those of the abdominal segments develop later, when the demarcation of the abdominal segments nears completion. While the invagination of the stomodaeal anlage occurs at an early stage of growth of the segmenting germ band, proctodaeal invagination is delayed until the majority of the abdominal segments are demarcated.

The margins of the embryonic primordium begin to fold over the embryonic anlage, finally closing mid-ventrally, thus forming the amniotic cavity (fig. 12.5d,e). The latter then segregates from the remainder of the extra-embryonic ectoderm, which surrounds the yolk mass and the germ band, and forms the serosa. In the case of blastokinesis the closure of the amnion is often delayed until the time when the corresponding region of the germ band submerges into the yolk. The amnion participates in the dorsal closure of the embryo while replacing the serosa during the dorsal overgrowth of the yolk mass. During the resorption and shrinkage of the yolk the amnion is replaced by the definitive embryonic ectoderm from back to front. The amnion and serosa fuse beneath the anterior part of the germ band and then rupture, so that the embryo becomes exposed again. During katatrepsis the serosa contracts into the dorsal organ.

The mesoderm segregates into paired somitic anlagen and a median haemocyte anlage. The hollow somites form anterior, posterior and ventrolateral outpocketings. The gonads develop within the abdominal splanchnic mesoderm. Except for a short unpaired vagina or ejaculatory duct, of ectodermal origin, the gonoducts arise from the abdominal somitic mesoderm. Germ cell origin and gonadal development will be dealt with separately (see below, p. 127 and p. 145).

Further information on the embryonic development of the Hemimetabola may be found in the following references: on the odonatan *Plathycnemis* (Seidel, 1929, 1934), on the dictyopteran *Blatella* (Tanaka, 1976), on the isopteran *Kalotermes* (Truckenbrodt, 1964), on the orthopterans *Gryllus* (Kanellis, 1952; Mahr, 1960*b*) and *Tachycines* (Krause, 1938*a,b*), on several embiopterans (Stefani, 1959), on the psocopteran *Liposcelis* (Goss, 1952, 1953), on the homopterans *Pyrilla* (Sander, 1956), *Apiomorpha* (Buchner, 1957) and *Euscelia* (Sander, 1959/60*a,b*) and in the general reviews by Anderson (1972*a*, and 1973, pp. 209–62).

Embryonic development of the Holometabola: descriptive data

The uncleaved eggs of the Holometabola are relatively small with less yolk than those of the Hemimetabola. In the former the chorion is less strongly developed than in the latter. Although the eggs are usually ovoid in shape, they nevertheless show a clear antero-posterior axis and a convex ventral and flat or concave dorsal surface. The eggs have a conspicuous reticular cytoplasm and a well-developed yolk-free periplasm, which is thicker posteriorly. In the Coleoptera, Diptera and Hymenoptera the so-called posterior pole plasm contains a mass of basophilic granular material which is often called the oosome or germinal plasm (see below, pp. 129–39).

The early development of the Holometabola has many features in common with that of the Hemimetabola, but also shows some characteristic differences. The cytoplasm of the cleavage energids originates, to a large extent, from the reticular cytoplasm of the egg; the energids are often connected through the reticulum. In the Holometabola the time of invasion of the periplasm by the cleavage energids also varies considerably, from the 128- to the 1024-nuclei stage (seventh to tenth cleavage). Although, in general, the invasion of the periplasm is synchronous, it may start locally at the equator, at the posterior pole, or at both poles, subsequently to extend over the entire egg. In all Holometabola a uniform syncytial blastoderm is formed initially. During the further nuclear divisions so-called 'secondary periplasm' deriving from the egg interior is added to the primary periplasm. The start of division asynchrony may vary from the ninth to the thirteenth cleavage cycle. When mitotic division of the peripheral nuclei ceases the nuclei become ovoid and form nucleoli. Cell walls then grow in from the egg surface, leading to the formation of a uniform, cuboidal, cellular blastoderm. In the majority of Holometabola a small number of cleavage energids remain in the yolk as 'primary' vitellophages. The latter divide but 'secondary' vitellophages are also split off from the syncytial or cellular blastoderm. Several Lepidoptera and some Diptera possess only 'secondary' vitellophages, while some Hymenoptera have 'tertiary' vitellophages which proliferate from the anterior and posterior midgut rudiments.

In the Holometabola the presumptive embryonic area comprises a large part of the uniform cellular blastoderm, only the mid-dorsal blastoderm region representing presumptive extra-embryonic ectoderm. The attenuation of the extra-embryonic ectoderm is therefore mainly a corollary of the gastrulation process. According to Krause's (1939) classification, eggs of the Holometabola belong to the 'semi-long' and 'long' germ types (fig. 12.6).

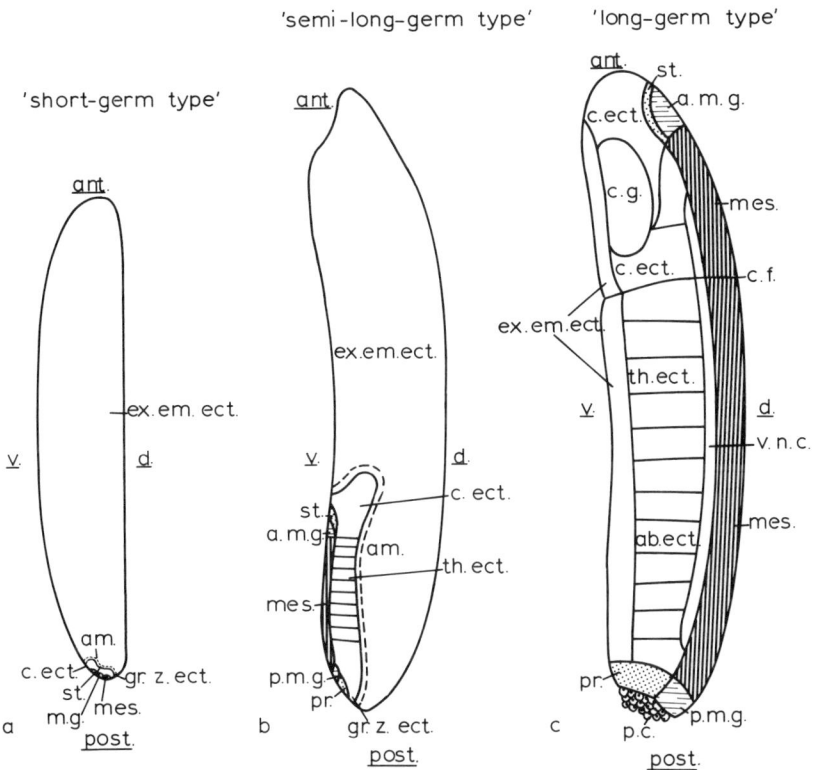

Fig. 12.6. Fate maps with blastoderm anlagen shown in lateral view. (a) 'short-germ type': orthopteran *Tachycines*; (b) 'semi-long-germ type': odonatan *Platycnemis*; (c) 'long-germ type': dipteran *Dacus*. ab.ect. = abdominal ectoderm; am. = amnion; a.m.g. = anterior midgut; ant. = anterior; c.ect. = cephalic ectoderm, c.f. = cephalic furrow; c.g. = cephalic ganglia; d. = dorsal; ex.em.ect. = extraembryonic ectoderm; gr.z.ect. = growth zone ectoderm; mes. = mesoderm; m.g. = midgut; p.c. = pole cells; p.m.g. = posterior midgut; post. = posterior; pr. = proctodaeum; st. = stomodaeum; th.ect. = thoracic ectoderm; v. = ventral; v.n.c. = ventral nerve cord. (After Krause, 1938b and Anderson, 1972a,b and 1973.)

Gastrulation through ventral groove formation starts when the germ anlage begins to elongate. Invagination is particularly pronounced at the anterior end of the groove, where the invaginating cells form an anterior cell mass situated between the future head lobes. The stomodaeum and anterior midgut anlagen subsequently invaginate at the anterior end of the gastral groove, the posterior midgut and proctodaeum anlagen at its posterior end.

The Lepidoptera have an aberrant form of blastokinesis. The embryo, surrounded by the amnion, becomes submerged in the yolk-filled amnioserosal space. Fluid originating from the latter is transferred from the amniotic cavity to the ventral haemocoel by suctorial action of the stomodaeum.

In the majority of the Holometabola, extra-embryonic membrane formation is similar to that in the Hemimetabola. However, in various Hymenoptera and Diptera, where the embryonic primordium comprises the far larger portion of the blastoderm, the embryonic membranes are reduced; although the serosa is complete, there is no amnion or only an amniotic vestige, while in some other Diptera even serosa formation is suppressed, the extra-embryonic ectoderm persisting as such.

In the Holometabola, organogenesis is essentially like that in the Hemimetabola. In Diptera and Hymenoptera, cephalic somite formation is suppressed. The gonads form in the splanchnic mesoderm of the fifth and sixth abdominal segments by the enclosure of the PGCs in a mesodermal sheath. Germ cell origin and gonad formation will be treated separately (see below, pp. 127 and 145).

It is evident that the fate maps of Hemi-and Holo-metabola (see figs 12.6a,b and 12.6c, 12.10 respectively), though varying in relative proportions, are essentially alike and show a distinct similarity to those of the Myriapoda and the Onychophora (see Anderson, 1973).

Further information on the embryonic development of the Holometabola is found in the following references: on the coleopterans *Tenebrio* (Ullmann, 1964), *Bruchidius* (Jung, 1966a,b), *Leptinotarsa* (Haget, 1970) and *Xyleborus* (Beeman & Norris, 1977a,b), on the neuropteran *Chrysopa* (Bock, 1939), on the mecopteran *Panorpa* (Ando, 1960), on the lepidopteran *Bombyx* (Kuwana & Takami, 1968), on the dipterans *Dacus* (Anderson, 1962), *Musca* (Formigoni, 1954; West, Cantwell & Shortino, 1968; Cantwell, Nappi & Stoffolano, 1976), *Aëdes* (Christophers, 1960), *Culex* (Idris, 1960a,b; Oelhafen, 1961; Davis, 1967), *Smittia* (Kalthoff & Sander, 1968; Zissler & Sander, 1973, 1977), *Wachtliella* (Wolf, 1977) and *Drosophila* (Rabinowitz, 1941; Mahowald, 1963a,b; Falk, Orevi & Menzl, 1973; Turner & Mahowald, 1976, 1977, 1979), on Diptera in general (Anderson, 1966c), on several Siphonaptera (Kessel, 1939), on the hymenopterans *Camponotus* (Reith, 1931), *Pimpla* (Bronskill, 1959), *Habrobracon* (Von Borstel, 1955; Amy, 1961), *Mesoleius* (Bronskill, 1964), *Lucilia* (Davis, 1967) and *Apis*

(DuPraw, 1967; Maul, 1967), and in the general reviews by Anderson (1972b, 1973, pp. 209–62).

Embryonic development of Hemi- and Holo-metabola: experimental data

General properties

For a proper understanding of insect development the following data, obtained from experimental analysis, must be considered.

Counce (1973) states that during early cleavage stages the cortical layer of the egg is a hostile environment for any nucleus. Nuclei remaining in this layer, like most of the maturation and sperm nuclei, stop developing, whereas the innermost maturation nucleus and one of the sperm nuclei, which are released from this developmental block by their inward migration, form the female and male pronuclei, respectively. Mutants show that several of these early developmental events are under genetic control.

The length of the cleavage cycle may vary from several hours or even days in the more primitive Hemimetabola (Anderson, 1972a) to only five to ten minutes in some higher Diptera (Anderson, 1972b).

Cytoplasmic streaming, amoeboid movement of energids, and locomotory activity of mitotic asters play an important role in effecting a more or less uniform distribution of the cleavage energids over the egg endoplasm (Sander, 1976b). Wolf (1975, 1978) suggests that competition for precursor molecules for aster microtubules may also play a role in this process.

Counce (1961) states that, in many insect eggs, the sixth cleavage cycle represents a critical or sensitive phase for a number of developmental events, such as migration of the cleavage energids towards the periphery, transition from a regulative to a mosaic type of development or vice versa, expression of several embryonic lethal mutants, chromosome elimination, etc.

At the stage of the syncytial blastoderm the cortical layer of the egg is apparently no longer hostile to nuclei, but on the contrary promotes nuclear development, so that a drastic change in its properties must have taken place since the initiation of development (Counce, 1973). This change apparently does not depend on nuclear information, since normal developmental changes have been observed in eggs of several species in the absence of nuclei, showing that the change must depend on a cytoplasmic developmental programme (for example Schnetter, 1965, in *Leptinotarsa*). However, pseudo-cleavage and pseudo-blastoderm formation is an exceptional event.

Nucleoli appear in the cleavage nuclei just prior to the time the energids enter the periplasm or during the subsequent syncytial blastoderm stage. There seems to be a close coincidence between the appearance of nucleoli in the blastoderm nuclei and the onset of RNA synthesis foreshadowing the beginning of differentiation. In this respect, Krause & Sander (1962) called attention to the fact that the merging of the cleavage energids with the periplasm also requires some nucleo-cytoplasmic interactions. This seems to hold equally for the merging of the secondary periplasm with the syncytial blastoderm layer in the Holometabola.

The formation of the germ anlage in the cellular blastoderm seems to depend both on a local contraction of the yolk system and on a fountain-like cytoplasmic streaming inside the yolk, and not on a local contraction of the blastoderm itself. The subsequent formation of an elongated germ band takes place by a reshuffling of cells rather than by directive growth through mitosis. In this reshuffling process giant locomotory organelles of the yolk system seem to play an important role (Sander, 1976b).

The gastrulation process, during which the mesodermal germ layer is formed through proliferation along the midventral groove and the subsequent lateral spreading of the invaginated cells, creates opportunities for interblastemic interaction. Defect experiments have demonstrated that the ectodermal germ layer possesses strong tendencies for regional self-differentiation in the absence of the mesoderm, whereas the mesodermal germ layer manifests hardly any differentiation tendencies in the absence of the ectodermal germ layer. The regional differentiation of the mesoderm therefore depends largely or entirely on an inductive influence emanating from the overlying ectoderm, which represents the leading germ layer. This was first demonstrated in the neuropteran *Chrysopa* by Bock (1942) and later confirmed by various authors in several species belonging to different groups: in the odonatan *Platycnemis* by Seidel (1952), in the coleopteran *Leptinotarsa* by Haget (1950, 1953b), in the orthopteran *Tachycines* by Krause (1953) and Krause & Krause (1957), in the hymenopteran *Apis* by Bertzbach (1960), in the dipteran *Calliphora* by Alléaume (1963), in the coleopteran *Dermestes* by Ede (1964). (See also Seidel, Bock & Krause, 1940; Seidel, 1961 and Counce, 1973.)

In blastokinesis the contractility of the yolk system plays an important role during anatrepsis (Sander, 1967; Vollmar, 1972), while the contraction of the serosa constitutes the driving force in katatrepsis (Mahr, 1961). (See also Krause & Sander, 1962 and Sander, 1976b.)

An active extension of the body flanks rather than a contraction of the embryonic covers seems to be responsible for dorsal closure of the embryo, as indicated by certain malformations which lead to eversion of the embryo (see Sander, 1976b).

Further information on morphogenetic movements and nucleo-cytoplasmic interactions may be found in the following references: on the orthopterans *Gryllus* (Mahr, 1957, 1960a,b; Schwalm, 1965; Heinig, 1967) and *Acheta* (Moser et al., 1970) on the coleopteran *Bruchidius* (Jung &

Krause, 1967), on the mecopteran *Panorpa* (Sander & Vollmar, 1967), on the dipterans *Calliphora* (Davis, Krause & Krause, 1968), *Wachtliella* (Wolf, 1973, 1977) and *Drosophila* (Howland, 1941; Yao, 1949, 1950; Ede & Counce, 1956; Fullilove & Jacobson, 1971; Zalokar, 1976), on the hymenopterans *Apis* (Müller, 1957/8) and *Pimpla* (Wolf & Krause, 1971), on general physiology (Agrell, 1963, 1964), and in the general review by Counce (1973).

From the beginning of experimental analysis of insect development two different egg types have been distinguished: the 'regulative' or 'indeterminate' type and the 'mosaic' or 'determinate' type (see Richards & Miller, 1937). The two egg types correspond in general to the Hemimetabola and Holometabola, respectively, and are characterised by Seidel (1952) as 'late' and 'early' differentiating egg types. According to Seidel, Bock & Krause (1940), Counce (1961) and Seidel (1961, 1966), the 'regulative' egg type, represented in, among others, the Orthoptera and Odonata, mostly coincides with the 'short-germ' egg type, with homogeneous periplasm and a germ anlage having a post-antennal growth zone and gradual organ formation, whereas the 'mosaic' egg type, represented in, among others, the Lepidoptera, Diptera, Hymenoptera and Neuroptera, mostly coincides with the 'long-germ' egg type, with early differentiated periplasm and a germ anlage with a very small or no growth zone and with *in situ* organ formation. The 'semi-long-germ' type, with a growth zone for at least the abdominal segments, occupies an intermediate position. It must, however, be emphasised that, in reality, a large range of intermediate types exists between the extreme 'regulative' and 'mosaic' egg types.

Although the problem of pattern formation is a central topic in the experimental analysis of insect embryogenesis, we consider it beyond the scope of this monograph. We therefore refer the reader to some of the relevant review articles, particularly the older reviews by Richards & Miller (1937) and Seidel, Bock & Krause (1940), the later reviews by Krause (1957, 1958, 1961), Seidel (1961, 1966), Counce (1961) and Krause & Sander (1962), and the more recent reviews by Lawrence (1971), Counce (1973), Lawrence & Morata (1976), Sander (1976*a*), Illmensee (1978) and Garcia-Bellido, Lawrence & Morata (1979).

The totipotentiality of the cleavage nuclei

Before discussing the origin, structure and fate of the germ cells, the totipotentiality of the cleavage nuclei generally must be discussed on the basis of the outcome of defect and transplantation experiments. Schnetter (1965) removed about two-thirds of the cleavage nuclei of the 32-nuclei stage or about one third of the 512-nuclei stage in embryos of the coleopteran *Leptinotarsa decemlineata*, and obtained initially retarded but ultimately normal development. In these embryos some additional mitoses of the remaining energids occurred, restoring a more or less normal number of nuclei before blastoderm formation was initiated. Geyer-Duszynska (1967) and Illmensee (1968) transplanted donor nuclei from young syncytial blastoderm stages of *Drosophila melanogaster* into recipient unfertilised eggs of virgin females and obtained diploid preblastoderm stages, demonstrating that these arbitrarily chosen nuclei could support normal early development. Schnetter (1967) transplanted cleavage nuclei and nuclei from large and small cellular blastoderm stages into enucleated eggs of *Leptinotarsa* and obtained normal development. The same result was obtained by Illmensee (1972), who performed nuclear transplantations from cleavage and syncytial blastoderm stages to unfertilised *Drosophila* eggs, and by Schubiger & Schneiderman (1971), who used donors and hosts of *Drosophila* of different genotypes. Zalokar (1973) performed similar transplantations using nuclei of pre-blastoderm and blastoderm stages of *a y w* strain of *Drosophila*, which he implanted into the posterior pole region of fertilised eggs of a *v; b w* strain, and obtained mosaics including gynandromorphs. Okada, Kleinman & Schneiderman (1974*b*) transplanted anterior energids into the posterior region of fertilised *Drosophila* eggs and obtained chimaeric adults, including one with a chimaeric germ cell population. Illmensee & Mahowald (1974) transplanted posterior pole plasm into the anterior region of recipient eggs which were in an early cleavage stage. The plasm was colonised by arbitrary energids of the host and formed ectopic pole cells. Reimplantation of the latter into the posterior region of genetically marked recipient eggs yielded four per cent donor progeny, showing that any energid entering posterior plasm, even in an ectopic position, can support normal pole cell and subsequent germ cell development. Santamaria (1975) transplanted nuclei between eggs of different *Drosophila* species and obtained viable adult chimaeras. Zalokar (1971) and Illmensee (1973) transplanted wild-type nuclei from different regions of early gastrulae of *Drosophila* into genetically marked, unfertilised eggs, showing that the percentage of initiation of development was the same for all regions. All these experiments demonstrate that, up to the syncytial and early cellular blastoderm stage, the various cleavage nuclei are fully isopotent, so that regional differentiation of the blastoderm, including pole cell differentiation, must be due to the regionally different composition and physiological properties of the periplasm. Irreversible nuclear changes seem to occur only at the differentiated blastoderm stage or later. (See further the reviews by Richards & Miller, 1937; Seidel, Bock & Krause, 1940; Seidel, 1961 and Counce, 1961, 1973.)

Origin of the germ cells in Hemi- and Holo-metabola

General survey

In the various groups of insects the time and place of origin of the germ cells vary considerably. Seidel (1924) stated that, in general, in the indeterminate egg type, which occurs predominantly in the Hemimetabola, the segregation of the germ cells occurs rather late in development, that is *after* blastoderm formation, while in the determinate egg type, occurring predominantly in the Holometabola, the germ cells segregate *before* blastoderm formation. Nelson (1934), reviewing the older literature, concluded that the time of germ cell formation in the pterygotes comprises the whole range from early segregation during blastoderm formation (in the form of pole cells) to post-embryonic segregation; he distinguished six different periods. Miya (1958) divided germ cell formation in apterygote and pterygote insects into three main classes according to the *site* of origin: (1) segregation at or near the posterior pole of the egg; (2) segregation on the ventral side of the egg, and (3) segregation in a non-distinct site. The first class was again subdivided into three categories: (a) with distinct polar granules, (b) without distinct polar granules, and (c) with delayed appearance of the germ cells.

In a recent review on the Hemimetabola, Anderson (1972a) distinguished three classes as follows: (1) very late segregation of the PGCs in the splanchnic wall of the first six abdominal segments, followed by fusion of the segmental anlagen into two continuous strands, which are then covered by splanchnic epithelium, thus forming the definitive gonads (for example in Orthoptera); (2) segregation of the PGCs during early gastrulation, and subsequent migration to the splanchnic wall of the anterior abdominal segments (for example in Dictyoptera, Embioptera (Stefani, 1961) and Heteroptera); (3) segregation of the PGCs during blastoderm formation in the form of a mass of internal cells, which becomes attached to the posterior end of the embryonic primordium in the differentiated blastoderm stage, later associating itself with the third and fourth abdominal segments during blastokinesis. The internal cell mass undergoes final segregation into two separate masses which give rise to the definitive gonads (for example in Dermaptera, Psocoptera and Homoptera).

In the Holometabola the PGCs are almost always recognisable beneath the posterior end of the embryonic primordium by the time the latter has been established and begins to elongate into the germ band. In the tenebrionid Coleoptera the PGCs are recognisable during gastrulation at the posterior end of the mesodermal layer. In the other Coleoptera and in the Megaloptera, Lepidoptera and symphytan Hymenoptera the PGCs appear as a postero-ventral group of cells beneath the posterior end of the embryonic primordium soon after completion of the uniform cellular blastoderm. In other Coleoptera again, and in the Diptera, Siphonaptera and parasitic Hymenoptera the PGCs segregate during the formation of the syncytial blastoderm in the form of posterior pole cells characterised by the presence of polar granules. The origin of the germ cells is still unknown in the Neuroptera, Mecoptera and Trichoptera (Anderson 1972*b*).

It is evident from the various reviews that, with the gradual improvement of histological techniques, PGCs have been recognised at ever earlier stages in several groups. At present the general conclusion seems to be that, in the great majority of insects, the germ cells originate in a particular region of the egg, which seems to be predetermined topographically but not necessarily functionally. Whereas an enormous literature exists on the higher insects, where pole cell formation is characterised by the presence of typical polar granules, little or no attention has been paid to those groups of insects which also show early and localised germ cell segregation, but *without* the presence of specific cytoplasmic structures which may be considered as germ cell determinants. It must be emphasised that these organisms of course also form typical germ cells.

There is a phenomenon which seems to argue against the presence of a well-defined germ cell lineage, that is the occurrence of polyembryony in some higher insects. Several forms of polyembryony have been distinguished. Vignau (1967) and Cavallin (1971) described that so-called 'substitutive' polyembryony occurring in the cheleutopteran *Carausius morosus*, where the initial embryonic anlage with typical germ cells sinks into the yolk and degenerates, to be replaced by an embryo formed by secondary germ band formation in the blastoderm. According to Cavallin, the *de novo* formation of germ cells in the replacement embryo points to a late and indeterminate germ cell origin. However, it is not known exactly what happens to the germ cells of the initial embryo, so that this conclusion may be incorrect. Cavallin (1976) could identify the PGCs in the very young unsegmented germ band by their ultrastructural features.

Ivanova-Kasas (1972) described two other forms of polyembryony in the parasitic Strepsiptera and Hymenoptera, respectively. In the former, polyembryony is associated with viviparity, the mother supplying the necessary nutrients, oxygen, etc. The rather yolk-poor eggs show total cleavage with segregation of the yolk, so that a coeloblastula is formed with a syncytial yolk ball attached to it. In the wall of the blastula many germ discs develop, each acquiring its own germ cells. Here the presence of a predetermined germinal anlage in the original egg seems out of the question, so that a more epigenetic mode of germ cell formation seems likely. In the parasitic Hymenoptera, which have small eggs, the male and female pronuclei, localised at the posterior end of the egg, form a synkaryon and separate, with the ooplasm and oosome surrounding them, from the peripheral ooplasm containing the polar body nuclei. The latter fuse and form a so-called paranucleus, which fragments amitotically and forms the trophamnion which surrounds the embryonic cell. The latter divides mitotically, the oosome being transferred to one of the daughter cells. The oosome disintegrates during the second division. During further cleavage the embryonic cell forms a large number of separate cell groups, in some species up to 180, in which embryo formation occurs by primary segregation of a group of central, presumptive midgut cells from a peripheral layer showing ventral groove formation. In each embryonic anlage germ cells appear. During embryonic development nutrients are supplied to the embryos through the trophamnion. As a consequence, the whole embryonic complex grows enormously, acquiring a volume which may be four million times the initial egg volume. One must assume either a rather unlikely redistribution of the original germ plasm of the egg among the many embryonic anlagen, or a *de novo* formation of germ cells in each of the embryonic anlagen. This intriguing problem still remains to be solved.

Pole cell formation and origin and structure of the polar granules

Metschnikoff (1866) and Balbiani (1882) presented the first descriptions of pole cells in the higher insects. Krause (1939) described the formation of pole cells with polar granules in chrysomelid Coleoptera, Diptera, Siphonaptera and Hymenoptera. Sonnenblick (1941) gave an extensive description of pole cell formation in the dipteran *Drosophila melanogaster*. After the eighth cleavage cycle, when 256 energids have been formed, a number of those reach the posterior pole region of the egg, which is characterised by a voluminous layer of periplasm and an abundance of polar granules. Subsequently, the posterior pole of the egg forms a varying number of protuberances into which cleavage energids enter, the nucleated outpocketings then being segregated from the syncytial blastoderm. Subsequently, the syncytial blastoderm forms a continuous blastoderm layer over the entire egg surface, including the posterior region, where it merges underneath the pole cells. Jura (1964a) described the same process in *Drosophila virilis*, where eight to twelve pole cells are formed initially. The latter divide a few times, forming a polar cap outside the embryo proper. The polar granules present in the original polar region of the egg are more or less evenly distributed among all the pole cells.

Ivanova-Kasas (1958) described the oosome at the posterior pole of the hymenopteran *Eurytoma aceculata* as being composed of many fine grains. During development it ends up in the pole cells. Tawfik (1957) found that the posterior pole region of the unfertilised egg of the hymenopteran *Apanteles* has a high alkaline phosphatase activity. According to Wolf (1967), the oosome of the egg of the dipteran *Wachtliella persicariae* is identical to the polar granules of the *Drosophila* egg. The oosome in the egg of the dipteran *Smittia* is a lens-shaped structure consisting of a three-dimensional network of electron-dense material, which has a fibrillar appearance but which is probably granular in nature (Zissler & Sander, 1973). It is not known, however, whether the oosome of the *Smittia* egg plays a role in germ cell formation. The oosome in the egg of the coleopteran *Leptinotarsa* contains polyribosomes, numerous ribosomes and sub-ribosome units (Haget, 1972).

In *Drosophila melanogaster*, Mahowald (1962) distinguished the pole cells from blastoderm cells by their less dense cytoplasm with annulate lamellae and polar granules, and by their hollow, spherical nucleoli. The polar granules are electron-dense granules without a limiting membrane, composed of ribosome-sized granules and smaller granular or fibrillar components. Ullmann (1965) noted that the pole cells of *Drosophila* contain so-called epsilon granules as well as polar granules. The former are ovoid in shape and have a delicate double limiting membrane, and seem to arise from convolutions of the cellular membrane in the mature oocyte. Mahowald (1968b) characterised the polar granules of *Miastor* as an inter-woven mat of fibrils. Schwalm, Simpson & Bender (1971) described the polar granules in the dipteran *Coelopa frigida* as being composed of dense granules of 2.5–3.5 nm diameter. Illmensee & Mahowald (1976) recently found that the pole cells in *Drosophila* are characterised, apart from polar granules, by round 'nuclear bodies'.

In *Drosophila melanogaster* the polar granules originate during early stages of vitellogenesis as small electron-dense bodies of 0.2 μm diameter, which become attached to mitochondria during the growth phase of the oocyte (fig. 12.7), a situation which lasts till the end of oogenesis (Mahowald, 1962, and the review of oogenesis by Mahowald, 1972). After fertilisation the association of polar granules with mitochondria is lost, however. Mahowald (1968b, 1975) states that, in the dipteran *Miastor*, the polar granules, which are attached to mitochondria, are first detectable during oogenesis both in the ooplasm and in the nurse cell cytoplasm prior to the breakdown of the nurse cells. Typical polar granules in the form of dense masses of amorphous material appear at the posterior pole of the oocyte.

Meng (1968, 1970) described the genesis of the oosome in the hymenopteran *Pimpla turionellae*. In the young oocyte a so-called 'chromidium' lies close against the nuclear membrane. This is later located in the lower, posterior part of the oocyte. At the beginning of vitellogenesis oosome material accumulates around the chromidium and the oosome is displaced to the posterior pole periplasm. Following the degeneration of the nurse cells trophocytic RNA accumulates in the oosome. During cleavage the oosome material ends up in the pole cells. Although the chromidium is proteinaceous in nature and does not contain RNA or DNA, the oosome is

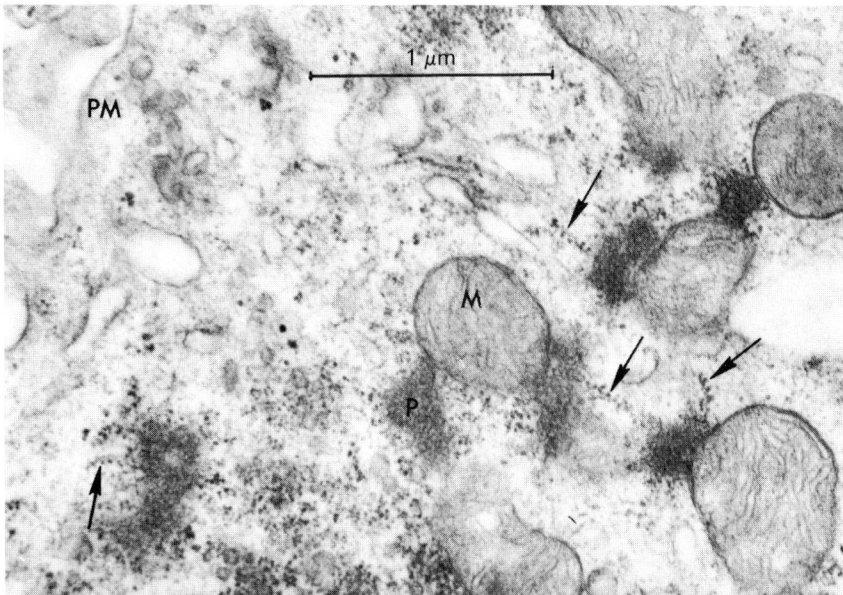

Fig. 12.7. Electron micrograph of polar granules at posterior tip of *Drosophila melanogaster* egg, 15 minutes after fertilisation. Polar granules (P) are often attached to mitochondria (M) and associated with helical polysomes (arrows). Plasma membrane (PM) at upper left of picture. (Courtesy A. P. Mahowald, 1971c.)

rich in RNA. Later protein, carbohydrate yolk, lipid droplets and glycogen particles are present as specific metabolites of the pole cells.

The structure of the polar granules seems to change during development. Counce (1959) stated that in nine *Drosophila* species the polar granules go through cycles of fragmentation and coalescence before they become included in the pole cells. She attributed changes in morphology and distribution to changes in the cortical plasm. Mahowald (1968a) observed that, in *Drosophila melanogaster*, during cleavage and pole cell formation the polar granules fragment and become associated with mitochondria and polysomes (fig 12.8). After pole cell formation the polar granules coalesce. During pole cell migration to the gonads (see below, p. 133), they again fragment into fibrous bodies which become attached to the outer nuclear membrane (Mahowald, 1971b). The latter configuration is maintained in the PGCs of the indifferent gonad and in the oogonia and spermatogonia. During oogenesis the fibrous bodies dissolve in the growing oocytes but are retained in the nurse cells until the end of vitellogenesis, possibly as a general store of mRNA. In *Miastor* the polar granules go through similar processes of association with mitochondria and the nuclear membrane, as well as fragmentation and reaggregation during the life-cycle (Mahowald, 1975).

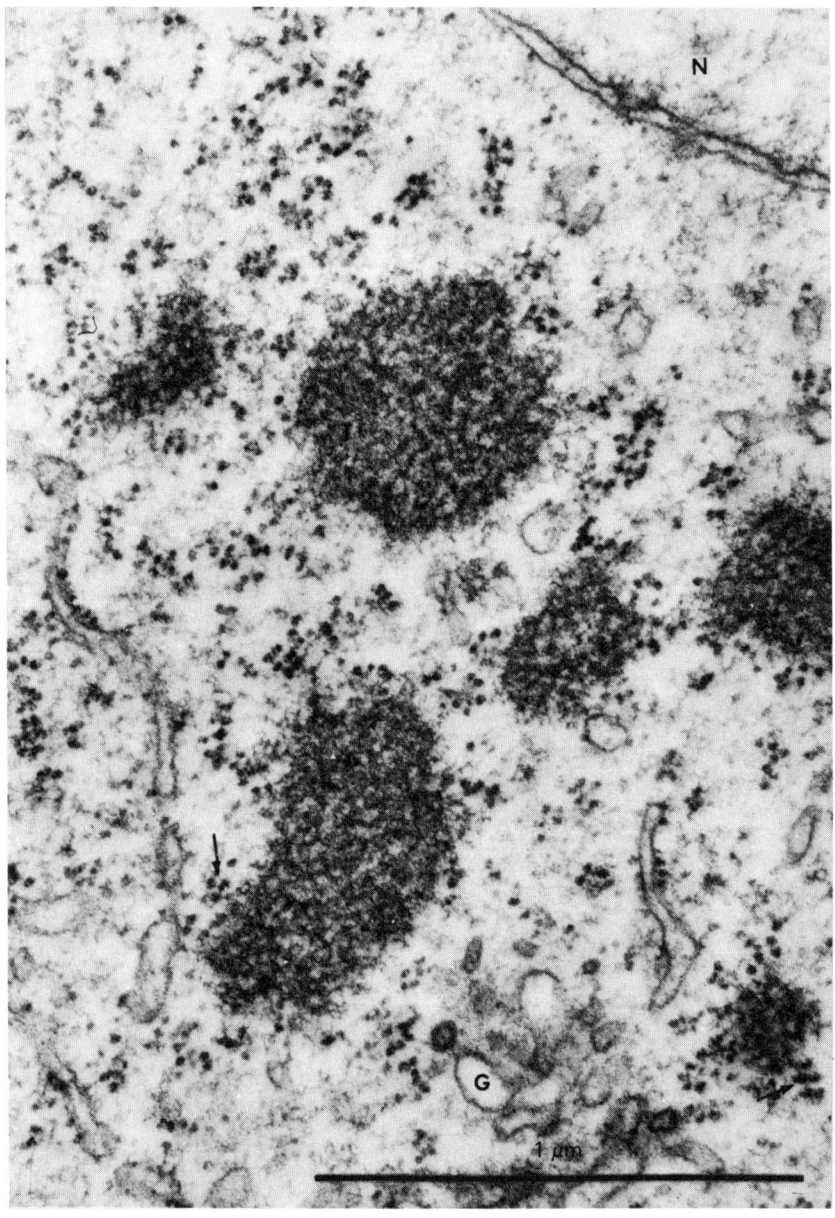

Fig. 12.8. High-magnification electron micrograph of polar granules of *Drosophila hydei* egg at stage when polar granules have started to reaggregate; polysomes (arrows) are still associated with the polar granules. G = Golgi complex; N = nucleus. Note fibrillo-granular appearance of polar granules. (Courtesy A. P. Mahowald, 1968a.)

According to Mahowald (1971c, 1975), association of the fragmented polar granules with ribosomes denotes a period of functional activity. In *Coelopa frigida* the polar granules also change during development (Schwalm, Simpson & Bender, 1971). From early cleavage to pre-blastoderm formation they are composed of globular subunits of 20–25 nm. At the pre-blastoderm stage the globular subunits are transformed into electron-dense, rod-like structures which are randomly oriented inside the polar granules, while during pole cell formation the polar granules aggregate into a granular complex. These changes were attributed to nucleo-cytoplasmic interactions until Schwalm (1974) demonstrated that in over-ripe unfertilised eggs of *Coelopa frigida* similar changes occur in the pole plasm in the absence of immigrating nuclei, so that they must be programmed in the cytoplasm (see the reviews by Mahowald, 1971a, 1972 and the general review by Eddy, 1975).

In addition to the polar granules, a naturally occurring symbiont ball in the posterior pole of the egg was described by Sander (1956, 1959/60a,b) in the homopterans *Pyrilla perpusella* and *Euscelis plebejus* and by Kalthoff & Sander (1968) in the dipteran *Smittia parthenogenetica*. This so-called 'mycetoma' contains bacteria, which seem to be involved in the formation of the mycetocytes (Sander, 1968). Brown & Bennett (1957) and Royer (1973) observed a similarly located symbiont ball in the homopterans *Pseudaulacaspis pentagona* and *Icerya purchasi*, respectively. Since the PGCs segregate in the neighbourhood of the posterior pole, the symbionts may be transferred through the germ cells to the next generation. This seems the more likely since Daniels & Ehrman (1974) and Ehrman & Daniels (1975) described mycoplasm-like symbionts in the pole cells of *Drosophila paulistorum*. These had been found previously in the larval and adult ovaries and testes and are transmitted via the egg cytoplasm from one generation to the next. In contrast to the homopterans, where the symbiont ball occurs naturally, those in *Drosophila paulistorum* are pathological, since their presence leads to hereditary sterility.

Developmental fate of the pole cells

Sonnenblick (1941) distinguished two periods of pole cell migration in *Drosophila*. The first phase occurs during cellular blastoderm formation, when pole cells pass between the columnar blastoderm cells and invade the yolk mass; this primary migration concerns close to half the total number of pole cells present in the polar cap. In the second phase the remaining pole cells are first passively moved inwards with the invagination of the midgut anlage (fig. 12.9) and subsequently move actively between the columnar cells of the midgut. In a *Drosophila* mutant in which the blastoderm remains syncytial, since it is unable to form cell walls, Rice & Garen

134 *Arthropoda/Hexapoda/Pterygota*

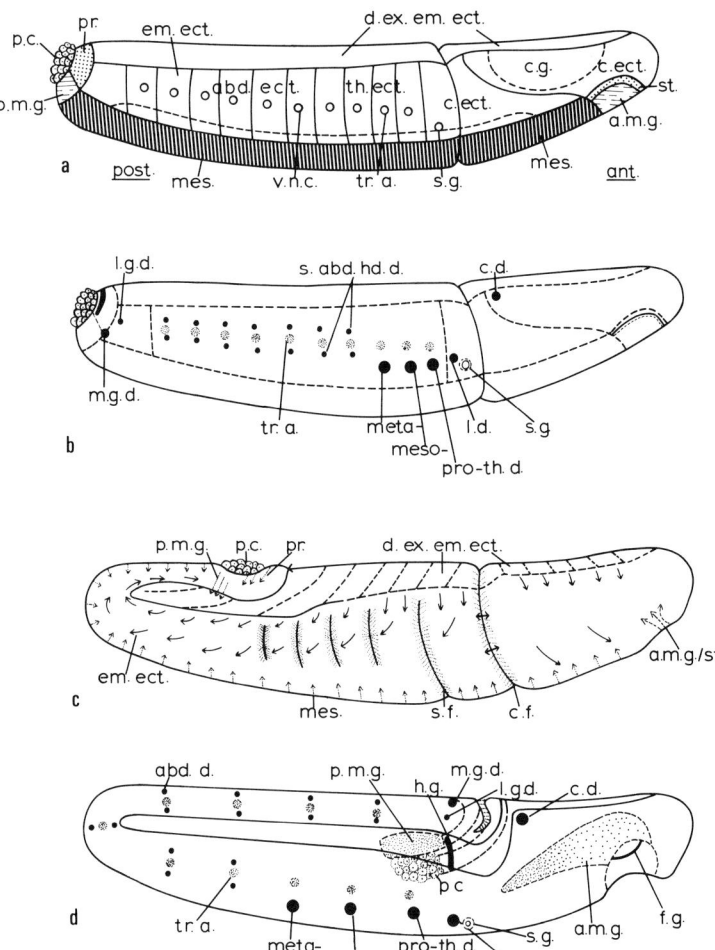

Fig. 12.9. Fate maps of larval and adult organs in *Dacus tryoni*. (a) and (b) before gastrulation; (c) during gastrulation, movements of embryonic ectoderm, points of resistance, and entry of rudiments into interior shown by arrows (single, double and broken, respectively); (d) location of adult rudiments at end of gastrulation. All figures show right lateral view. Note displacement of pole cells with invaginating midgut anlage. abd.d. = abdominal discs; abd.ect. = abdominal ectoderm; a.m.g. = anterior midgut; **ant**. = anterior; c.d. = cephalic disc; c.ect. = cephalic ectoderm; c.f. = cephalic furrow; c.g. = cephalic ganglia; d.ex.em.ect. = dorsal extraembryonic ectoderm; em. ect. = embryonic ectoderm; f.g. = foregut; h.g. = hindgut; l.g.d. = lateral genital disc; l.d. = labial disc; mes. = mesoderm; m.g.d. = median genital disc; p.c. = pole cells; p.m.g. = posterior midgut; **post**. = posterior; pr. = proctodaeum; s.abd.hd.d. = segmental abdominal hypodermal discs; s.f. = secondary furrows; s.g. = salivary gland; st. = stomodaeum; (meso-, meta-, pro-) th.d. = thoracic discs; th. ect. = thoracic ectoderm; tr.a. = tracheal anlagen; v.n.c. = ventral nerve cord. (After Anderson, 1966c.)

(1975) described passive displacement of the pole cells by means of cytoplasmic streaming in the syncytial blastoderm, as well as subsequent active amoeboid movements of the pole cells.

Although several pole cells reach the body cavity, many seem to remain in the gut. Those reaching the body cavity migrate towards their final destination in the gonad-forming area of the abdominal segments, where they are shunted into two groups and surrounded by small mesodermal cells forming the definitive gonads. Sonnenblick pointed out that, although there are approximately 90 pole cells in the polar cap, only five to seven PGCs are found in each of the minute female gonads, and nine to thirteen in the more voluminous male gonads. These observations were essentially confirmed by Counce (1963), although she assumed the migration of the pole cells to be more or less continuous (see also Prudhommeau & Laugé, 1972). The fact that only a fraction of the original pole cells reaches the gonads raises the question of the fate of the remaining ones.

Sonnenblick (1950) stated that, in the higher insects, the pole cells which do not reach the gonads take part both in midgut formation and in the formation of secondary vitellophages (Rabinowitz, 1941). According to Aboïm (1945) and Poulson (1947, 1950) the early migration of pole cells gives rise to PGCs, the late migration to midgut cells and vitellophages. This conclusion was supported by Counce & Ede (1957) and Poulson & Waterhouse (1960). However, according to Hathaway & Selman (1961) and Jura (1964b), the situation is just the converse: the early migrating pole cells form midgut cells and vitellophages, while those involved in the secondary migration form PGCs. The pole cells which become incorporated into the midgut would give rise to the larval cuprophilic cells or calycocytes, which degenerate during metamorphosis. Aboïm (1945) and Poulson (1947) stated that only the pole cells which enter the lateral mesoderm represent true germ cells. Recently, Heming (1979) confirmed the differential fate of the pole cells in the thysanopteran *Haplothrips*, where the pole cells become germ cells when reaching the gonad but differentiate into vitellophages when dispersed in the yolk. However, Underwood *et al.* (1980) conclude from morphological and autoradiographic studies in *Drosophila melanogaster* that those pole cells which do not reach the gonads degenerate instead of taking part in gut formation or differentiating into vitellophages. (See the reviews by Counce, 1961, 1973; Krause & Sander, 1962; Charniaux-Cotton, 1964; Anderson, 1966c; Beams & Kessel, 1974; Smith & Williams, 1975 and Mahowald *et al.*, 1979.)

There are a few other experimental facts which bear on the question of the differential fate of the pole cells. Counce & Ede (1957) analysed the female sterility mutant f_s *nos A* in *Drosophila melanogaster*, in which no secondary pole cell migration occurs but fertile gonad formation is, nevertheless, observed. According to Poulson & Waterhouse (1960), a reduction in the number of cuprophilic cells in the midgut is observed after UV irradiation of the posterior region of *Drosophila* and *Lucilia* eggs. On the other hand, Fielding (1967) found that the female sterility mutant *grandchildless* of *Drosophila subobscura* has no pole cells but normal midgut development. In his pole cell transplantation experiments, Illmensee (1976) found donor cells in the host midgut but could not determine whether they were actually cuprophilic cells or only represented ectopic germ cells.

Counce (1963) stated that, in *Drosophila*, there is no difference in polar granule morphology between pole cells which become PGCs, form midgut cells or are included in the yolk. However, the possibility that pole cells found in the midgut and the yolk represent ectopic germ cells, which would degenerate sooner or later, seems very unlikely in the light of the following observations. Counce & Selman (1955) observed ectopic gonad formation around displaced germ cells after ultrasonic treatment of pre-blastoderm stages. Illmensee & Mahowald (1974, 1976) transplanted posterior pole plasm into the anterior pole region of eggs at early cleavage stages and obtained ectopic pole cell formation. Reimplantation of these pole cells into the posterior or ventral region of genetically marked recipient eggs yielded four per cent donor progeny, demonstrating that posterior pole plasm retains the capacity to support the formation of viable pPGCs in an ectopic position. Similar results were obtained with posterior pole plasm of unfertilised eggs and of oocytes of stages 13–14 (stages after King, 1970), but not with pole plasm of oocytes of stages 10–12. In these younger oocytes the posterior pole plasm is apparently not yet functional, but it certainly becomes functional prior to egg maturation (Illmensee, Mahowald & Loomis, 1976). Similar results were obtained in interspecific transplantations of polar plasm leading to the formation of 'hybrid' pole cells (Mahowald, Illmensee & Turner, 1976). All these experiments demonstrate that ectopic pole cells can survive. Moreover, it is evident that the posterior pole plasm must contain specific cytoplasmic factors for pole cell formation. (See the reviews by Gehring, 1973; Illmensee, 1976 and Mahowald *et al.*, 1979.)

It is evident from many experiments on eggs of Coleoptera, Diptera and Hymenoptera that destruction of the pole cells leads to sterility of the larva and hence to sterile adults. Hegner (1908) was the first to try to remove the posterior pole plasm or the later pole cells by pricking the posterior pole of eggs of the coleopteran *Leptinotarsa*; he obtained incomplete or complete absence of PGCs, while similar pricking of the anterior pole had no such effect. Hegner (1911) performed the same operation with a hot needle. In *Drosophila*, Geigy (1931) destroyed the posterior pole region at the blastoderm stage by local UV irradiation and obtained absence of pole cells and subsequently of PGCs in the larva. Cornman (1943) caused reversible pole cell regression as a result of high hydrostatic pressure. Repeated and prolonged treatment led to irreversible suppression of pole cell formation.

Unfortunately, the ultimate effect on PGC formation is unknown. Counce & Selman (1955) succeeded in preventing the internalisation of the pole cells by ultrasonic treatment, resulting in gonad formation around ectopic cells, for instance in the anterior region behind the mouth. Poulson & Waterhouse (1960) applied UV to the posterior pole region of *Drosophila* and *Lucilia* eggs and obtained the most effective sterilisation at the time of pole cell formation. Hathaway & Selman (1961) used a UV microbeam to destroy the pole cells in *Drosophila melanogaster*, while Jura (1964b) applied UV to the posterior pole region of the egg of *Drosophila virilis* at various stages. Hathaway & Selman and Jura obtained a strong reduction of the number of PGCs or even their complete absence in the larvae. In *D. melanogaster*, Warn (1972) obtained 90 per cent sterile adults by UV irradiation of the posterior pole of the egg, using a shielding device; he obtained similar results by micropuncture of the posterior pole region at pre-blastoderm stages. Pole cell formation could be restored by injection of non-irradiated posterior pole plasm showing that the posterior pole plasm contains a factor essential for pole cell formation (Warn, 1975). This confirmed the results of Okada, Kleinman & Schneiderman (1974a), who further observed that UV irradiation delays the migration of the cleavage nuclei to the posterior periplasm and prevents the formation of protoplasmic protrusions at the posterior pole. Levin *et al.* (1974), by using UV light of 313 nm wavelength, showed that pole cell destruction can also be obtained by interfering with protein synthesis, particularly at the cleavage stages, when the polar granules are associated with ribosomes. Graziosi & Micali (1974) used UV light of 235.7 nm wavelength, which specifically affects RNA, and found that the highest incidence of sterility following irradiation occurred either shortly before 55 minutes after oviposition, or after nuclei have entered the polar region. Micali *et al.* (1978) found a high UV sensitivity immediately after oviposition, followed by a period of decreased sensitivity lasting until the migrating nuclei approach the egg surface and thus become exposed to the UV beam. This recent work was all done on *Drosophila melanogaster*.

There is still another argument in favour of the thesis that, in the Diptera, the pole cells are the precursors of the PGCs. Of the rather large number of female sterility mutants in *Drosophila* species the sex-linked, recessive, temperature-sensitive mutants *grandchildless* in *D. subobscura* (Fielding, 1967) and *grandchildless-like* in *D. melanogaster* (Thierry-Mieg, 1976) are of particular interest, since they cause sterility by affecting the pole plasm, thus preventing the cleavage energids from entering it at the normal stage of development. A much delayed nuclear invasion of the posterior pole plasm does occur but no pole cells are formed, notwithstanding the presence of polar granules (Fielding, 1967; Thierry-Mieg, Masson & Gans, 1972). Gehring (1973, 1976a) assumes that, in these mutants, the polar granules are affected, rendering them inactive in promoting pole cell formation.

Mahowald, Caulton & Gehring (1979) found that the earliest defect in oogenesis involves the anterior and posterior tips of the oocyte, which are devoid of ribosomes, mitochondria and other cell organelles; this apparently causes retardation of the movement of the cleavage nuclei towards both the anterior and the posterior egg cortex. As a consequence, no cleavage energids enter the posterior pole plasm, although a more or less normal syncytical blastoderm forms underneath the non-nucleated posterior pole plasm. (See Gehring, 1973, 1976a; Mahowald, 1979; Mahowald et al., 1979 and Warn, 1979 for a description of various mutants affecting male and female fertility.)

The method of genetic fate mapping in *Drosophila melanogaster*, in which use is made of an unstable ring-X chromosome, so that XX–XO gynandromorphs are formed with random boundaries between the two components (Janning, 1974), confirms that the presumptive PGCs are indeed located in the most posterior region of the egg, while the mesodermal component of the gonad has its primordium in a medio-ventral location (Gehring, Wieschaus & Holliger, 1976; Nissani, 1977; Zalokar, Erk & Santamaria, 1980). Lohs-Schardin et al. (1979) corroborated the genetic fate map, particularly for adult structures including the gonads, by means of very accurate UV laser beam irradiation. Underwood, Turner & Mahowald (1980) recently verified the genetic fate map by the application of very small local defects. Schüpbach, Wieschaus & Nöthiger (1978) and Wieschaus & Szabad (1979) concluded from gynandromorphs and from X-ray-induced somatic recombination, respectively, that only about a dozen pole cells situated at the surface of the embryo actually form germ cells. (See also Janning, 1978 and Janning, Pfreundt & Tiemann, 1979.)

Some observations on Lepidoptera and Homoptera throw some doubt on the general validity of the thesis that the pole cells with their germ plasm are invariably the precursors of the PGCs. Miya (1955, 1957, 1958) demonstrated that, in the lepidopteran *Bombyx mori*, partial or complete cauterisation of the presumptive genital region which corresponds to the localisation of the germinal cytoplasm, does not lead to a corresponding reduction in the number of germ cells; however, a regulatory mechanism of the genital region might be responsible for this result. Achtelig & Krause (1971) and Günther (1971) found that, in the homopteran *Pimpla*, prevention of pole cell formation, or pole cell destruction by X-irradiation did not prevent the later appearance of PGCs in the fully segmented embryo, from which they concluded that the pole cells are not invariably the precursors of the PGCs and that, at least in *Pimpla*, the oosome is not a germ cell determinant. Since these observations are contradictory to the great majority of studies on the possible role of the polar granules, the experiments should, in our opinion, be repeated.

The possible role of the germinal granules in pole cell and germ cell formation

When centrifuging *Drosophila* eggs at 4250 g with the posterior pole up, Jazdowska-Zagrodziňska (1966) found that the polar granules can be displaced and completely removed from the pole plasm, which itself is not affected. The presence of pole plasm without polar granules does not lead to detachment of pole cells, while ectopic polar granules do not induce the detachment of other cells as pole cells, so that both pole plasm and polar granules seem to be required for pole cell formation (see also Counce, 1963). Wolf (1967) was able, by centrifugation, to displace the basophilic material of the posterior pole of the egg of *Wachtliella* to an equatorial region, where it prevented chromosome elimination in adjacent nuclei. However, in *Wachtliella* the non-displaced posterior pole plasm supported the formation of pole cells, although it could no longer prevent chromosome elimination in these cells. Evidently two separate factors are involved, one supporting pole cell formation and one preventing chromosome elimination (see also Chen, 1971). From the fact that only a fraction of the original pole cells become PGCs, Gehring (1976b) concluded that the germ cell determinant must consist of two components, one involved in pole cell formation and one in germ cell determination (see below).

Allis, Waring & Mahowald (1977) devised a method for mass isolation of pole cells in *Drosophila*. Mahowald (1977), who considers the polar granules to be the unique organelle of the posterior pole plasm and later of the pole cells and the PGCs, has made a purified polar granule fraction from isolated pole cells. Waring, Allis & Mahowald (1978) found that this fraction contains one major basic protein of about 95 000 daltons molecular weight. Recently, Allis *et al.* (1979) have succeeded in culturing *Drosophila* pole cells *in vitro*, where they follow their normal developmental programme and remain capable of forming functional germ cells after being re-introduced into a recipient embryo. However, the crucial experiment still has to be done that will enable us to decide whether the germ cell determinant is a particular subcellular organelle, a specific molecular component of that organelle, or a component of the surrounding pole plasm.

In our opinion, the fact that the pole cells, as potential germ cells, may enter other pathways of cellular differentiation when present in particular environments, for example the midgut or the yolk, is much more essential than the still unresolved question of whether the PGCs result from the primary or the secondary migration of the pole cells. The deviant differentiation of many pole cells actually touches the crucial question of their state of determination. We shall return to this question in the general discussion on p. 176.

Chromosome elimination

We have already encountered the phenomenon of chromatin diminution in the nematode *Ascaris*. The related phenomenon of chromosome elimination is found in several dipteran families: the nematoceran families of the Sciaridae, Cecidomyidae, Mycetophilidae and Chironomidae. Moreover, in the nematoceran dipteran *Tipula*, the siphonapteran *Pulex* and the coleopteran *Dytiscus* so-called 'nucleic bodies' (*Nukleinkörper*) are found attached to the long arms of one or both X chromosomes in the germ cells, but not in the somatic cells (Bayreuther, 1952, 1956). Although Charniaux-Cotton (1964) considers chromosome elimination to be relatively rare in the animal kingdom, W. Beermann (1956) points out that nuclear differences between germ cells and somatic cells are less exceptional.

In the nematoceran Diptera, chromosome elimination occurs during early cleavage and involves all the cleavage nuclei except those which happen to lie in the posterior pole plasm. The latter contains particular electron-dense, proteinaceous granules without a limiting membrane which are rich in RNA, are often associated with mitochondria and are considered to be identical to the polar granules of, for example, *Drosophila* (Wolf, 1967, 1969). Nuclei which enter the posterior pole plasm temporarily stop dividing. Their number may be one, two or more, depending on the number of preceding cleavage divisions occurring in the different species. During cleavage a number of chromosomes in the somatic nuclei stay behind in the equatorial plane of the cleavage spindle or become scattered throughout the mitotic apparatus; as a result these are not included in the daughter nuclei and degenerate in the cytoplasm (fig. 12.10).

The number of chromosomes involved in chromosome elimination, called E or L chromosomes, varies enormously among the different genera and species. It ranges between one and three in various *Sciara* species with a relatively low number of chromosomes – 7–9 in the male and 8–10 in the female germ cells (Du Bois, 1933; Metz, 1938) – to about 76 chromosomes in certain *Chironomus* species, where the germ cells have about 80 chromosomes, as against the somatic cells with only two pairs of autosomes (Bauer & Beermann, 1952).

In several species the elimination of a smaller or larger number of E chromosomes in the somatic cells is complicated by additional elimination of chromosomes in the germ cells during the sixth to ninth cleavage cycles, as a consequence of sex differentiation or aberrant chromosome behaviour. In *Sciara coprophila* this elimination involves the two large E chromosomes and one X chromosome in the male (Du Bois, 1932, 1933) (fig. 12.11). Chromosome behaviour during this elimination may vary considerably, for example in the primitive Cecidomyid *Mycophila speyeri* (Nicklas, 1960). There may also be a second round of chromosome elimination in the somatic

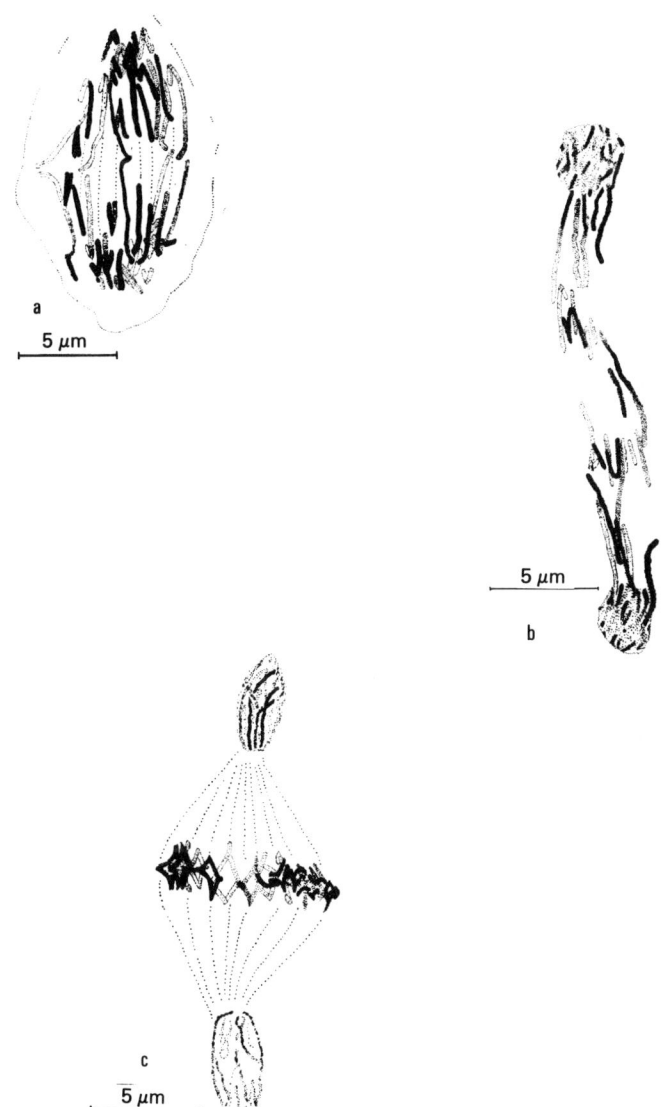

Fig. 12.10. Chromosome elimination in somantic cells of the dipteran *Miastor* sp. during third or fourth cleavage cycle; (a) mid-anaphase; (b) early telophase; (c) late telophase. (After Nicklas, 1959.)

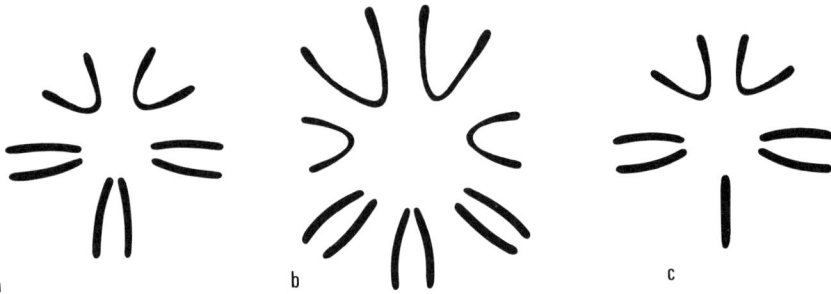

Fig. 12.11. Diagram showing chromosome pattern in *Sciara coprophila*. (a) In somatic cells of female; (b) in germ cells of both sexes; (c) in somatic cells of male. (After Du Bois, 1933.)

cells during the sixth to eighth cleavage cycles, for example in *Heteropeza pygmaea* (Hauschteck, 1962).

The time at which chromosome elimination occurs also varies among the different genera and species. In *Oligarces paradoxus* it occurs during the third cleavage, the three anterior nuclei showing chromosome elimination, the fourth, most posterior one dividing normally (Reitberger, 1939/40). In *Heteropeza pygmaea* the phenomenon takes place stepwise during the third, fourth and fifth cleavage cycles (Camenzind, 1966), in *Sciara* species during the fifth or sixth cycle (Metz, 1938).

Experimental procedures, such as UV irradiation of the pole plasm, temporary ligation, centrifugation leading to displacement of nuclei and dispersion of polar granules, etc., have shown that, whenever cleavage nuclei are prevented from reaching the intact pole plasm at the time that is normal for the species concerned, this leads to chromosome elimination in all nuclei, while the temporary arrest of mitosis characteristic of pole cells does not occur. This was found in *Miastor* species by Nicklas (1959); in *Mayetiola destructor* by Bantock (1961); in *Wachtliella persicariae* by Geyer-Duszynska (1959, 1966) and Wolf (1967, 1969) and in *Sciara coprophila* by Du Bois (1933). Wolf (1967) centrifuged *Wachtliella* eggs in a forced position and observed that the basophilic granules were displaced to an equatorial region, leaving the pole plasm otherwise unaffected. Nuclei located in close proximity to the displaced polar granules did not show chromosome elimination. Apparently, chromosome elimination is a largely 'autonomous' process resulting from changes occurring in the egg endoplasm at a particular stage of development. It can be prevented only by the close proximity of polar granules (Du Bois, 1933; Metz, 1957; Geyer-Duszynska, 1959, 1961; Nicklas, 1959; Counce 1961, and Wolf, 1967). Fux (1975) concluded that, in *Heteropeza pygmaea*, prevention of chromosome elimination is not restricted to the area containing the polar granules, but must be under the control of a gradient of some inhibitory factor originating from this area.

When, under experimental conditions, chromosome elimination takes place in all the nuclei, pole cells are, nevertheless, formed, but they have a reduced chromosome number. PGC development is normal during embryonic and larval life but, at the pupal stage, gametogenesis is arrested. The E chromosomes are apparently required for gametogenesis, and particulary for germ cell differentiation; in other words, development of germ cells is at least partially under genetic control of these 'supernumerary' chromosomes. This was found in *Mayetiola destructor* by Bantock (1961, 1970) and in *Wachtliella persicariae* by Geyer-Duszynska (1959, 1966).

Reitberger (1939/40) and White (1946) considered the large number of chromosomes in the Cecidomyidae to be the result of polyploidy, so that chromosome elimination would serve only to restore the 'normal' chromosome number. White (1950) later began to doubt this conclusion. W. Beermann (1956) called E chromosomes 'supernumerary' to the 'normal' somatic set of chromosomes, the so-called S chromosomes. However, according to Painter (1966), the eliminated E chromosomes are not morphologically different from the S chromosomes. The extra E chromosomes would serve to increase the polysome-forming capacity of the nurse cells in the egg follicles, a need being particularly great in species showing paedogenesis, where embryos develop parthenogenetically at larval stages. On the other hand, Nicklas (1959) stated that the majority of the E chromosomes are not homologous with the S chromosomes, so that there is no question of polyploidy in the germ cells. Normal DNA synthesis occurs in all the chromosomes prior to chromosome elimination. This was confirmed by Rieffel & Crouse (1966) by means of ^{3}H thymidine administration. Camenzind (1966) concluded that, due to the aberrant chromosome behaviour during gametogenesis in *Heteropeza pygmaea* (see below, p. 144), few if any germ line chromosomes can occur in the somatic cells. However, Bantock (1970) still assumes that the high chromosome number in Cecidomyidae may be due to polyploidy. The chromosomal banding pattern in the germ line of *Miastor* also suggests the occurrence of polyploidy (Bregman, 1975). Kunz & Eckhardt (1974) found that highly repetitive satellite DNA sequences without ribosomal cistrons are present in equivalent amounts in S and E chromosomes; they comprise only 15 per cent of the total DNA, so that the rest of the DNA is obviously very different in the two types of chromosome (see the review by W. Beermann, 1956 and the general review by Beams & Kessel, 1974).

Jazdowska-Zagrodzińska & Matuszewski (1978) described extensively a complex system of nuclear lamellae occurring in several Cecidomyidae; they form intranuclear compartments separating the E and S chromosomes from each other during gametogenesis, a phenomenon which may explain their different behaviour in male and female embryos (Kunz, 1970; White, 1973). Similar nuclear lamellae were described by Mahowald (1975) in the pole cells of *Miastor*, by Panelius (1968) in the PGCs of *Heteropeza*, and by Kunz (1970) in the oocytes of several Cecidomyidae.

On the basis of our present knowledge we must conclude that chromosome elimination in somatic cells does indeed represent an irreversible loss of genetic material. The phenomenon itself must, however, be considered as exceptional.

Paedogenesis and parthenogenesis

In the cecidomyid genera *Miastor* and *Heteropeza*, development is paedogenetic and involves apomictic parthenogenesis (Nicklas, 1959). In *Heteropeza pygmaea*, where the choice between bisexual and parthenogenetic development depends on environmental factors, the female-producing eggs undergo only one maturation division and remain diploid, while the male-producing eggs undergo two maturation divisions and become haploid. Thus, after chromosome elimination the somatic nuclei of the resulting females have ten chromosomes and those of the males only five (Hauschteck, 1962).

Parthenogenesis also occurs in the embiopteran *Haploembia solieri*, which has parthenogenetic and bisexual races. Here, parthenogenesis is again accomplished by the suppression of one maturation division, the pronucleus remaining diploid (Stefani, 1959). Parthenogenetic development in the isopteran *Kalotermis flavicollis* is preceded by normal maturation, so that the parthenogenetic egg is haploid. Although sporadic diploid nuclei may appear from the 64-nuclei stage onwards, the embryo remains essentially haploid and the nuclei have a single nucleolus, whereas embryos originating from fertilised eggs are diploid and their nuclei have two nucleoli (Truckenbrodt, 1964). The origin of diploid parthenogenesis in the lepidopteran *Solenobia triquetrella* is not yet fully understood. In parthenogenetic eggs maturation seems normal and identical to that in bisexual eggs. Nevertheless, the parthenogenetic embryo is diploid. It seems that a diploid nucleus is formed by the fusion of two polar body nuclei (Von Borstel, 1957). It is assumed that this fusion nucleus prevents the egg nucleus from developing further; in the bisexual egg a fusion nucleus is also present but there it is blocked by the zygote nucleus (Seiler, 1959, 1960). In the lepidopteran *Bombyx mori* double fertilisation may be responsible for the production of gynandromorphs as a result of fusion of two sperm nuclei into a diploid synkaryon (Astaurov & Ostriakova-Varshaver, 1957).

Gonad development, sex determination and related topics

In the coleopteran *Leptinotarsa* gonad development has been described by Richard-Mercier (1972) and others. The indifferent gonads are formed on the fourth day of development and consist of a group of PGCs surrounded by a sheath of somatic cells. With sexual differentiation the female gonad forms a number of ovarioles and the male gonad a number of testicular follicles. The testis contains relatively more somatic cells than the ovary. The apical testicular tissue and the terminal filament of the ovary are formed from dorsal splanchnic mesoderm. Brien (1965) described the entire genital system of the coleopteran *Lampyris noctiluca*. In the male the two testes, containing testicular follicles, ampullae and an evacuation canal, lead into two spermioducts. These fuse into a single ejaculatory canal with a vesicula seminalis, glands and a terminal penis. In the female the two ovaries, containing a varying number of ovarioles, lead into two oviducts, which fuse into a single uterus and vagina.

In the Apterygota, which are sometimes incorrectly called the Ametabola, at least some of the adult structures, for example the genital apparatus, develop from distinct embryonic anlagen which are called 'imaginal discs'. Imaginal disc formation is more pronounced in the hemimetabolous Pterygota, where for instance the wings and the genital apparatus develop from embryonic anlagen, while most of the remaining adult structures develop gradually from the corresponding larval structures. In the holometabolous Pterygota the majority of the imaginal structures arise from individual imaginal discs, which originate in the embryo and grow as embryonic anlagen during larval life, replacing the larval structures during metamorphosis (see fig. 12.9 on p. 134 and fig. 12.12). Here, the genital apparatus of both male and female is formed from the single, bilaterally symmetrical disc.

Many investigations have shown that destruction of the posterior pole plasm or of the later pole cells leads to the absence of PGCs but does not impair 'normal' gonad development. This destruction was done by means of surgical extirpation in *Leptinotarsa* by Schnetter (1965); cauterisation in *Leptinotarsa* by Haget (1953a) and Richard-Mercier (1977) and in *Calliphora* by Alléaume & Haget (1962); UV irradiation in *Drosophila* by Geigy (1931), Geigy & Aboïm (1944), Poulson & Waterhouse (1960), Hathaway & Selman (1961), Jura (1964b), Laugé & Prudhommeau (1971) and Levin *et al.* (1974) and in *Lucilia* by Poulson & Waterhouse (1960). In the complete absence of germ cells the agametic, mesodermal gonads arise at the normal site and time and go through a normal process of sexual differentiation, forming either a normal testis containing testicular follicles, or a normal ovary containing ovarioles without follicle epithelium (see also Buchner,

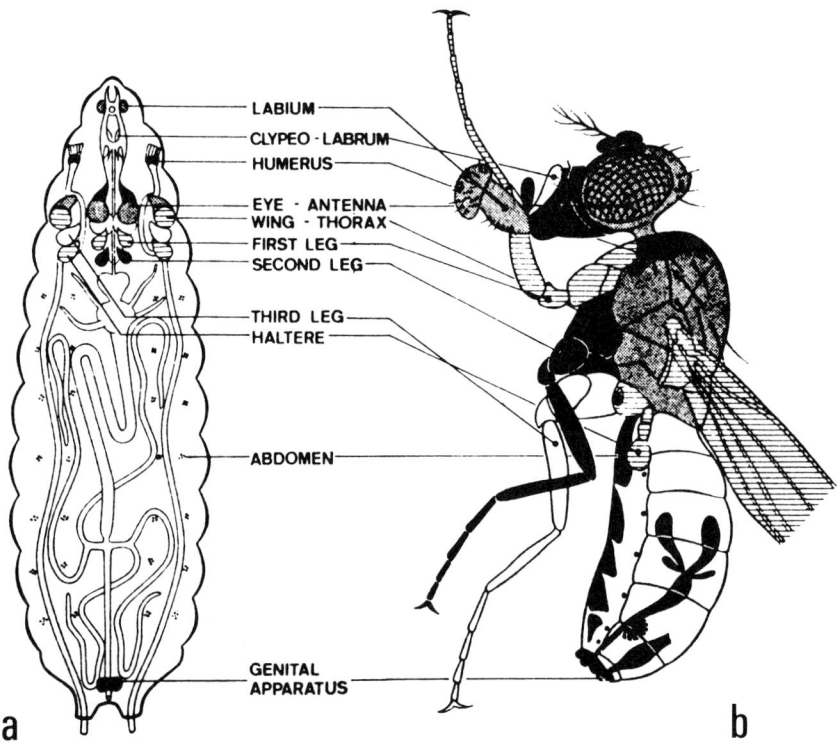

Fig. 12.12. Diagram of (a) larval imaginal discs, and (b) their derivatives in the adult fruit fly *Drosophila melanogaster*. Corresponding parts are indicated with same hatching density. (After Wildermuth, 1970.)

1957). On the other hand, destruction of the mid-abdominal mesoderm suppresses gonad formation (Haget, 1953a). When, in *Leptinotarsa*, the PGCs are prevented from reaching the gonadal anlagen by a local burn between the germ cell and gonadal anlagen, sterile gonads develop, while the ectopic PGCs degenerate (Haget, 1969). Therefore, gonadal development apparently does not require the presence of PGCs. The latter migrate towards the independently formed, indifferent gonadal anlagen and colonise them.

Miya (1955) analysed the mutant *new additional crescent*, E^n, of *Bombyx mori* and found that in the homozygous condition the development of the gonadal anlagen is disturbed although normal PGC multiplication takes place, showing again that germ cell and gonadal development are independent processes. This conclusion was corroborated by genetic mapping in gynandromorphs of *Drosophila* by Gehring, Wieschaus & Holliger (1976), who found separate locations for the anlage of the germ cells at the posterior pole and that of the mesodermal gonadal tissues on the postero-ventral side of the egg. (See Charniaux-Cotton, 1964, and Anderson, 1966c, for reviews.)

Counce & Selman (1955), applying ultrasonic treatment to *Drosophila* embryos, observed that, in cases showing posterior damage, the posterior midgut invagination failed to occur and the pole cells were not internalised normally. They claimed that the ectopic germ cells 'induced' more or less normal but ectopic gonads with a gonadal sheath and interstitial cells. However, the ultrasonic treatment may have displaced the gonadal anlagen to the ectopic sites where the PGCs were ultimately found. From centrifugation experiments in *Lucilia* and *Culex*, Davis (1970) concluded likewise that pole cells are required for gonad sheath formation, but similar objections can be raised against his interpretation.

In insects sex is determined by the number of X chromosomes in the somatic tissues, that is XX or XX' in the female and X in the male, irrespective of the genetic constitution of the germ cells (Metz, 1938). *Miastor* lacks X chromosomes; here, sex is determined by the number of chromosomes eliminated from the somatic cells, that is 36 in the female and 32 in the male, so that sex determination is again somatic (Nicklas, 1959). In *Oligarces paradoxus* egg dimorphism exists, the 'male' eggs being large and the 'female' eggs small (Reitberger, 1939/40).

Seasonal dimorphism in the hymenopteran *Formica rufa rufa-pratensis minor* was described by Bier (1952, 1954). Large eggs (36 μm) with much polar plasm are formed in the period from November to May, giving rise to sexual individuals, whereas small eggs (28 μm) with very little polar plasm are formed in the period from June to September, when sterile workers develop. Seasonal dimorphism can be experimentally controlled by changing the temperature and food supply. Sexual dimorphism in the hymenopteran *Apis mellifera* was analysed by Reinhardt (1960); drones develop from haploid, unfertilised eggs, and worker bees and queens from diploid, fertilised eggs. Although in haploid drone development cleavage is initially delayed in comparison with diploid worker eggs, drone embryos ultimately form the double number of smaller cleavage energids. Artificially fertilised drone eggs develop entirely as worker eggs. In the homopterans *Pseudaulacaspis pentagona*, *Aspidiotus simulans* and *A. destructor* sexual dimorphism manifests itself: in the male the germ cells undergo normal meiosis and become haploid and in the female undergo only simple mitosis and remain diploid. After normal mating the females first produce diploid, coral-coloured female embryos and subsequently haploid, pinkish-white male embryos, the latter undergoing elimination of the paternal chromosome set during early embryogenesis (Brown & Bennett, 1957; Brown & De Lotto, 1959).

In several Dipteran groups highly aberrant types of chromosome behaviour are encountered. In the Orthocladiinae, nematoceran Diptera, which have a set of 'ordinary' chromosomes and a large number of E chromosomes, the zygote is diploid but the PGCs become haploid by partial chromosome elimination. The germ cells actually go through reduction and duplication twice. During the last gonial differential mitosis unequal chromosome distribution results in one of the daughter cells becoming temporarily diploid. It then becomes haploid again during final meiosis. The other daughter cell, which has a reduced chromosome number, becomes a nurse cell in the female and degenerates in the male (Bauer & Beermann, 1952). In the cecidomyid *Sciara ocellaris* one paternal sex chromosome and one L chromosome are eliminated from the PGCs by moving intact through the nuclear membrane and subsequently degenerating in the cytoplasm. In other *Sciara* species there is sexual dimorphism in addition; in *Sciara coprophila* one X chromosome is eliminated in the male during the seventh or eighth cleavage cycle (Du Bois, 1933; Berry, 1941; Rieffel & Crouse, 1966).

Hagan (1951) distinguished four different types of 'viviparity' in the insects, viz. 'ovoviviparity' without special nutritional adaptations (found among the Thysanoptera, Blattodea, Corrodentia, Anoplura, Plectoptera, Homoptera, Lepidoptera, Diptera, Coleoptera and Hymenoptera); 'adenotrophic viviparity', with uterine nutrition of the larva after hatching (found among the Diptera); 'haemocoelous viviparity', with embryonic nutrition by means of a trophamnion, trophoserosa or trophochorion (found among the Diptera and Strepsiptera), and 'pseudo-placental viviparity', characterised by the presence of a pseudo-placenta in the genital tract of the female (found among the Dermaptera, Blattodea, Corrodentia, Homoptera and Hemiptera).

CRUSTACEA

General characterisation

Although the Crustacea are typical representatives of the Arthropoda, with a distinctive type of development, they form a highly diversified group with an immense variation of details. They have a highly variable number of metameric segments, ranging from 13 to 42, including three preoral metameres. The head usually comprises the first six metameres. All segments bear a pair of appendages, each of which branches into two. The head appendages comprise the antennulae, antennae, mandibulae, maxillulae and maxillae. The trunk segments may or may not bear legs. The body ends in a telson. (See Anderson, 1973, pp. 263–364.)

Embryonic development

Although various groups of crustaceans retain a recognisable spiral cleavage pattern, their fate maps (fig. 12.13) do not seem to bear any relation to those of the spiralian annelids. According to Anderson (1969, 1973) they therefore do not constitute a group intermediate between the annelids and the higher arthropods.

There is an immense diversity in cleavage pattern among the Crustacea. In his review Anderson (1973) emphasises the great influence the amount of yolk has on the cleavage pattern. The balanomorph cirripede crustaceans, which have relatively small eggs with densely packed yolk, show a bilaterally modified spiral cleavage, with the various organ anlagen already segregated at the 33-cell stage. At this stage the yolk is confined to the 4D cell, while the mesoderm is represented by the yolk-free 3A, 3B and 3C blastomeres, which together with the yolk-free ectoderm cells almost completely enclose the large 4D cell except for its postero-ventral surface, thus expressing the bilateral symmetry of the blastula. The 4D blastomere divides into two large anterior and two small posterior midgut cells, forming the stomach and intestine respectively. After moving inwards and posteriorly the blastomeres 3A, 3B and 3C first form a single mesodermal mass, which subsequently divides into two masses giving rise to paired mesodermal bands. The stomodaeum is a midventral invagination of the ectoderm, which later connects up with the midgut anlage. The ectoderm becomes subdivided into protocerebral, antennular, antennal, mandibular and post-naupliar embryonic ectoderm areas as well as a dorsal extra-embryonic ectoderm

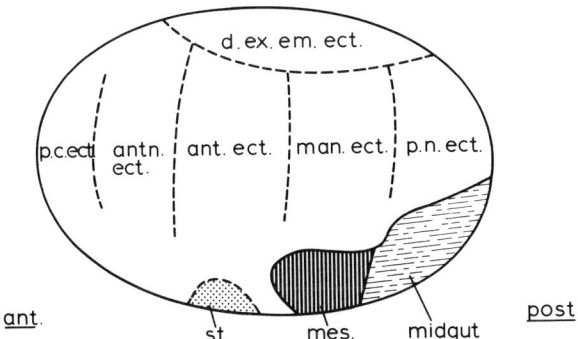

Fig. 12.13. Fate map of a cirripede crustacean in left lateral view. **ant.** = anterior; ant.ect. = antennal ectoderm; antn.ect. = antennulary ectoderm; d.ex.em.ect. = dorsal extraembryonic ectoderm; man.ect. = mandibular ectoderm; mes. = mesoderm; p.c.ect. = protocerebral ectoderm; p.n.ect. = post-naupliar ectoderm; **post.** = posterior; st. = stomodaeum. (After Anderson, 1969 and 1973.)

region (see fig. 12.13). The latter may vary in size from almost non-existent to covering nine tenths or more of the blastular surface. The central nervous system develops from ectoderm situated behind the labrum and stomodaeum anlagen. (See Anderson, 1973.)

In the small eggs of some cladocerans, copepods and penacid malacostracan decapods cleavage is nearly equal and of a radial type, with *Polyphemus* and *Cyclops* as well-known examples. Here, the A, B and C quadrants form mesoderm, 2d the endoderm and 2D the single presumptive PGC (see fig. 12.14, p. 151). The branchiopod *Artemia* shows a very pronounced form of radial cleavage leading to a hollow blastula of 512 cells, still without indications of embryonic axes (Ivanova-Kasas, 1977a,b; Yang, 1977; Zilch, 1978). In the ostracod *Cyprideis*, which shows clear evidence of spiral cleavage, the egg is so rich in yolk that the cleavage furrows penetrate only superficially. The yolky eggs found in some anostracans and astracodans (see Ivanova-Kasas, 1977a,b) and in some malacostracan decapods (see Fioroni, 1969) may show the so-called 'intermediate' cleavage type, which has features of both total and superficial cleavage. Initially, cleavage is of the total, equal type, leading to a solid blastula consisting of pyramidal cells. Subsequently, the peripheral, cytoplasmic portions of the cells, containing the nuclei, separate from the central, yolky parts. The 'yolk pyramids' may fuse into a central yolk mass, while the peripheral cells form a blastoderm. In some cladocerans cytokinesis starts only at the 8-nuclei stage, the yolk being evenly distributed among the octants. 'Yolk pyramid' formation then takes place and a blastoderm is formed. The majority of the malacostracans, which have large, yolky eggs, show intralecithal cleavage initially, after which the cleavage energids move to the surface. Superficial blastomeres are formed, which may cover only a small part of the egg surface at first but ultimately form a complete blastoderm.

Weygoldt (1963) distinguished four different cleavage types, viz. the 'spiral cleavage' type of the Ostracoda, the *'Lepas'* type, the *'Polyphemus'* type, and the 'superficial cleavage' type of the higher Malacostraca. Anderson (1973) has the various cleavage patterns of the Crustacea evolving from that of the Cirripedia; he does not believe in a possible annelid ancestry of the Crustacea. Ivanova-Kasas (1977a,b) distinguishes three different evolutionary lines among the Crustacea; starting from the primitive isolecithal Anostraca and Ostracoda, showing equal spiral cleavage and possibly having annelid ancestry, these would lead to (a) the telolecithal Cirripedia, with unequal cleavage but spiral spindle orientation, (b) the isolecithal Cladocera and Copepoda, with radial cleavage, and (c) the more highly evolved centrolecithal Malacostraca, with superficial cleavage. In our opinion it should be emphasised that not all crustacean groups are characterised by a single type of cleavage; many groups show several of the above diversifications among their representatives, which makes a comparison of the various groups on the basis of egg and cleavage type a rather hazardous undertaking.

Many Crustacea have a free larval stage, the nauplius larva, which may be compared to the 'short-germ' type of the insects, since the nauplius forms only the cephalic segments with the antennular, antennal and mandibular buds. It contains a so-called 'post-naupliar' growth zone which, during a number of moults, will furnish the maxillular and maxillar head segments as well as a large number of trunk segments with or without appendages. The growth zone is usually characterised by the presence of ecto- and mesoteloblasts. Crustacean development shows yet another type of diversity in that the larval stage sometimes is partially or completely eliminated and the embryo develops more or less directly into a juvenile adult, as in *Branchiura*.

Origin of the germ cells and development of the gonads

In the Crustacea the site and time of the first cytoplasmic differentiation of the PGCs is highly variable. Later, mesodermal cells form the gonadal anlage around the PGCs.

In the Branchiopoda, Copepoda, Ostracoda and pericaridan Malacostraca the PGCs segregate early from a definite region of the egg. The presumptive PGC segregates as the 2D cell of the early blastula (fig. 12.14), for example in the cladocerans *Holopedium* and *Polyphemus* and the copepod *Cyclops*. After division of the 2D cell the pPGCs become associated with the post-naupliar growth zone, each of them dividing into a large true PGC and a small gonoduct cell (Anderson, 1969). In the Isopoda, Amphipoda and Tanaidacea the PGCs differentiate during gastrulation as a group of cells situated in the ventral midline between the presumptive mesoderm and the midgut. They become associated with the mesoteloblasts until the first pair of thoracic somites is formed. In the anostracan and conchastracan Branchiopoda the PGCs are first found on the median face of the first pair

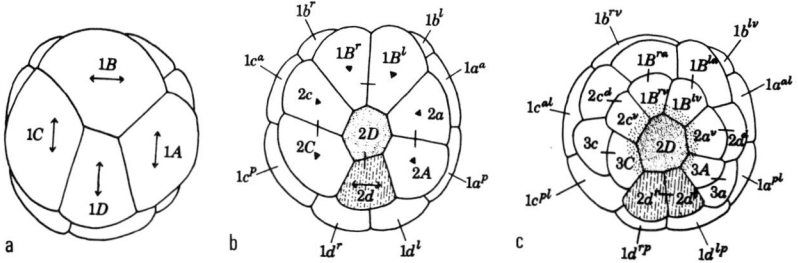

Fig. 12.14. Third to fifth cleavage in cirripede crustacean *Polyphemus* sp., seen in ventral view. (a) 8-cell stage; (b) and (c) 16- and 32-cell stages, with 2D as the presumptive primordial germ cell; stippled area in (c) indicates inner contour. (After Anderson, 1969.)

Arthropoda/Chelicerata

of trunk somites. In the Malacostraca, such as *Hemimysis*, *Nebalia* and *Anaspides*, a pair of PGCs is first distinguishable in the ventral wall of each pair of segmental coelomic cavities, near the pericardial septum. Later, the gonads are formed as mesodermal cells from the pericardial floor surround the PGCs (Manton, 1928, 1934, and Anderson, 1973).

In the egg of the copepod *Cyclops* so-called 'ectosomes' (specific granules found near one of the poles of the first cleavage spindle) are transferred through five differential divisions to the 2D blastomere, that is the presumptive PGC. During the sixth cleavage they are evenly distributed between the two daughter PGCs (see Charniaux-Cotton, 1964).

Zaffagnini & Lucchi (1970) described the ultrastructure of the so-called germ cell determinant found in the eggs of the copepods *Daphnia pulex* and *D. magna*. It is extruded from the oocyte nucleus through the nuclear pores in the form of nuage material, and consists of a coarse network of fine granular and fibrillar material. Moreover, it contains vesicles of different sizes, some dense material and glycogen particles, and is associated with mitochondria and multivesicular bodies. Initially lying in the perinuclear cytoplasm, it later migrates to the cortical ooplasm.

S. Beermann (1959) observed chromatin diminution in *Cyclops strenuus*. In contrast to the PGCs, the somatic cells show a loss of heterochromatin from the ends of the chromosomes during the prophase of the fifth and sixth cleavages. In 1966 and 1977 Beermann observed similar phenomena in *C. divulsus* and *C. furcifer*. The chromosome ends remain in the equatorial plane, while the central regions of the polycentric chromosomes divide and move towards the poles of the cleavage spindle (fig. 12.15). The process shows much resemblance to that in the nematode *Ascaris*. (See also Terpitowska, 1976, and the general review by Beams & Kessel, 1974.)

In the framework of his studies on gonadal differentiation, Brien (1965) observed complete sex reversal of genetic females into functional males, and vice versa, under the influence of male and female hormones, respectively. Reversal involved both the generative and the somatic sexual characteristics. Whereas oocytes differentiate in the absence of hormones, spermatogenesis requires the presence of an androgenic hormone (see also Charniaux-Cotton, 1964).

CHELICERATA (Arachnoidea)

General characterisation

In the Chelicerata the head and the thorax are unified into the cephalothorax. This bears six pairs of appendages, the first two pairs, the preoral cheliceres and the postoral pedipalps (which are not homologous to the

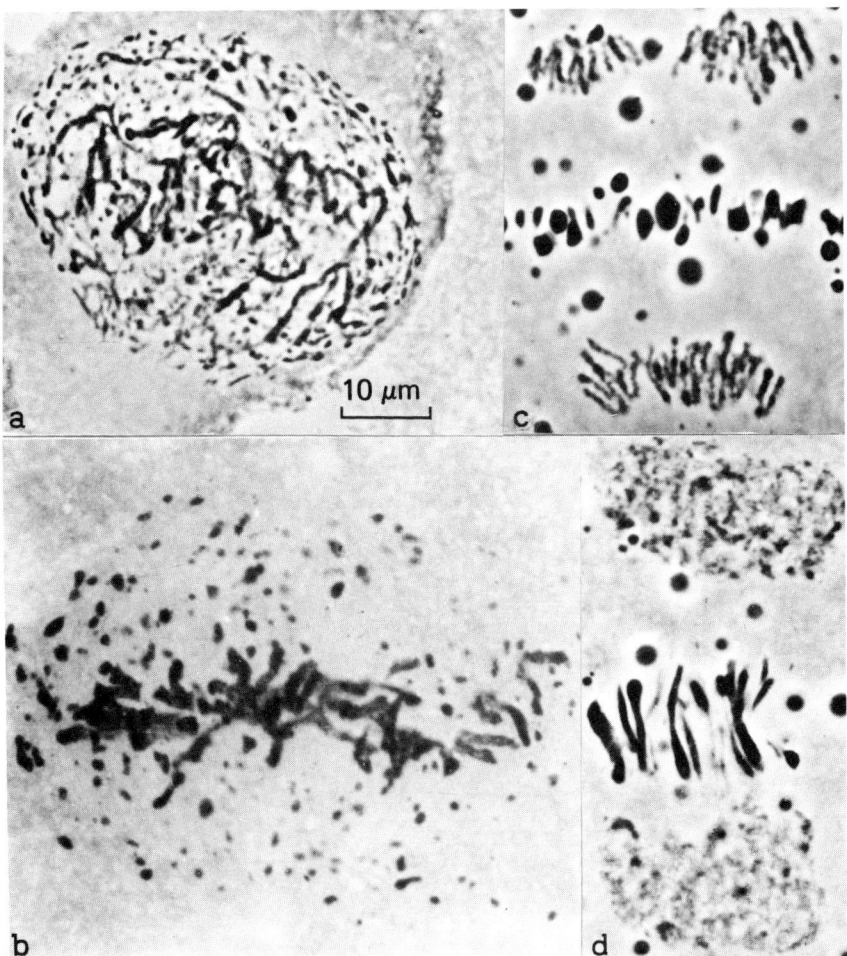

Fig. 12.15. Successive steps in chromosome elimination in somatic cells of the copepod *Cyclops strenuus*. (a) Prophase; (b) metaphase; (c) anaphase; (d) telophase. (Courtesy S. Beermann, 1977.)

antennulae and antennae of other arthropods) serving for feeding and defence, while the remaining four pairs serve for locomotion. The abdomen bears no legs. In addition to the chitinous exoskeleton, a sort of endoskeleton has developed, on which the muscles insert. The nervous system consists of a double chain of segmental ganglia and a pair of more complex cerebral ganglia in the precheliceral lobes. There are several sets of single, well-developed eyes. The respiratory organs are either book-lungs or tracheae. The second, so-called 'opisthosomal', segment is the gonadal segment. The gonads develop from coelomic sacs and the gonoducts are coelomoducts. All Chelicerata have direct development. They emerge from the eggs as juveniles of adult bodyform having the full complement of segments (see Anderson, 1973, pp. 365–451).

The Chelicerata comprise, *inter alia*, the scorpions (with a poison gland at the tip of the large postabdomen), the spiders (with spinning-glands), and the primitive mites and ticks (with fused cephalothorax and abdomen).

Embryonic development

The eggs are usually rounded or ovoid and rich in yolk. They have a central nucleus, a reticular inner cytoplasm and a thin periplasm. The cleavage pattern of the Xiphosura, for example *Limulus*, and the Acarina (mites and ticks) is more primitive than that of the other arachnoids. Although the first nuclear divisions are intralecithal, furrowing begins at the animal pole and ultimately leads to a columnar blastoderm surrounding the undivided central yolk. In other arachnoids the first three nuclear divisions occur at first without cytokinesis, but the cytoplasm then becomes divided into eight 'pyramids' surrounding a central blastocoel. Subsequently, a yolk-free blastoderm is formed while the 'yolk pyramids' fuse into the central yolk mass. Some blastoderm cells invade the yolk and become vitellophages, which will later form the midgut epithelium. The eggs of the ovoviviparous scorpions have a polar cap of yolk-free cytoplasm and show discoidal cleavage, while the small eggs of the viviparous scorpions show total cleavage and coeloblastula formation. In the pseudoscorpions cleavage is at first total, leading to eight equal blastomeres; later the yolk-free micromeres segregate from the yolk-laden macromeres, the former giving rise to a peripheral blastoderm and the latter forming an inner yolk mass.

In the arachnoids the germ anlage is formed by accumulation of blastoderm cells on the ventral side accompanied by attenuation on the dorsal side. A gastral groove appears in the midline of the embryonic primordium, followed by elongation and subsequent segmentation of the germ band. A broad cephalic lobe with precheliceral and cheliceral segments, a pedipalpal segment, four ambulatory segments of the prosomal, and a growth zone for the opisthosomal segments can be distinguished. The midventral gastral groove may extend through all segments or it may only be short, shifting posteriorly during germ band formation, as in ticks and mites. The extent of primary and secondary segmentation varies considerably among groups. In the scorpions the growth zone is situated directly behind the cephalic lobe, while in other arachnoids it produces only the opisthosomal segments.

Apart from the dorsal extra-embryonic ectoderm, in many forms a wide ventral sulcus separates the left and right segmental anlagen, forming a zone of ventral extra-embryonic ectoderm (compare fig. 12.2 on p. 111). In the scorpions the mesoderm and the midgut usually delaminate from the ectoderm without gastral groove formation. In embryos with a gastral groove its anterior end becomes the stomodaeum and mouth, while the proctodaeum forms much later as an independent invagination of the future telson. In the scorpions, during the early elongation of the germ band a serosal layer is formed as well as an amnion.

In the scorpions the germ anlage releases vitellophages, which surround the yolk mass. In the other arachnoids the midgut develops in two parts; the major portion is formed by the vitellophages of the yolk mass, while the posterior part develops by proliferation from the growth zone. In Acarina the growth zone produces the major portion of the midgut. The mesoderm proliferates along the ventral midline of the germ band, forming paired somites and an unsegmented terminal mesodermal mass. In limb-bearing segments the somites form an appendicular lobe.

There is a rather fundamental divergence between the scorpions and the other arachnoid orders. Nevertheless, the embryonic development of the Chelicerata as a whole follows a distinctive pattern. Their fate map (fig. 12.16, p. 156) has nothing in common with those of the Crustacea or the Onychophora, Myriapoda and Hexapoda. Chelicerate development is, moreover, devoid of any reminiscence of spiral cleavage. (For individual references see the general review by Anderson, 1973, pp. 365–451.)

Origin of the germ cells

Very little is known about the origin of the germ cells in the Chelicerata. In the scorpions the PGCs segregate more precociously than in the other groups. A compact group of PGCs proliferates from the centre of the germ disc. The cells remain located beneath the growth zone during the formation of the opisthosomal segments, and later migrate towards the gonadal coelom (Moritz, 1957). In the spiders the PGCs are recognisable only at a late stage of development as large cells in the walls of the third to sixth pairs of opisthosomal coelomic sacs (Kautzsch, 1910).

General conclusion

The Arthropoda comprise, on the one hand, the Myriapoda and Hexapoda, which form a natural assemblage with the Onychophora, and, on the other hand, the Crustacea and Chelicerata. The latter two groups stand so far apart from each other and from the Onychophora – Myriapoda – Hexapoda assemblage that the Arthropoda must be considered to be polyphyletic in origin.

In the Myriapoda the germ cells originate at a relatively late stage of

156 *Arthropoda*

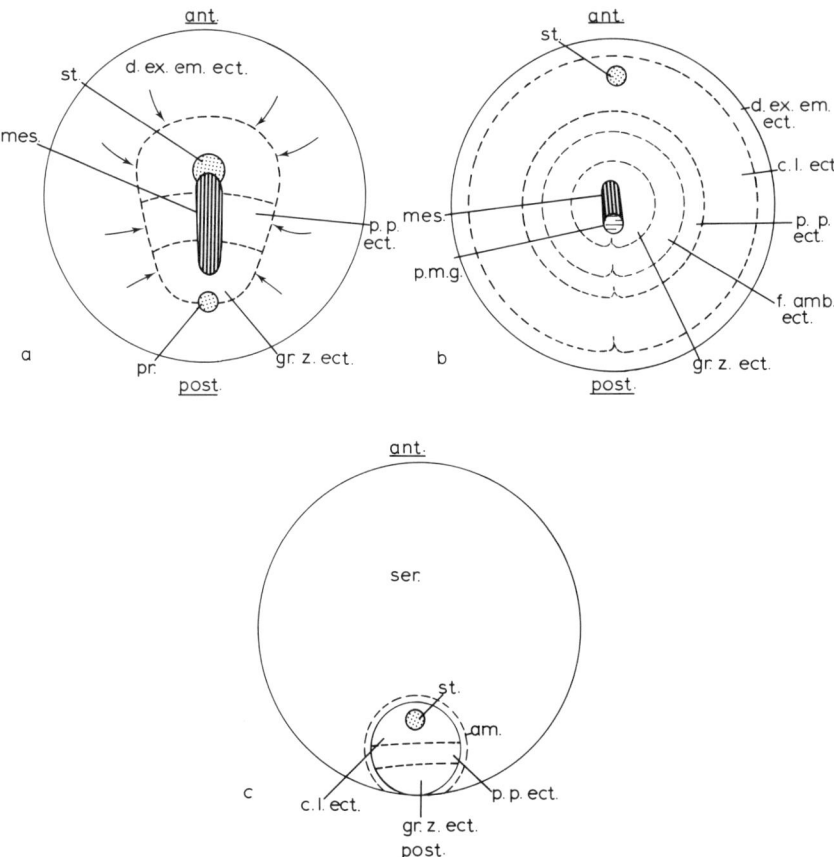

Fig. 12.16. Fate maps of chelicerate blastoderms. (a) *Xiphosura*, arrows indicate extension of dorsal extraembryonic ectoderm; (b) spiders; (c) scorpions. am. = amnion; **ant.** = anterior; c.l.ect. = cephalic lobe ectoderm; d.ex.em.ect. = dorsal extraembryonic ectoderm; f.amb.ect. = first ambulatory segment ectoderm; gr.z.ect. = growth zone ectoderm; mes. = mesoderm; **post.** = posterior; p.m.g. = posterior midgut; p.p.ect. = pedipalpal ectoderm; pr. = proctodaeum; ser. = serosa; st. = stomodaeum. (After Anderson, 1973.)

development in the medio-ventral lobes of somites of the genital segments. Germ cell origin in this group resembles that in the Onychophora, although there they may arise during gastrulation.

Among the apterygote Hexapoda, in the Collembola the PGCs segregate during intralecithal cleavage, but in the Thysanura only during blastokinesis.

In the pterygote Hexapoda both the time and site of origin of the germ cells vary considerably. In the Hemimetabola the time varies from as late as the advanced embryo, when the germ cells arise in the mesoderm of the abdominal segments, to as early as the germ anlage, where they arise at its posterior end. In the Holometabola the time of origin varies from as late as germ band formation to as early as the syncytial blastoderm stage, when the presumptive germ cells segregate in the form of posterior pole cells. The latter situation obtains in many Coleoptera, Diptera, Siphonaptera and Hymenoptera. Here germ cell origin is generally considered to be of the typically preformistic type. Among the hexapods the site of germ cell origin may vary from a ventral via a 'non-distinct' to a posterior location.

The dipteran pole cells, which are characterised by the presence of germ cell-specific polar granules, can, nonetheless, be regarded only as potential germ cells, since most of them never reach the gonadal anlagen but differentiate into midgut cells and vitellophages. Germ cell determination is therefore a stepwise process, which is completed only after the PGCs have reached the gonads. Although many authors regard the polar granules as the germ cell determinant, their role in pole cell and germ cell formation has not yet been fully elucidated. Moreover, in many insects germ cell formation is not characterised by the presence of germinal granules.

Chromosome elimination is encountered in several dipteran families. It occurs in all somatic cells during early cleavage and probably involves a partial loss of unique genetic information. Chromosome elimination does not occur in the germ cells, where it seems to be prevented by the presence of the polar granules.

In the Crustacea the time of origin of the germ cells is highly variable, ranging from early cleavage to late segregation within the walls of the mesodermal coelomic sacs. Chromatin diminution has been observed in the copepod *Cyclops*, while germ plasm has been described in the germ cells of the copepod *Daphnia*.

In the Chelicerata the PGCs appear either early in the germ anlage, or late in the walls of the mesodermal coelomic sacs.

Surveying the entire phylum of the Arthropoda, it is evident that germ cell origin generally has pronounced preformistic features. However, several groups in which the germ cells appear rather late also show epigenetic trends. Consequently, germ cell formation in the Arthropoda as a whole exhibits both the typically *preformistic* and the so-called *intermediate* modes.

Gonad development in the Arthropoda seems to be independent of the presence of germ cells. The site of origin of the gonadal anlagen is usually different from that of the germ cells, so that the latter must be actively or passively displaced to the gonadal anlagen.

13

Remaining phyla

There are quite a number of phyla of which little or nothing is known as to germ cell origin; indeed, even their embryonic development has been studied only superficially. That is why we have placed all these phyla in a separate chapter at the end of the systematic part. We shall discuss only those phyla of which at least something is known regarding their germ cell origin. In analogy with the preceding chapters we shall present a short characterisation of each phylum and a brief discussion of its embryonic development and the possible origin of the germ cells.

Eleven phyla will not be discussed since no relevant data are available on germ cell origin. They are the acoelomate Acanthocephala, the pseudocoelomate Kinorhyncha and Nematomorpha, and the eucoelomate Brachiopoda, Phoronida, Sipunculida, Priapulida, Tardigrada, Pentostomida, Pogonophora and Hemichordata. The remaining phyla will be arranged according to the taxonomic classification proposed in the introduction (see pp. 5–6).

CTENOPHORA

Ctenophora are biradially symmetrical Radiata with an oral–aboral axis and an aboral sensory region. They consist of an outer epidermal layer bearing eight meridional rows of ciliary comb plates, a voluminous ectomesoderm containing connective tissue, muscle fibres and amoebocytes, and a rather complex gastrovascular system. The latter consists of an ectodermal stomodaeum or pharynx, a central infundibulum, eight meridional canals underlying the ciliary bands and two tentacular canals. The ctenophores are hermaphrodites with separate ovaries and testes situated along the meridional digestive canals. Adult ctenophores are able to regenerate lost parts. They may even regenerate completely from small fragments, leading to a form of asexual reproduction.

The Ctenophora were originally classified under the Coelenterata. Due to their different spatial organisation and completely aberrant development they are now placed in a separate phylum. (See Spek, 1926; Hyman, 1940, vol. 1, pp. 662–96; and the recent reviews by Uchida & Yamada, 1968b, and Pianka, 1974.)

The oral–aboral axis of the egg as well as the localisation of the comb-plate-forming potential are established at the time of first cleavage (Freeman, 1977). The egg shows a strictly determinate cleavage pattern, with splitting-off of primary micromeres from the initially formed eight more-or-less equal blastomeres, arranged in two tiers of four cells (fig. 13.1a–c). The primary micromeres, containing the greater part of the original ectoplasm of the egg, form the future ectoderm, while the remaining ectoplasm is split off from the macromeres later in the form of secondary micromeres (fig. 13.1i). Ortolani (1963a,b; 1964) concluded from marking experiments that these represent the presumptive ectomesoderm. According to Siewing (1977), the ctenophores have no endomesoderm but only internalised myoepithelial cells. Gastrulation takes place by epibolic extension of the dividing primary micromeres around the invaginating large endodermal macromeres (fig. 13.1d–h). After differentiation of the various organ anlagen a free-swimming 'cydippid' larva is formed.

The fact that gonad formation occurs along the meridional digestive canals suggests an endodermal origin of the germ cells late in development. It is not known whether *de novo* formation of germ cells occurs during regeneration.

NEMERTINI (Rhynchocoela)

Nemertini are cylindrical or more-or-less flattened acoelomate Bilateralia with a nerve-less cephalic lobe. The body is not segmented, although the internal organs may show pseudometamerism. Underneath the ciliated epidermis lies a muscular layer consisting of longitudinal and circular muscle sheaths. The mouth is situated ventro-cranially. An evertible proboscis is enclosed in a tubular cavity, the rhynchocoel. The digestive tract is composed of various regions and ends in an anal opening at the posterior end of the worm. There is a closed circulatory system with one dorsal and two lateral blood vessels. The excretory organs are represented by one to many pairs of protonephridia. The well-developed paired cerebral ganglia are connected to a pair of longitudinal nerve cords. Ectodermal ocelli and a pair of cerebral organs are innervated by the brain. The dioecious gonads are simple sacs opening to the exterior. They are paired and arranged in pseudometameric order (see Hyman, 1951a, vol. 2, pp. 459–528).

The eggs show determinate, spiral cleavage with three, four, or sometimes five quartets of large micromeres. The fourth quartet contains a larger 4d

160 *Remaining phyla*

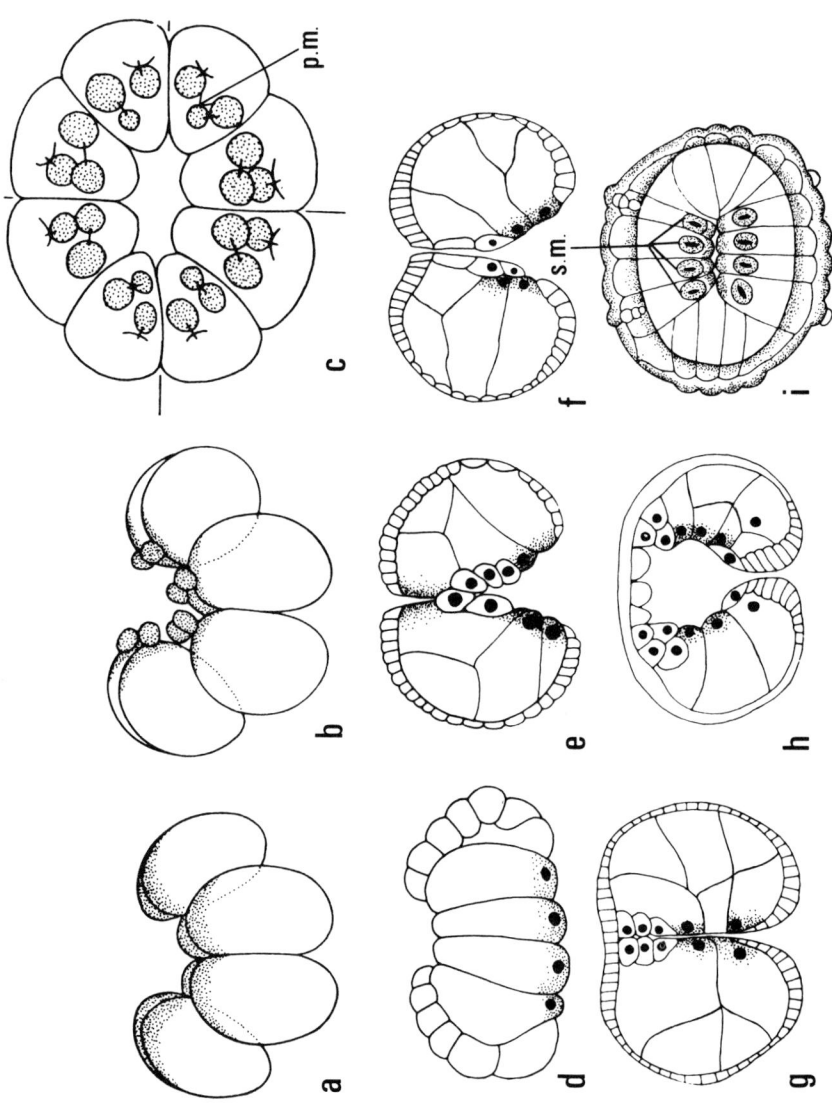

cell which will form the endomesoderm. The ectomesoderm is formed by the second and third quartets. Either a hollow coeloblastula or a solid stereoblastula develops, the former showing invagination and the latter polar ingression of macromeres and mesodermal micromeres. The mesodermal cells become mesenchymatous, fill up the blastocoel and differentiate partly into muscle fibres.

Development is usually indirect via a free-living ciliated larva, called the 'pilidium' (Gontcharoff, 1960). Some *Lineus* species lack a free-living larval stage (Schmidt, 1964). During metamorphosis several ectodermal discs are formed, which will give rise to adult organs: there is a single proboscis disc and paired brain and cerebral organ discs. At the original blastoporal site the stomodaeum with the proboscis and the surrounding rhynchocoel develops, while the proctodaeum is formed at the posterior end.

Hörstadius (1937) found that fragments of *Cerebratulus* oocytes form normal dwarf embryos, provided they contain a nucleus and are fertilised. Isolated blastomeres of the two- to four-cell stage also form dwarf embryos (fig. 13.2), but animal and vegetal isolates of the eight-cell stage and beyond form partial embryos corresponding to the prospective significance of the isolated region, so that development is determinative beyond the four-cell stage.

The nemertines have considerable regenerative capacities. Both the front end with the proboscis and the posterior end of the worm can regenerate. In

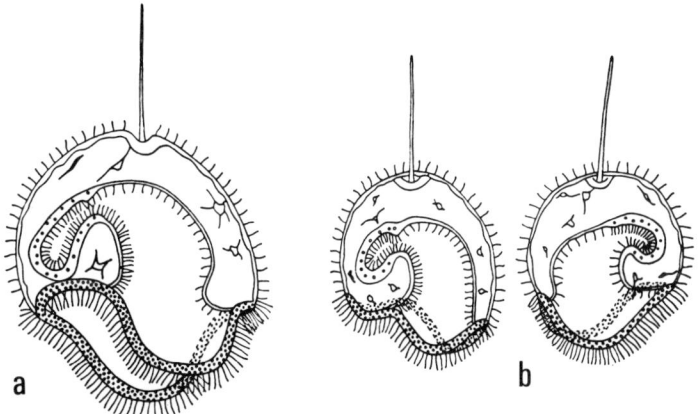

Fig. 13.2. (a) Normal pilidium larva of the nemertine *Cerebratulus lacteus*. (b) Mirror-image dwarf-larvae after isolation of blastomeres at 2-cell stage. (After Hörstadius, 1937.)

Fig. 13.1. (a–c) Cleavage and formation of primary micromeres (p.m.) in the ctenophore *Beroe ovata*; (a) 4-cell stage; (b) 8-cell stage; (c) 32-cell stage. (d–h) Epiboly of primary micromeres and invagination of macromeres. (i) Formation of secondary micromeres (s.m.). (After Siewing, 1977.)

some species with very high regenerative capacity a form of asexual reproduction occurs through local constriction and subsequent regeneration of the missing parts (scissiparity) (see Stolte, 1936 and Berrill, 1952c). After wounding or fragmentation the wound is closed by epithelial cells migrating from the basal layer of the epidermis. Beneath the wound epidermis a typical blastema is formed by neoblasts present in the parenchyma. Several organs such as the musculature and proboscis are formed from the blastemal mesenchyme, while the epidermis gives rise to the brain, the cerebral organs, the mouth and the linings of fore- and hind-guts (see Hermann, 1978/9).

It seems likely that, as in the planarians, during regeneration or asexual reproduction, the germ cells develop *epigenetically* from undifferentiated cells, which may either persist in the parenchyma or may be formed *de novo* by dedifferentiation (Dawydoff, 1928). Unfortunately, nothing is known about germ cell formation during embryogenesis.

MESOZOA

The term 'Mesozoa', proposed by Van Beneden (1876), indicates that this would be a very primitive group, systematically intermediate between the Protozoa and Metazoa. However, this point of view has now been abandoned by most authors. The Mesozoa are now considered as some form of highly degenerate acoelomate flatworm related to the digenetic trematodes or the cestodes (Stunkard, 1937). As we will see, they also have some characteristics in common with the nematodes.

The Mesozoa, which belong to the acoelomate Bilateralia, are endoparasites living in the cephalopod kidney. They are characterised by cell constancy for all cell types except the 'germinal' cells. The animals are solid in structure and consist of an outer layer of somatic cells and one or more inner, so-called axial, cells. The latter have a germinative function, producing agametes. The nucleus of the axial cell undergoes a number of endogenous divisions, as a result of which the cell becomes filled with agametes, which are nourished from the remains of the axial cell.

The life cycle, which consists of an alternation of sexual and asexual generations, has been studied extensively by Nouvel (1947, 1948) and by McConnaughey (1951). A ciliated larva with two or three axial cells containing agametes infects a young cephalopod and grows out into a 'stem nematogen'. Inside the latter the agametes multiply and form 'primary nematogens' or vermiform larvae, consisting of only a few ciliated cells and a single axial cell. Several generations of primary nematogens may be formed by endogenous multiplication. Modified nematogens, the so-called 'rhombogens', are formed at sexual maturity of the cephalopod, when the renal organs of the latter have become heavily infected. The agametes of the rhombogens constitute the sexual generation of 'infusorigens', very simple organisms whose outer and inner cells form egg cells and spermatozoa, respectively. Inside the infusorigen autofertilisation occurs. The zygotes of the sexual generation form still other, more complex, individuals, the 'infusoriforms'. The latter leave the axial cell of the grandparents and become free-living. However, the infusoriform is not the infectious form. Here, a part of the life cycle is still unknown. Rhombogens may occasionally form secondary nematogens (McConnaughey, 1951), or infusorigens may stop forming infusoriforms and start forming secondary nematogens by polyembryony (Nouvel, 1948). In this way the continuous infection of the host is guaranteed. (See Hyman, 1940, vol. 1, pp. 233–47, and also Grassé, 1961, and Austin, 1964.)

In the sexual infusorigen the axial cell splits into a small cell giving rise to egg cells, and a large cell forming spermatozoa. McConnaughey (1951) observed normal chromosome reduction only during spermatogenesis, but Short & Damian (1967) found that, in *Dicyema aegira*, chromosome reduction occurs both in oogenesis and in spermatogenesis, leading to haploid gametes and a diploid zygote. In the asexual generations a similar unequal division occurs, leading to the formation of a small cell giving rise to the epithelial, somatic cells and a large cell forming the agametic axoblast(s). Lameere (1919) was the first to observe chromatin diminution in the somatic cells, a process which occurs neither in the agametic axoblasts nor in the male and female germ cells. This was confirmed by Goldschmidt & Lin (1947) and by McConnaughey (1951). The axoblasts must therefore be considered as true germinal cells. In both the sexual and the asexual generation a very early germ or germinal cell formation can be distinguished, which differs clearly from the formation of the somatic cells (fig. 13.3), which is characterised by chromatin diminution.

The cell constancy of the somatic cells and the sharp and early distinction between germinal or germ cells on the one hand and somatic cells on the other characterises the development of the Mesoza as *highly preformistic*; moreover, it shows distinct parallels with the development of the nematodes.

ROTIFERA

The very small, sessile, pseudocoelomate Rotifera are characterised by a round or slightly flattened body with a terminal anterior mouth surrounded by a double spiral of cilia, a more-or-less straight intestinal tube, and an anal opening located at the base of a short posterior glandular stalk, by which the animal is attached to the substrate. There are protonephridia, which

164 *Remaining phyla*

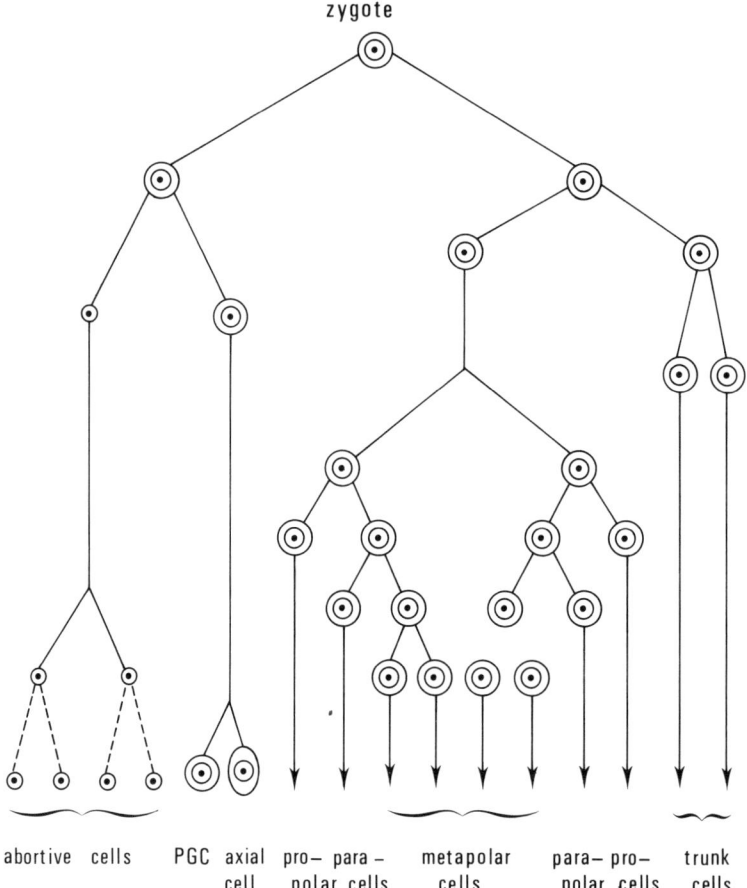

Fig. 13.3. Cell lineage of primary nematogen of the mesozoan *Microcyema vespa*. (After Grassé, 1961.)

debouch together with the gonoduct into the cloaca. The animals show sexual dimorphism. Although parthenogenesis is common among the Bdelloïdes, of which only female animals are known, *Asplanchna* species are either parthenogenetic or bisexual and, moreover, viviparous. Parthenogenetic eggs have a thick shell and require a resting period for further development, while amphimixic eggs have a thin shell and do not require such a period. Cleavage is unequal and basically of the spiral type. The larger D cell becomes enclosed by the descendants of the A, B and C blastomeres. The D cell divides into a so-called primordial yolk cell, which degenerates, and a primordial germ cell characterised by the presence of a 'germinal determinant' (Nachtwey, 1925).

It is evident that the Rotifera are characterised by very early and *preformistic* germ cell formation.

GASTROTRICHA

The pseudocoelomate Gastrotricha are small, non-sessile freshwater and marine animals with an anterior mouth surrounded by rings of cilia, and a dorso-posterior anus situated close to the posterior attachment gland. The straight intestinal tube has a muscular pharynx. There are protonephridia. As in the Rotifera, parthenogenesis is common; for instance, no male individuals are known of the freshwater gastrotrichs.

The slightly elongated egg divides nearly equally, forming first a single and then a double quartet of blastomeres. The cephalic quartet of slightly smaller cells turns 45° about the longitudinal egg axis such that a medioventral row of two cells is formed; these divide transversely and then sink in. The cells bordering the blastoporal groove thus formed also divide transversely and form the stomodaeum anlage. Another posterior depression develops into the proctodaeum. On either side of the latter, two PGCs can soon be distinguished (Sacks, 1955). In the Gastrotricha the PGCs evidently segregate rather early in development.

ENTOPROCTA

The Entoprocta are tiny pseudocoelomate Bilateralia. They are solitary or colonial, stalked, sessile animals with a distal, bowl-shaped calyx provided with a circlet of ciliated tentacles; this surrounds the atrium or vestibule, which encloses the mouth and the anus, hence the name Entoprocta. The mouth and anus are connected by a U-shaped alimentary tract. In the solitary forms the stalk is directly attached to the substrate by means of a foot plate; in the colonial forms it is a branch of the stolon. The stalk possesses basal and intercalary muscular nodes. While the boundary between calyx and stalk is marked by a constriction, the stolon also shows regular constrictions and consists of so-called 'fertile' and 'sterile' sections, the former with, the latter without calyx (fig. 13.4). A subenteric ganglion and two gonads lie in the concavity of the U-shaped digestive tract. A straight gonadal duct opens into the atrium or vestibule. The internal organs are embedded in a loose mesenchyme, which is called the pseudocoelomic cavity. The animals are dioecious or hermaphroditic; it is possible, however, that the dioecious animals are actually protogynous or protandrous hermaphroditic animals. The paired, sac-like ovaries are connected to the atrial

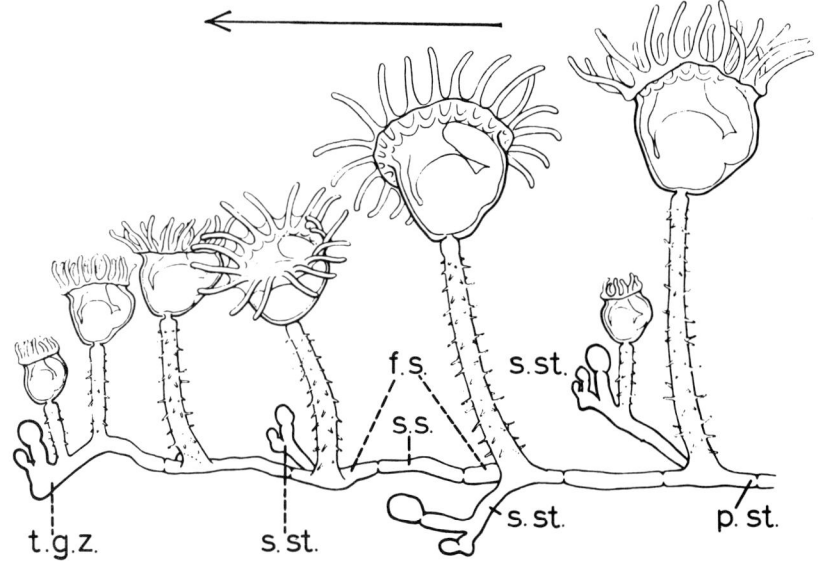

Fig. 13.4. Budding in the endoproct *Pedicellina cornua*. f.s. = fertile segments; p.st. = primary stolon; s.s. = sterile segment; s.st. = secondary stolon formation; t.g.z. = terminal growth zone of primary stolon. Arrow indicates direction of growth. (After Brien, 1956.)

cavity by means of a straight oviduct containing gland cells, which secrete the vitelline membrane surrounding the eggs. The paired testes are also connected to the atrial cavity. In hermaphroditic animals the ovaries and testes have a common gonoduct (Mariscal, 1975). (See Hyman, 1951*b*, vol. 3, pp. 521–44.)

The phylum comprises three families, the Loxosomatidae, the Pedicellinidae and the Urnatellidae, which together number about 120 species.

The main mode of propagation is asexual reproduction. Depending on the species, budding may occur on the calyx, the stalk or the stolon (Brien, 1956). Budding involves both the ectodermal layer and the underlying mesenchyme. Local cellular proliferation leads to the formation of an ectodermal thickening, which sinks into the mesenchyme, forming a solid ectodermal ball. Within the latter two interconnected cavities are formed. The distal vesicle develops into the atrial cavity, the proximal one into the digestive tract. The persisting opening between the two cavities represents the future mouth and stomodaeum, while the anus opens secondarily into the atrial cavity through a proctodaeal funnel. The ectoderm of the atrial cavity forms finger-like outgrowths, the tentacles, which are first directed inwards but later, when the atrial wall is perforated, turn outwards. The subenteric ganglion originates from the atrial floor, while the muscles and the gonadal anlagen and gonadal ducts are formed from mesenchymal cells.

In *Barentsia discreta* three types of regeneration can be distinguished, viz. from the calyx, the stalk or the stolon. Double malformations of the calyx result from calyx regeneration without retention of the original calyx. Stalks can regenerate a calyx both apically and basally, but muscular nodes only apically. The stolon can regenerate distally but not proximally from 'fertile' as well as 'sterile' sections. The stolon and the muscular node therefore show a clear proximo-distal polarity (Mukai & Makioka, 1978). Under unfavourable conditions calices degenerate, while hibernacula are formed as thickened finger-like outgrowths of the base of the stolon. Favourable conditions lead to calyx regeneration and new bud formation (Mariscal, 1975).

Sexual reproduction seems to be favoured by higher temperatures. In the majority of species the eggs are brooded in the bowl-shaped atrial cavity, where they become attached to the so-called embryophore by means of the vitelline membrane. The embryos are released only when fully developed.

The yolky egg shows total, unequal cleavage of the spiral, determinate type. Five quartets of rather large micromeres and a sixth quartet of macromeres are formed. The first three quartets form the presumptive ectoderm, the other three the endoderm, while the 4d cell gives rise to the endomesoderm. A hollow coeloblastula develops. The endodermal cells sink in first, followed by the 4d cell. The initially round blastopore becomes slit-shaped and then closes. A stomodaeum is formed nearby as an invagination of the gastrula wall. From the stomodaeum anlage, cells are pinched off to form mesenchyme. A proctodaeum anlage then invaginates, followed by the formation of a groove on the ventral surface of the embryo between the mouth and anus, the future atrial cavity. Apical plate cells form a ciliated apical organ. The subenteric ganglion and the protonephridia arise from the ectoderm of the atrial floor, while a girdle of locomotory cilia develops from trochoblast cells around the equator of the larva. The free-swimming larva has often been characterised as a modified trochophore larva but actually is a miniature adult (Mariscal, 1965, 1975).

The free-swimming larva of *Pedicellina* attaches itself with its oral surface to the substrate, thus sealing off the atrial cavity. The larva divides into three portions; the lower portion forms the attachment disc with the foot gland and stalk, the middle portion degenerates, while the upper portion containing the U-shaped gut rotates through 180°. As a result the mouth and anus become directed distally and form the calyx of the adult, while tentacles grow out from the atrial wall.

Certain species show larval budding, in which miniature adults develop from internal or external buds. The former are set free by rupture of the dorsal surface of the larva, which then perishes (Jägersten, 1964; Nielson, 1966). Nielson (1967) has described how, in *Loxosomella*, the larva attaches itself to the substratum with the frontal or preoral side. The ciliary girdle then contracts, sealing off the larval vestibule. No rotation occurs and the stalk grows out from the attached preoral side. Finally the vestibule reopens and tentacles grow out (Mariscal, 1975).

Although no special studies are available on the origin of the germ cells in the Entoprocta, several authors think that the PGCs derive from common mesodermal cells at a late stage of development. Although embryonic development is determinate, with typical spiral cleavage and a 4d cell representing the future endomesoderm, later development is highly regulative.

In the Entoprocta the mode of germ cell formation is probably *epigenetic*, as in all animal groups showing active asexual reproduction.

ECHIURIDA

The Echiurida are non-metameric eucoelomate Bilateralia. They consist of a sack-shaped trunk region and a gutter-shaped proboscis. Underneath the cuticle lie circular and longitudinal muscle layers. The digestive tract has a terminal mouth at the base of the proboscis and consists of a long, coiled intestine ending in a terminal posterior anus. There is a closed circulatory system and an unsegmented ventral nerve cord, but no cerebral ganglia. The animals are dioecious with unpaired gonads in the ventral coelomic wall (Dawydoff, 1959).

Development is of the annelid type. Spiral cleavage leads to a coelo- or stereo-blastula, which gives rise to a trochophore larva containing ecto- and meso-teloblasts. During metamorphosis the teloblasts form 11 to 15 segments, each with a ventral ganglion, but the metamery disappears later, probably in connection with sessile life.

According to Loosli (1935), in *Bonellia viridis*, germ cells are first recognisable in the mesoderm of the undifferentiated larva, while Dawydoff (1959) states that the ovary appears very late in development. There is a very pronounced sexual dimorphism in Bonellia, which has large female and only minute male animals (Dawydoff, 1959). (See also Gould-Somero, 1975.)

CHAETOGNATHA

The eucoelomate Chaetognatha are bilaterally symmetrical, small, planktonic animals with head, trunk and tail regions. They have horizontal fins both on trunk and tail. The head has a pair of simple eyes and an olfactory organ. The nervous system consists of a dorsal cerebral and a large ventral trunk ganglion, with circumenteric connectives. The digestive system consists of an anterior mouth flanked by two groups of hooks, and a straight digestive tube which ends in an anal opening at the ventral extremity of the trunk. A circulatory system and excretory organs are lacking. The Chaetognatha are hermaphroditic; the paired ovaries, each with its own oviduct, are situated in the trunk region, while the paired testes, each with a short vas deferens, lie in the tail region. There are left and right coelomic cavities in the head, trunk and tail regions, separated by median mesenteries. (See Hyman, 1959, vol. 5, pp. 1–71.)

Development is direct. Cleavage is equal and leads to a coeloblastula. Gastrulation occurs by invagination of the vegetal endo- and meso-derm. The archenteron forms two lateral diverticula representing the future coelomic cavities. After closure of the blastopore a mouth is formed by perforation of the archenteron and the overlying ectoderm on the side opposite the blastopore, so that the Chaetognatha belong to the so-called Deuterostomia.

Ghirardelli (1953) observed a highly stainable, roundish body at the vegetal pole of the uncleaved egg of *Spadella cephaloptera*. During cleavage it is transferred to one of the blastomeres of the 32-cell stage. According to Ghirardelli, this blastomere divides into a small cell containing the so-called 'germinal determinant' and representing the primordial germ cell, and a larger cell, the primordial endoderm cell. After a first division the two PGCs, initially situated in the tip of the endo- and meso-dermal archenteron, leave the archenteric wall and move to the coelomic out-pocketings. One of the daughter cells of each PGC will form the ovary anlage and the other the testis anlage, after moving to the trunk and tail regions respectively.

The 'germinal determinant' consists of a clew of more-or-less ramified threads, large grains and little plaques. Grains and plaques are composed of agglomerates of very fine grains, some of which may be glycogen and some RNA. Part of the grains tend to form a network around the nucleus (Ghirardelli, 1966a,b) (fig. 13.5a,b).

It is evident that the Chaetognatha show a very early segregation of germ cells, which are, moreover, characterised by the presence of a specific cytoplasmic structure. Although experimental evidence is entirely lacking, the mode of PGC formation in the Chaetognatha seems to be typically *preformistic*, showing much resemblance with that in anuran amphibians (see Eddy, 1975).

170 *Remaining phyla*

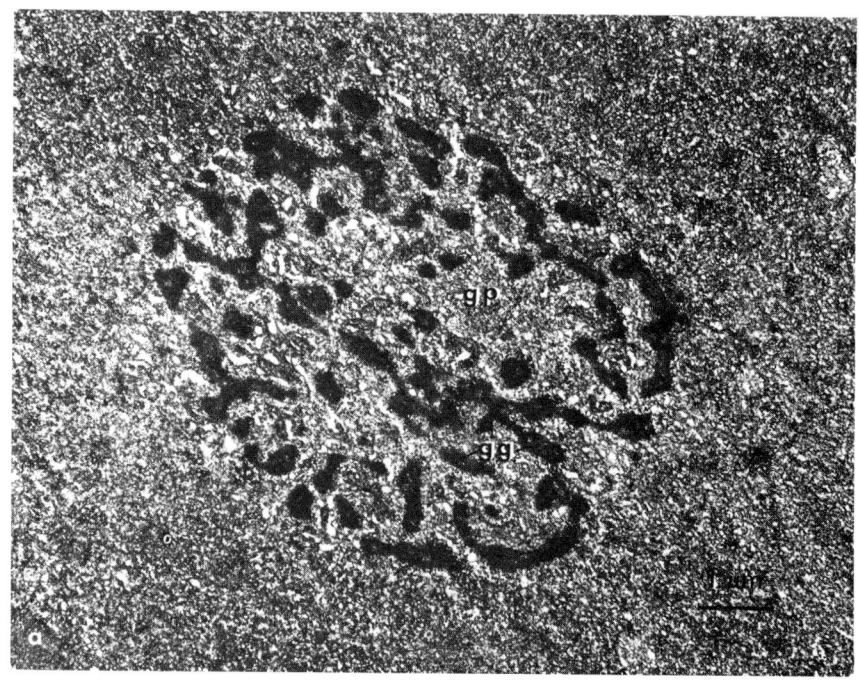

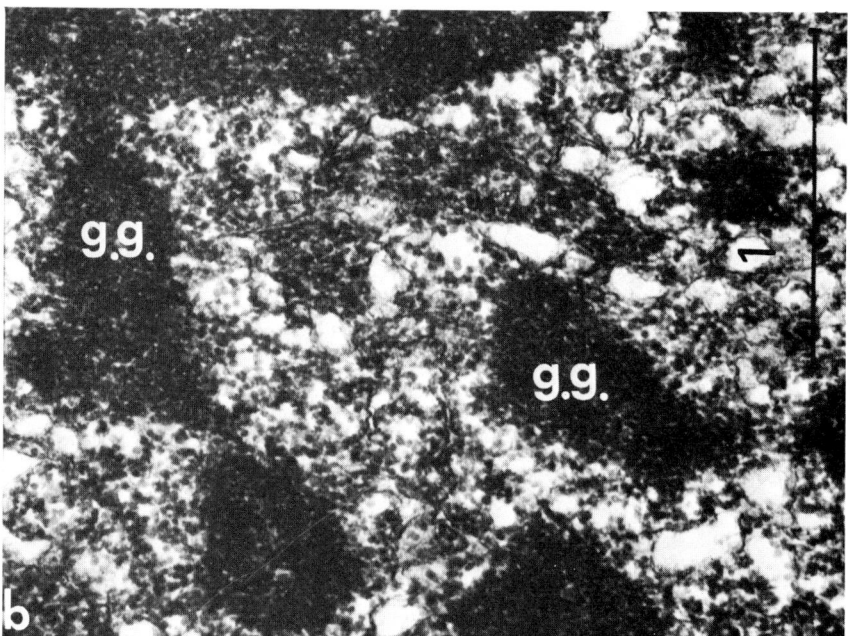

Conclusions

The Ctenophora, belonging to the Radiata, show highly determinate spiral cleavage, although the adults possess some regenerative capacity. The germ cells have a late, probably endodermal origin.

The acoelomate Nemertini show determinate spiral cleavage but considerable regenerative capacity in later life, bound up in some forms with asexual reproduction. The germ cells probably develop epigenetically from undifferentiated parenchymal cells, as in the planarians.

The acoelomate endoparasitic Mesozoa, which are probably related to the digenetic trematodes or the cestodes, are characterised by cell constancy in the somatic tissues. During early development chromatin diminution occurs in the somatic nuclei, as in the nematodes. They have a highly preformistic mode of germ cell formation.

The pseudocoelomate Rotifera show very early segregation of the germ cells. These are characterised by the presence of a 'germinal determinant', indicating a highly preformistic mode of germ cell formation.

The pseudocoelomate Gastrotricha show early segregation of the germ cells.

The pseudocoelomate Entoprocta, which have a typical determinate spiral cleavage, are active in asexual reproduction and manifest a typically epigenetic mode of germ cell formation, both in late embryonic development and during budding.

The eucoelomate Echiurida show an 'annelid' type of development with a typical spiral cleavage pattern. The germ cells probably have a late, mesodermal origin.

The eucoelomate Chaetognatha are characterised by an early segregation of the germ cells, which contain a typical germ plasm, so that the mode of germ cell formation is highly preformistic.

Fig. 13.5. (a) Electron micrograph of cytoplasmic 'germinal determinant' at vegetal pole of 2-cell stage of the chaetognathe *Spadella cephaloptera*. (b) High-magnification electron micrograph of part of 'germinal determinant', showing granular structure of main component as well as distribution of vacuoles and granules in interjacent cytoplasm. g.g. = germinal granules; g.p. = germ plasm. (Courtesy E. Ghirardelli, 1966*b*.)

14

General discussion

Embryonic development in the animal kingdom

A few words must be said about what we believe to be a fundamental aspect of normal development in the entire animal kingdom, as exemplified by the echinoderms and the chordates. The initial development of these groups is governed by an interaction of the first two segregated moieties of the embryo, that is a totipotent animal moiety and an already more-or-less determined vegetal moiety, the interaction of which leads to the formation of a third moiety. The subsequent interactions between the three moieties lead to a stepwise further increase in the complexity of the developing organism.

In the echinoderms the vegetal moiety forms the primary and secondary mesenchyme, representing the mesodermal germ layer, while the animal moiety gives rise only to epidermis. The interaction of these two moieties leads to the appearance of the endoderm, formed from the animal moiety under an inductive influence of the mesenchyme, thus demonstrating the typically epigenetic character of early development.

Comparing the situation in the echinoderms with that in the chordates as described in our previous book (Nieuwkoop & Sutasurya, 1979) an interesting parallel manifests itself. As in the echinoderms, the totipotent animal moiety of the chordate embryo, which is likewise capable of epidermal differentiation only, is partially transformed into a third moiety under an inductive influence emanating from the vegetal region of the embryo. However, here, the already-determined vegetal region represents the endodermal component of the embryo and induces the entire mesoderm and even a part of the endoderm from the totipotent animal moiety. It must therefore be concluded that, although the inductive action in both groups emanates from the vegetal region of the embryo, in the echinoderms the leading germ layer is the mesoderm, whereas in the chordates it is the endoderm.

Can this line of thought be extended to other groups of invertebrates? In our opinion it may indeed also hold for the arthropods. Although we do not know whether the mesoderm of the insect embryo is actually formed as a result of an epigenetic interaction between the peripheral ectodermal blastoderm and the central yolk mass, nor whether an inductive action emanates from the yolk mass or the ectoderm, it looks as if the leading germ layer in the arthropods is the ectoderm, since the ectoderm determines the regional pattern of the underlying mesoderm and probably also that of the endoderm. Nature apparently uses the same general principle to initiate embryonic development, that is the segregation of only two different moieties along the primary animal–vegetal (or inside–outside) axis of the egg, and the epigenetic formation of a third moiety forms a basic theme; this shows considerable variation in its realisation, however.

Weismann's *Keimplasma* theory and the phenomenon of chromosome elimination

In the introduction we arrived at the conclusion that Weismann's *Keimplasma* theory is no longer tenable in the light of our present understanding of nucleo-cytoplasmic interactions during embryonic development. His theoretically highly interesting concept – a differential distribution of *Determinanten* (genes) among the different somatic cell types during development, with conservation of the full complement of *Determinanten* in the germ cells as forerunners of the next generation – is not in agreement with the totipotency of all somatic and germ cell nuclei shown by nuclear transplantation in both vertebrates and invertebrates. Weismann's hypothesis must be replaced by the concept of the differential activation of genes in the various cell types of the organism. According to the now widely accepted hypothesis of Davidson (1976) and others, derepression and repression of genes in the nucleus are due to nucleo-cytoplasmic interaction. Which genes are activated depends essentially upon the composition of the cytoplasm. Development is the response of the developing system to pre-existing spatial differences in the cytoplasm of the egg. These may be in the form either of qualitative differences resulting from cytoplasmic segregation, or of purely quantitative differences in the form of gradients. Both lead to a differential activation of genes and the subsequent formation of different cytoplasmic machineries in the different cell types. This concept of differentiation is considered to be universal and intrinsic to all developmental processes. It takes differentiation to be a stepwise process, which progresses until a final equilibrium state is reached.

Does the phenomenon of chromatin diminution or chromosome elimination, representing a 'genome reduction', constitute an argument against the hypothesis of differential gene activation as the universal mechanism of cellular differentiation, and is it therefore an argument in favour of Weismann's classical *Keimplasma* theory? First of all, genome reduction is certainly not a universal mechanism but a rather rare and exceptional occurrence. Secondly, it is restricted to the distinction between germ cells and somatic cells and plays no role in the development of different somatic cell types. As far as we know, all the somatic cells acquire the same reduced chromosome complement. It certainly does not, therefore, constitute an argument in favour of Weismann's theory. Neither is it an argument against the hypothesis of differential gene activation, since it cannot be made responsible for differential somatic differentiation.

Although this has been suggested by some, it does not seem likely that genome reduction simply compensates for polyploidy or gene amplification in the germ cells, or that the somatic chromosome set represents the normal and essentially complete set of chromosomes of the species. We have seen that a temporary prevention of the association of cleavage nuclei with polar plasm leads to chromosome elimination in all nuclei. Although in *Wachtliella* this does not prevent PGC formation, it nevertheless interferes with normal gametogenesis. This shows that the eliminated E or L chromosomes are indispensable for normal gametogenesis, so that the eliminated chromosomes must contain information not present in the chromosomes of the somatic cells. Genome reduction must therefore be considered a special adaptation in certain groups of animals.

It has also been suggested that genome reduction may represent an adaptation to a parasitic mode of life in a special and constant environment. This may seem to hold for the endoparasitic nematodes but is less likely for the Cecidomyidae, the larval development of which occurs in the special environment of plant galls, though they have normal free-living imagos. The same is true for the Mycetophilidae. The suggestion certainly does not hold for the Sciaridae, the free-living larvae of which live on detritus, for the Chironomidae, which have aquatic larvae, nor for the free-living copepods such as *Cyclops*.

The germ line concept

We should consider here the 'germ line' concept, which still plays such an important role in the literature on germ cell formation. We must ask ourselves whether the distinction between a separate and in principle continuous, immortal germ line and the mortal somatic tissues of the organism is still valid, or whether it is an artificial distinction which has merely been retained in the literature as a remnant of Weismann's *Keimplasma* theory. We have already seen that Weismann's theory is no longer in agreement with our present understanding of the nucleo-cytoplasmic interactions which govern cellular differentiation. It is therefore necessary to investigate whether the concept of a totipotent germ line, which is a consequence of Weismann's *Keimplasma* theory, should not be abandoned altogether, now that the theory itself has lost its fundamental validity.

In our discussion of the various phyla and classes of the invertebrates and vertebrates we have come across arguments which are relevant to this question, and which actually plead against a fundamental distinction between *germen* and *soma*. As we shall see, this holds equally for the typically 'epigenetic', the so-called 'intermediate' and the typically 'preformistic' modes of germ cell formation.

In the lower invertebrates, with their typically 'epigenetic' mode of germ cell formation, separate phases of sexual and asexual reproduction alternate during the life cycle of the individual. All these groups are characterised by the presence of undifferentiated cells, such as the gonidia of the colonial flagellates, the archaeocytes of the sponges, the interstitial cells of the coelenterates and the neoblasts of various groups of platyhelminthes. Depending upon external environmental factors such as light, temperature and oxygen supply, these undifferentiated cells may either differentiate into various somatic cell types during growth, regeneration and asexual reproduction, or be switched into the pathway of germ cell formation during sexual reproduction. Here it is perfectly evident that germ cell development represents a special pathway of cellular differentiation. The mature germ cells are highly differentiated and adapted for particular functions: the self-propelling flagellated spermatozoon, designed to reach the mature egg by chemotaxis and to penetrate its envelopes and cell membrane, and the egg itself, endowed with complex machineries for rapid nuclear multiplication and cell membrane formation and provided with a large quantity of reserve substances and molecular building blocks needed to carry out these initial functions as well as functions which ultimately lead to the development of a new individual. A special characteristic of germ cell differentiation is that it does not lead to a temporary or permanent loss of developmental potentialities; consequently, cellular differentiation and totipotentiality are not necessarily mutually exclusive, as is so often postulated.

In the preceding chapters, in which the origin of the germ cells was placed in the context of embryonic development, regeneration and asexual reproduction, the presence of undifferentiated cells, whether constantly present as an embryonic reserve or formed *de novo* by dedifferentiation, was seen not to be restricted to the 'primitive' invertebrate phyla, but to occur also in more highly evolved groups like the annelids and the ectoprocts, and perhaps even in the colonial ascidians belonging to the cephalochordates. In all these groups a continuous germ line does not exist, so that the germ line concept certainly is not valid here.

In the phyla of the echinoderms and the molluscs germ cell formation has been classified as 'intermediate' between the typically epigenetic and the typically preformistic modes, since no alternation of sexual and asexual reproduction occurs, nor is there early segregation of the germ cells. In these groups the germ cells are recognisable only at a late stage of embryonic development, often as late as after metamorphosis. Although we do not know the actual time of determination of the germ cells in these groups, and although determination may markedly precede cellular differentiation, it seems plausible to assume that the future germ cells simply form part of the somatic tissues of the embryo during a rather long period of development. This is the more likely since undifferentiated cells do not seem to exist during embryonic development in these groups, which have only weak powers of regeneration. Here, the concept of a distinct germ line is certainly not applicable to the initial phases of embryonic development, and is therefore fundamentally invalid.

The typically 'preformistic' mode of germ cell formation is exemplified by the highly evolved holometabolous insects, where, from the very beginning of embryonic development, a posterior pole plasm characterises the precursors of the future germ cells, and where, at the same time, a nearly uninterrupted germ line seems to be present throughout the entire life cycle. However, in these groups we have encountered the well-documented observation that, in normal development, only a small fraction of the pole cells, all of which are endowed with germ cell-specific polar granules, actually do form germ cells, whereas the majority contribute to various somatic tissues. Even in these groups, therefore, the distinction between a germ line and the somatic cell line does not seem to be an absolute one, although germ cells do not develop from cells other than pole cells. Despite the fact that the pole cells must be regarded as potential germ cells, they can apparently still form various other cell types, such as cuprophilic cells of the midgut and vitellophages. We must therefore conclude that the pole cells are still in a labile state of determination and that germ cell differentiation requires an additional inductive influence, probably from the gonadal anlage, so that germ cell determination is a stepwise process. This throws considerable doubt on the validity of the germ line concept even in groups with a typically preformistic mode of germ cell formation.

Taking all these arguments together we cannot avoid the conclusion that the fundamental distinction between *germen* and *soma*, implying the postulate of an uninterrupted germ line extending from one generation to the next, is no longer adequate and valid, since it is no longer supported by our present insight into germ cell development. We believe that it is time for this obsolete concept to be abandoned and to be replaced by the idea that germ cell formation represents just another type of cellular differentiation. It differs from other types only in that, during this particular differentiation process, the full complement of developmental potentialities is preserved. But even here a restriction must be made: contrary to oogenesis, where the developmental potentialities are preserved in the cytoplasm as well as the nucleus, in spermatogenesis only the nuclear potentialities seem to be preserved.

In our previous book on chordate development (Nieuwkoop & Sutasurya, 1979) we still maintained the germ line concept, notwithstanding the fact that we encountered at least two thoroughly different principles of germ cell development, these being the preformistic mode of germ cell formation from the endoderm in the anuran amphibians (and possibly also in the birds), and the epigenetic mode of germ cell formation from the mesoderm in the urodele amphibians (and possibly also in the mammals). In discussing germ cell formation in the invertebrates, in particular in the lower phyla, it has become clear that the germ line concept is essentially inadequate. Since there is no reason to assume that any fundamental difference in germ cell formation exists between the invertebrates and the vertebrates or chordates, the germ line concept should also be abandoned in the vertebrate and chordate literature and replaced by the concept of germ cell formation as a manifestation of cellular differentiation. Although germ cells behave initially as rather independent elements in the organism, which are in a sense 'foreign' to the tissues through which they migrate to reach the gonadal anlagen (the latter being the only adequate tissue environment for their ultimate differentiation), it must be concluded that the differentiation of germ cells and of somatic, or rather non-germinal, cells is not fundamentally different and should therefore no longer be considered as such.

Epigenesis and preformation in germ cell development; a continuous variable in the animal kingdom

In the introduction we distinguished three main modes of germ cell formation: (a) the 'epigenetic', (b) the so-called 'intermediate', and (c) the 'preformistic' modes.

(a) In the 'epigenetic' mode sexual and asexual forms of reproduction alternate under the influence of environmental factors. While during sexual reproduction germ cells are formed from undifferentiated or dedifferentiated embryonic cells, the same embryonic cells give rise to various somatic cell types during asexual reproduction. Their development can be switched from one pathway into the other.

(b) In the 'intermediate' mode the germ cells, once formed, are in general no longer replaceable by other cells, but they are formed at a rather late stage of development from cells which must have passed through a shorter or longer phase of somatic development. Asexual reproduction, though rather exceptional, still occurs, but it is not clear whether *de novo* formation of germ cells takes place.

(c) In the 'preformistic' mode the germ cells segregate from the somatic cells at a very early stage of embryonic development. They are often 'predetermined' by the presence of a germ cell-specific germ plasm, so that either presumptive or true germ cells can be distinguished during the entire or almost the entire life cycle.

The more primitive groups of multicellular invertebrates, that is the colonial flagellates, the sponges, the coelenterates, the Turbellaria and the Cestodes, belong to the first, epigenetic category. Some of the more highly evolved invertebrate groups, however, also show a typically epigenetic mode of germ cell development, for instance the Entoprocta and Ectoprocta. Asexual reproduction even occurs in the colonial tunicates, which belong to the chordate phylum. The solitary tunicates have been extensively discussed in our previous book (Nieuwkoop & Sutasurya, 1979). Their germ cells are assumed to derive from lymphocytes. It is not clear, however, whether, in the colonial forms, germ cells develop *de novo* from the somatic tissues of the bud (Berrill, 1975) or are transported by the bloodstream from the mother animal to the differentiating buds, so that no new formation of germ cells occurs (Sabbadin & Zaniolo, 1979). (See further the reviews by Brien, 1964, 1966.)

A preformistic mode of germ cell formation is characteristic in the highly evolved arthropods, particularly the holometabolous insects. However, this mode is not restricted to the highly evolved invertebrate groups and is also encountered in some rather primitive groups, such as the ctenophores and the nematodes. On the other hand, not all highly evolved groups show a typically preformistic mode of germ cell formation. The molluscs, for example, which certainly belong to the most highly evolved invertebrate groups and reach a level of organisation comparable to that of the vertebrates, typically belong to the intermediate category.

In our previous book we tried to show that, as far as germ cell formation is concerned, the phylum of the Chordata cannot be treated as a uniform group, but actually falls apart into two main groups, one with a preformistic mode of germ cell development (the anuran amphibians and possibly the birds) and the other with a much more epigenetic mode of germ cell formation (the urodele amphibians and possibly the mammals). This pronounced divergence has induced us to propose an early dichotomy in the evolution of the vertebrates. A similar divergence exists in the phylum of the Platyhelminthes, although from a comparative–anatomical point of view it is a rather homogeneous group. The Turbellaria show a typically epigenetic, the Trematoda a much more preformistic, and the Cestodes again an epigenetic mode of germ cell formation. It is perfectly clear that taxonomic classification and mode of germ cell formation do not run parallel. In table 14.1 we have tried to arrange the various invertebrate phyla and classes in a more-or-less continuous series, using as a criterion the three main modes of germ cell formation as distinguished in the introduction. We hope that this will bring out the pronounced divergence in the mode of germ cell formation among, and even within, the various invertebrate phyla.

It is evident that the various chordate classes can also be inserted into this series. They will find their places in the intermediate and preformistic

Table 14.1. *Arrangement of invertebrate phyla and classes according to decreasing importance of asexual reproduction (−) and regenerative capacity (- - -) on the left, and increasing importance of early segregation of the pPGCs on the right*

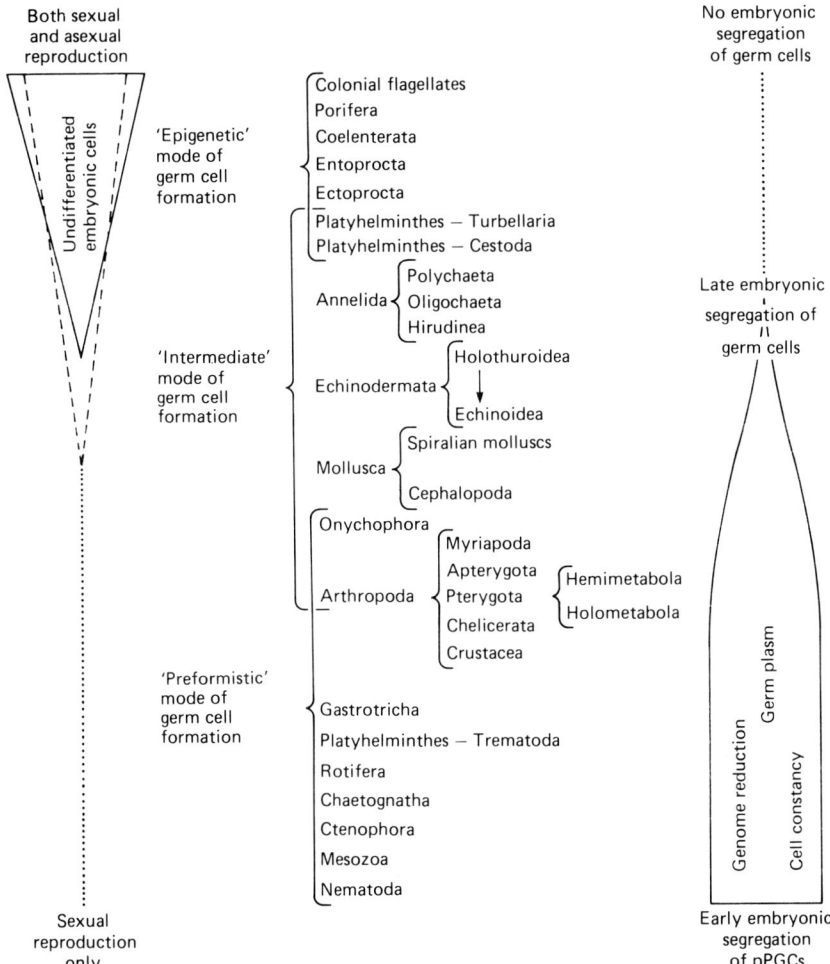

categories, since the vertebrates do not show the extreme epigenetic mode of germ cell formation of the lower invertebrate phyla, which are characterised by high regenerative capacity and asexual reproduction.

In the introduction we have asked whether the epigenetic and preformistic modes of germ cell formation are mutually exclusive, or whether they represent simply the extremes of a continuous range of gradations. The fact that we have been able to arrange the various phyla and classes in a

180 *General discussion*

more-or-less continuous series, starting from the most epigenetic and ending with the most preformistic mode of germ cell formation, obviously supports our personal belief in the second alternative. This is, of course, by no means a proof in itself. We must even concede that it is, at present, rather difficult to prove or disprove the idea, since our knowledge, particularly of the mode of germ cell formation, is still very scanty in many groups. Such groups of which the actual mode of germ cell formation is not well known may have been placed in the intermediate category, erroneously suggesting a bridge between the better-known extremes. In our opinion, however, the fact that undoubtedly related groups, like the various classes of the Platyhelminthes or the various orders of the Insecta, show much variation in time and mode of germ cell formation renders the first alternative very unlikely. We shall return to this question when discussing the essential features of germ cell formation (see p. 186).

One critical remark must be made. When we speak of more 'primitive' or more 'evolved' groups and apply this distinction to phyla and classes, we must bear in mind that the most primitive groups are assumed to be the oldest groups, which must therefore have the longest evolutionary history, while the most highly evolved groups are supposed to be phylogenetically the youngest. Each present-day group, of course, represents an end-branch of the evolutionary tree. The fact that we know so little about the actual evolution of the various invertebrate phyla and classes has made it impossible to consider germ cell development in the invertebrates in the light of evolution, as we tried to do in our previous book on the chordates, where rather extensive palaeontological data are available. This impossibility of an evolutionary approach constitutes one of the reasons for our more pragmatic approach here of treating the subject solely on the basis of the various modes of germ cell formation. The other reason lies in the fact that our knowledge of germ cell formation in the individual groups is so very unequal: some groups have been studied in great detail, for instance the higher insects, whereas other groups are still completely unknown. A more systematic approach to the problem is simply impossible. The frontier of research on germ cell formation shows a very irregular contour, with advanced sectors of extensively studied problems or groups and deep recesses of almost completely untouched areas. We shall return to this aspect when mentioning possibilities for further research (see p. 187).

Stability of cellular differentiation in epigenetic and preformistic development

How do the concepts of epigenetic and preformistic development relate to the concept of differential gene activation as a universal mechanism of cellular differentiation? First of all, it must be emphasised that all these concepts refer to the cell as a whole, that is the assembly of nucleus, cytoplasm and plasma membrane. The concepts of epigenetic and preformistic development refer to different degrees of *stability* of the process of cellular differentiation. Although, under certain conditions, nuclei may undergo irreversible changes during development, as in the case of chromosome elimination, it is now generally agreed, though not yet fully proven, that differential gene inactivation and activation are essentially *reversible* processes. However, during development a particular state of the cytoplasm will activate a specific set of genes for a certain pathway of differentiation, which, in their turn, will act back on the state of the cytoplasm. This feedback will lead to a progressive stabilisation of the state of differentiation of the cell. It is therefore understandable that cellular differentiation is generally a stable process. The question which concerns us here is how rigid the process of cellular differentiation is. In other words, can it still be partially or completely reversed under particular circumstances? For such reversal a breakdown of the cellular machinery characteristic of a particular cell type is required, a process which is called dedifferentiation. We must therefore ask ourselves how widespread the phenomenon of dedifferentiation is. We have seen that it is a common phenomenon in regeneration and asexual reproduction, processes which are rather widespread among the invertebrates.

Various levels of dedifferentiation must be distinguished, however. Dedifferentiation may be restricted to a restoration of mitotic activity in certain cell types, in which case the pathway of differentiation is not affected, the daughter cells redifferentiating like the parent cell. In other instances dedifferentiation may result in the dedifferentiated cell or its descendants being switched into a related pathway of differentiation. This is often called germ layer-specific* regeneration and is encountered, for instance, in various annelid groups. Finally, dedifferentiation may go so far that the cell reacquires an 'embryonic' state and may subsequently redifferentiate in any other direction, including that of germ cells. This phenomenon is encountered particularly in the alternation of asexual and sexual reproduction. In budding or fission non-germinal tissues partially reacquire the capacity to redifferentiate into germinal elements. In the sponges, coelenterates and turbellarians embryonic cells are preserved during embryonic development or are formed *de novo* by dedifferentiation under special circumstances. Brien (1953) concluded that the archaeocytes of the sponges, the interstitial cells of the coelenterates and the neoblasts of the turbellarians are the true undifferentiated, totipotent cells which can be switched either into various pathways of somatic differentiation or into that of germ cell formation. However, as we have seen, asexual reproduction is

*The term 'germ layer' must not be taken too strictly, since many exceptions plead against the classical germ layer concept.

not restricted to the primitive invertebrates, but is also found in various much more highly evolved groups.

Germ cell determination

In the introduction we concluded that the concept of totipotency is rather inadequate. In a strict sense *functional* totipotency holds only for the fertilised egg, which can autonomously express its developmental potentialities. It holds neither for the oocyte (the mature unfertilised egg) nor for the undifferentiated embryonic cell. These may only be called *potentially* totipotent. The undifferentiated embryonic cell must first be switched into the pathway of germinal differentiation before it can develop into a *functionally* totipotent cell. When we speak of totipotent cells, therefore, this implies only that the cell *could* develop to a state in which it could express the full range of developmental potentialities of the species. (See the distinction made in the introduction, p. 3, between nuclear potentialities and cellular expression.) Only in the germinal pathway of differentiation does a cell retain this capacity. It would therefore be preferable to use the term '*totipotential*' rather than 'totipotent'. It should, moreover, be emphasised that a totipotential cell is not necessarily a presumptive germ cell which has already entered the pathway of germinal differentiation. The concept of totipotential cells is much broader and includes all cells which *could* develop into germ cells or, in an asexually reproducing animal, could give rise to the entire range of somatic cells. It is evident that, in animals with an epigenetic mode of germ cell formation, the number of totipotential cells is very large. Any cell which dedifferentiates to an undifferentiated embryonic state is a totipotential cell. In the case of preformistic germ cell formation the number of toti-, better multi-, potential cells is quite restricted. In the higher insects, for instance, it comprises only the pole cells, since no other cells can develop into germ cells. The pole cells, moreover, represent the first cells to be segregated in the embryo. It is extremely interesting that, in *Drosophila* and other Diptera, only a small fraction of the multipotential pole cells actually develop into germ cells and therefore represent presumptive PGCs. There is convincing evidence that the majority of the pole cells ultimately realises other pathways of differentiation, for instance that of cuprophilic cells of the midgut or of secondary vitellophages. This means that the pole cells have been determined only to the state of multipotential cells and not yet to that of germ cells. This second determination step probably occurs only in the gonads, when the PGCs become surrounded by the mesodermal gonadal sheath and the interstitial cells.

Something may now be said about the question raised in the introduction: do the various epigenetic or preformistic modes of germ cell formation imply different basic mechanisms of germ cell formation or is the actual mechanism the same everywhere? In the case of 'epigenetic' germ cell formation the mechanisms involved in certain invertebrate and vertebrate groups may be at least partially different. For instance, in the urodele amphibians germ cell development seems to depend upon successive inductive actions emanating from the endoderm, whereas in the colonial flagellates, sponges and coelenterates external environmental factors release gametogenesis from totipotential embryonic cells. Moreover, in the coelenterates cells in early stages of gametogenesis can be reconverted into totipotential embryonic cells when asexual reproduction is re-established, a phenomenon never encountered in the vertebrates. Germ cell formation in the Turbellaria and Cestoda, which also starts from totipotential embryonic cells, may be slightly different from that in the coelenterates, since there are no indications that the process, once started, can then be reversed. In the case of typically 'preformistic' germ cell formation the mechanism in invertebrate and vertebrate groups seems much more similar. For instance, the structure and function of the germ cell-specific germ plasm seems to be exactly the same in the higher insects and the anuran amphibians. The only significant difference may lie in the fact that the pole cells of the higher insects, although potential PGCs, may give rise to other, somatic cell types, a phenomenon never encountered in the vertebrates.

Determinancy in cleavage, embryonic development and germ cell formation

We must now return to the concepts of epigenesis and preformation. These concepts were originally formulated to characterise embryonic development. They are equivalent to the terms 'indeterminate' or 'regulative', and 'determinate' or 'mosaic' development, respectively.

A clear distinction must first be made between 'determinate cleavage' and 'determinate embryonic development'. Determinate cleavage refers only to an accurately programmed subdivision of the egg into a number of equal or unequal blastomeres in a particular spatial configuration. The best-known example is spiral cleavage in the Spiralia. The latter comprise a rather large group of different phyla, some with more determinate, others with much more indeterminate, embryonic development, so that the two concepts by no means coincide. Another example of the disparity of the two concepts are the echinoderms, which show a fairly determinate cleavage pattern but exhibit the most pronounced type of epigenetic embryonic development.

A clear distinction must also be made between the concepts of 'indeterminate *vs* determinate embryonic development' and of 'epigenetic *vs* preformistic germ cell formation'. The two concepts may or may not coincide in any one group, but there is certainly no overall correlation between them.

Among the vertebrates the anuran amphibians show typically preformistic germ cell formation, although their embryonic development is highly epigenetic. Even the highly preformistic mode of pole cell formation in the higher insects is not necessarily correlated with a determinate type of embryonic development. The cecidomyid and chironomid Diptera show a typically preformistic mode of germ cell formation characterised by early pole cell segregation and, in addition, a pronounced germinal *vs* somatic nuclear differentiation in the form of chromosome elimination in all somatic cells. Nevertheless, embryonic development is rather regulative and epigenetic, which is expressed in, among other things, the possibility of reduplication of posterior segments in place of anterior ones (Price, 1958, in *Wyeomyia*; Yajima, 1960, 1964, in *Chironomus*, and Kalthoff & Sander, 1968, in *Smittia*). Other forms, such as *Drosophila*, show a pronounced mosaic type of blastoderm formation, with accurate localisation of all larval and adult organ systems in the egg and virtually complete absence of regulatory processes. They also have a typically preformistic mode of germ cell development, so that here the two concepts do coincide.

Possible role of the germinal granules in germ cell formation

We now come to a critical discussion of the possible role of the germinal granules in germ cell determination. When discussing germ cell formation in the vertebrates in our previous book (Nieuwkoop & Sutasurya, 1979) we expressed doubt about the validity of the idea that the characteristic germinal granules of the anuran amphibians really act as a germ cell determinant. This doubt was based particularly on the non-universality of the germinal granules among the vertebrate groups and on the fact that, in the urodeles, the PGCs acquire germinal granules only long after primary germ cell determination must have taken place. In the invertebrates similar objections can be raised: germinal granules, though found in various invertebrate groups, are certainly not a universal characteristic of germ cells. The counter-argument that germinal granules may not have been detected in many invertebrate groups because they have not been searched for sufficiently or adequately, cannot be rejected, but in our opinion this is rather improbable. Even in the insects germ cell formation does not seem to be necessarily associated with the presence of germinal granules, since many groups do without them. There is, however, still another argument of an entirely different nature against the unique character and role of the germinal granules in germ cell development. 'Germinal granules' have recently been found in somatic cells of the coelenterates and the turbellarians. In *Hydra*, Noda & Kanai (1977) found that interstitial cells contain typical 'germinal granules', which are also found in developing cnidoblasts during asexual reproduction, although they gradually diminish in number and finally disappear during the differentiation process. 'Germinal granules' are likewise found in cells of the regeneration blastema in *Planaria* (Le Moigne, 1967a; Spiegelman & Dudley, 1973; Hay & Coward, 1975). In sexually reproducing *Hydra* the germ cells contain markedly more germinal granules than the interstitial cells from which they are derived. Consequently, as far as the germinal plasm is concerned, the distinction between somatic differentiation in asexually reproducing animals and germinal differentiation in sexually reproducing animals appears to be merely *quantitative*, not qualitative in nature. This may also hold for the regenerative elements in planarians, the neoblasts, which, like the interstitial cells of the coelenterates, represent totipotential embryonic cells. In these and other lower groups the undifferentiated embryonic cells may either have been preserved during embryonic development or have been formed *de novo* by dedifferentiation of somatic cells, viz. from the choanocytes in the sponges, the gastrodermal and/or epithelio-muscular cells in the hydroids and the syncytial parenchyma in the planarians. The presence of germinal granules is therefore not a unique feature of germ cells but also of undifferentiated embryonic cells which are still capable of being switched into either the germinal or the somatic differentiation pathway.

Although Mahowald and co-workers have presented convincing evidence for the role of the posterior pole plasm in pole cell formation, it is still unclear whether it is the polar granules or the pole plasm itself that plays the essential role in germ cell determination. The crucial experiment, in which typical pole cell and germ cell formation is induced by an injection of purified germinal granules or of a purified protein fraction isolated from either pole plasm or germinal granules, still has to be performed. Graziosi and co-workers are no longer convinced that the polar granules represent the germ cell determinant (personal communication).

Displacement of the polar granules but not the pole plasm by forced centrifugation of *Drosophila* eggs (Jazdowska-Zagrodzińska, 1966) interferes with pole cell formation, so that here both pole plasm and polar granules seem to be indispensable for pole cell formation. However, the same experiment made in *Wachtliella* by Wolf (1967) led to a partially different but very interesting result: pole cell formation occurred normally in the posterior pole region of the egg in the absence of polar granules and with the participation of nuclei which had gone through the process of chromosome elimination. These 'reduced' pole cells could still develop into germ cells, but could not give rise to normal gametes. In the equatorial region of the egg, where the displaced polar granules had ended up, only those nuclei which happened to be in direct proximity to the polar granules failed to undergo chromosome elimination, so that the polar granules seemed to be responsible for preventing chromosome elimination. We feel that both experiments, which are so crucial to our understanding of the role of the germinal granules, and which are partially contradictory, should be repeated.

What, then, may be the essential role of the germinal granules in germ cell development? On the one hand, it is difficult to accept their purported role in germ cell determination, since germinal granules are not, as far as we know, a universal feature of germ cells, and while, moreover, the ultimate determination of germ cells may occur only after their arrival in the gonadal anlagen. On the other hand, it is also difficult to deny them any function at all in germ cell development, as they are such a characteristic feature of germ cells in many different groups of animals. In the light of these considerations we want to propose as a working hypothesis that the germinal granules, rather than being concerned with the *determination* of the germ cells, are involved merely in the *preservation* of the full complement of developmental potentialities of the species. In other words, they may be a necessary component of the cellular machinery that is required for germ cell *differentiation*.

Fundamental features of germ cells

In our opinion, then, germ cell formation must be considered as a specific type of cellular differentiation. In *Hydra* only oogonial and spermatogonial cells can still dedifferentiate into 'embryonic' interstitial cells when the animal switches back from sexual to asexual reproduction. Cells in later stages of gametogenesis cannot revert, which shows that they represent a more advanced state of cellular differentiation.

It should be realised that both the epigenetic and the preformistic modes of germ cell formation result in typical germ cells, so that the two modes of germ cell formation must have some essential features in common. It is conceptionally very difficult to envisage how following essentially different, mutually exclusive pathways the same end result could be achieved. In our opinion the essential feature of germ cells is best defined as the capacity to maintain a state which guarantees the expression of the full complement of developmental potentialities of the species. This state is apparently independent of the particular type of cellular differentiation which germ cells undergo. Viewed in the light of the hypothesis of differential gene activation this means that the cytoplasm of the germ cell must contain factors which prevent any irreversible, or not easily reversible, repression of essential genetic information. Although cellular differentiation is usually attended by the formation of a particular cellular machinery that leads to loss of functional totipotentiality, germ cell differentiation apparently does not fall under this rule. It would therefore be of great interest to compare germ cell differentiation with somatic cellular differentiation, both ultrastructurally and biochemically. An important clue in this analysis may be the pole cell-specific basic protein isolated by Waring, Allis & Mahowald (1978).

We have already mentioned that germ cell determination in the holometabolous insects seems to be completed only after the settling of the germ cells in the gonadal anlagen. This may also hold for germ cell development in other groups. This implies that it would be preferable to speak of *potential* primordial germ cells as long as they are still able to form other cell types.

During the initial phases of embryonic development the PGCs are characterised by a particular behaviour: they behave as 'foreign' elements among the somatic cells of the embryo, and may in a sense be compared to benign tumour cells. On what morphological and physiological features is this 'foreign' character of the PGCs based? Under the light microscope a well-defined cell boundary constitutes one of the characteristics of the PGCs. Ultrastructurally this has been shown to be due to the presence of rather large intercellular spaces surrounding the PGCs and separating them from surrounding somatic cells (see Nieuwkoop & Sutasurya, 1979, p. 62). This absence of intimate contacts with surrounding somatic cells may indeed play an important role in the preservation of the full complement of development potentialities, since interaction with surrounding cells is a prerequisite for many cellular differentiation processes. In the case of PGC development such cellular interactions are apparently precluded.

Although PGCs may temporarily associate themselves with endodermal cells, probably for nutritional purposes, the PGCs behave as 'foreign' elements as long as they have not become intimately associated with somatic cells of the gonadal anlagen. Their migratory activity seems to become suppressed as soon as they enter the gonadal anlagen. The definitive determination of the germ cells may be associated with this change in behaviour.

Suggestions for further analysis

In our opinion there are at least five lines of research on germ cell development that seem particularly promising.

(1) A comparison of germ cell differentiation with somatic cellular differentiation, for example in lower forms such as *Hydra*, where the interstitial cells may be switched experimentally into one or the other pathway.

(2) A comparison of germ cell characteristics, both ultrastructural and biochemical, in species *with* and *without* germinal granules. In our opinion attention has been focussed far too exclusively on insects with pole cell development and on animals containing germinal granules.

(3) A biochemical analysis of the increase in the number of germinal granules and possibly other cell organelles during germ cell formation in sexually reproducing *Hydra*, and of the reverse phenomenon during somatic differentiation in asexually reproducing animals.

(4) A descriptive and experimental analysis of 'late' germ cell formation in representatives of the so-called 'intermediate' groups, for example the echinoderms and the molluscs. In these groups the study of germ cell formation has been almost entirely neglected and may yield important new information.

(5) A careful study of the membrane properties of PGCs during early embryonic development, as well as of their possible changes after arrival of the PGCs in the gonadal anlagen.

We hope that this survey of the present state of knowledge of germ cell formation in the various groups of invertebrates may stimulate research particularly in those areas where progress has been slow. We have tried to be as objective as possible, emphasising that one should be very careful in making generalisations, as these tend to impede rather than advance understanding.

References

Reviews marked with *.

Aboïm, A. N. (1945). Développement embryonnaire et post-embryonnaire des gonades normales et agamétiques de *Drosophila melanogaster*. *Rev. Suisse Zool.* **52**, 53–154.

Achtelig, M. & Krause, G. (1971). Experimente am ungefurchten Ei von *Pimpla turionellae* L. (Hymenoptera) zur Funktionsanalyse des Oosombereichs. *Wilhelm Roux' Arch. Develop. Biol.* **167**, 164–82.

Agrell, I. (1963). Mitotic gradients in the early insect embryo. *Ark. Zool.* **15**, 143–8.

Agrell, I. (1964*). Physiological and biochemical changes during insect development. In *The Physiology of Insecta*, ed. M. Rockstein, vol. 1, pp. 91–148. Academic Press, London & New York.

Alléaume, N. (1963). Etude expérimentale des capacités d'auto-différenciation de l'ectoderme et de l'endomésoderme présomptifs, au stade blastule, chez *Calliphora erythrocephala* Meig (note préliminaire). *Procés-Verbaux Soc. Sci. Phys. Natur. Bordeaux*, Ann. 1962–3, 58–62.

Alléaume, N. & Haget, A. (1962). Développement de 'gonades mésodermiques' après microcautérisation des initiales germinales au stade blastule, chez *Calliphora erythrocephala* Meig. *Procés-Verbaux Soc. Sci. Phys. Nat. Bordeaux, Ann.* 1961–2, 92–6.

Allis, C. D., Underwood, E. M., Caulton, J. H. & Mahowald, A. P. (1979). Pole cells of *Drosophila melanogaster* in culture. Normal metabolism, ultrastructure and functional capabilities. *Develop. Biol.* **69**, 451–65.

Allis, C. D., Waring, G. L. & Mahowald, A. P. (1977). Mass isolation of pole cells from *Drosophila melanogaster*. *Develop. Biol.* **56**, 372–81.

Amy, R. L. (1961). The embryology of *Habrobracon juglandis* (Ashmead). *J. Morphol.* **109**, 199–217.

Anderson, D. T. (1962). The embryology of *Dacus tryoni* (Frogg.) Diptera, Trypetidae (=Tephritidae), the Queensland fruitfly. *J. Embryol. Exp. Morphol.* **10**, 248–92.

Anderson, D. T. (1966a*). The comparative embryology of the Polychaeta. *Acta Zool.* **47**, 1–42.

Anderson, D. T. (1966b*). The comparative early embryology of the Oligochaeta, Hirudinea and Onychophora. *Proc. Linn. Soc. New South Wales* **91**, 10–43.

Anderson, D. T. (1966c*). The comparative embryology of the Diptera. *Ann. Rev. Entomol.* **11**, 23–45.

Anderson, D. T. (1969). On the embryology of the cirrepede crustaceans *Tetraclita rosea* (Krauss), *Tetraclita purpurascens* (Wood), *Chthamalus antennatus* (Darwin) and *Chamaesipho columna* (Splengler) and some considerations of crustacean phylogenetic relationships. *Phil. Trans. Roy. Soc., ser. B* **256**, 183–235.

Anderson, D. T. (1972a*). The development of hemimetabolous insects. In *Developmental Systems: Insects*, eds S. J. Counce & C. H. Waddington, vol. 1, pp. 95–163. Academic Press, London & New York.

Anderson, D. T. (1972b*). The development of holometabolous insects. In *Developmental Systems: Insects*, eds S. J. Counce & C. H. Waddington, vol. 1, pp. 165–242. Academic Press, London & New York.

Anderson, D. T. (1973*). *Embryology and Phylogeny in Annelids and Arthropods*, 495 pp. Pergamon Press, Oxford & New York.

Ando, H. (1960). Studies on the early embryonic development of a scorpion fly, *Panorpa pryeri* MacLachlan (Mecoptera, Panorpidae). *Sci. Rep. Tokyo Kyoiku Daigaku, section B* **9**, 227–42.

André, F. & Davant, N. (1977). La régénération de l'appareil reproducteur et la différenciation d'éléments germinaux extragonadiques chez le lombricien *Eisenia foetida* Sav. *Wilhelm Roux' Arch. Develop. Biol.* **182**, 189–201.

André, F. & Davant, N. (1979). Influence du chlorure de zinc sur l'inversion des sexes et sur le développement et la différenciation d'éléments germinaux supplémentaires chez *Eisenia foetida* Sav. (Oligochète, Lombricidae). *Wilhelm Roux' Arch. Develop. Biol.* **186**, 115–28.

Arnold, J. M. (1961). Observations on the mechanism of cellulation of the egg of *Loligo pealii*. *Biol. Bull.* **121**, 380–1.

Arnold, J. M. (1963). Developmental analysis of the cephalopod embryo. *Proc. XVI Int. Congr. Zool. Washington, D. C., August 20–7, 1963*, vol. **1**, p. 76.

Arnold, J. M. (1965a). Normal embryonic stages of the squid, *Loligo pealii* (Lesueur). *Biol. Bull.* **128**, 24–32.

Arnold, J. M. (1965b). The inductive role of the yolk epithelium in the development of the squid, *Loligo pealii* (Lesueur). *Biol. Bull.* **129**, 72–8.

Arnold, J. M. (1968a). The role of the egg cortex in cephalopod development. *Develop. Biol.* **18**, 180–97.

Arnold, J. M. (1968b). An analysis of cleavage furrow formation in the egg of *Loligo pealii*. *Biol. Bull.* **135**, 413.

Arnold, J. M. (1971*). Cephalopods. In *Experimental Embryology of Marine and Freshwater Invertebrates*, ed. G. Reverberi, pp. 265–311. North-Holland, Amsterdam.

Arnold, J. M. & Williams-Arnold, L. D. (1974). Cortical–nuclear interactions in cephalopod development: cytochalasin B effects on the informational pattern in the cell surface. *J. Embryol. Exp. Morphol.* **31**, 1–25.

Arnold, J. M. & Williams-Arnold, L. D. (1976). The egg cortex problem as seen through the squid eye. *Amer. Zool.* **16**, 421–46.

Arnold, J. M. & Williams-Arnold, L. D. (1977*). Cephalopoda: Decapoda. In *Reproduction of Marine Invertebrates*, eds A. C. Giese & J. S. Pearse, vol. 4, pp. 243–90. Academic Press, London & New York.

Astaurov, B. L. & Ostriakova-Varshaver, V. P. (1957). Complete heterospermic androgenesis in silkworms as a means for experimental analysis of the nucleus–cytoplasm problem. *J. Embryol. Exp. Morphol.* **5**, 449–62.

Atkinson, J. W. (1971). Organogenesis in normal and lobeless embryos of the marine prosobranch gastropod *Ilyanassa obsoleta* *J. Morphol* **133**, 339–52.

Austin, C. R. (1964). Gametogenesis and fertilization in the mesozoan *Dicyema aegira*. *Parasitology* **54**, 597–600.

Balbiani, E. G. (1882). Sur la signification des cellules polaires des insectes. *C. R. Acad. Sci., Paris* **95**, 927–9.

Banchetti, R. & Gremigni, V. (1973). Indirect evidence for neoblasts migration and

for gametogonia dedifferentiation in ex-fissiparous specimens of *Dugesia gonocephala* S. L. *Accad. Naz. Lincei* **55**, 107–15.
Bantock, C. (1961). Chromosome elimination in Cecidomyidae. *Nature, Lond.* **190**, 466–7.
Bantock, C. R. (1970). Experiments on chromosome elimination in the gall midge *Mayetiola destructor*. *J. Embryol. Exp. Morphol.* **24**, 257–86.
Barnes, R. D. (1963*). *Invertebrate Zoology*, 632 pp. W. B. Saunders, Philadelphia.
Bauer, H. & Beermann, W. (1952). Der Chromosomencyclus der Orthocladiinen (Nematocera, Diptera). *Z. Naturforsch.* **7b**, 557–63.
Bautz, A. (1978). Jeûne et régénération chez la planaire *Dendrocoelum lacteum*. *103ᵉ Congrès nat. Soc. Savants, Nancy* 123–35.
Bayreuther, K. (1952). Extrachromosomale feulgenpositive Körper (Nukleinkörper) in der Oogenese der Tipuliden. *Naturwiss.* **39**, 71.
Bayreuther, K. (1956). Die Oogenese der Tipuliden. *Chromosoma* **7**, 508–57.
Bazitov, A. A. & Lapkalo, A. V. (1977). Fission and gastrulation in *Microsomacanthus paramicrosoma* (Cyclophyllidea, Cestoda). *Parasitology* **11**, 104–12. (Russian with English summary.)
Beams, H. W. & King, R. L. (1937). The suppression of cleavage in *Ascaris* eggs by ultracentrifuging. *Biol. Bull.* **73**, 99–111.
Beams, H. W. & Kessel, R. G. (1974*). The problem of cell determinants. *Int. Rev. Cytol.* **39**, 413–79.
Bednarz, S. (1962). The developmental cycle of germ-cells in *Fasciola hepatica* L. 1758 (Trematoda, Digenea). *Zool. Pol.* **12**, 439–66.
Bednarz, S. (1973). The developmental cycle of the germ cells in several representatives of Trematoda (Digenea). *Zool. Pol.* **23**, 279–326.
Beeman, R. D. (1977*). Gastropoda: Opisthobranchia. In *Reproduction of Marine Invertebrates*, eds A. C. Giese & J. S. Pearse, vol. 4, pp. 115–79. Academic Press, London & New York.
Beeman, S. L. & Norris, D. M. (1977a). Embryogenesis of *Xyleborus ferrugineus* (Fabr.) (Coleoptera, Scolytidae). I. External morphogenesis of male and female embryos. *J. Morphol.* **152**, 177–220.
Beeman, S. L. & Norris, D. M. (1977b). Embryogenesis of *Xyleborus ferrugineus* (Fabr.) (Coleptera, Scolytidae). II. Developmental rates of male and female embryos. *J. Morphol.* **152**, 221–7.
Beermann, S. (1959). Chromatin-Diminution bei Copepoden. *Chromosoma* **10**, 504–14.
Beermann, S. (1966). A quantitative study of chromatin diminution in embryonic mitosis of *Cyclops furcifer*. *Genetics* **54**, 567–76.
Beermann, S. (1977). The elimination of heterochromatic chromosomal segments in *Cyclops* (Crustacea, Copepoda). *Chromosoma* **60**, 297–344.
Beermann, W. (1956*). Nuclear differentiation and functional morphology of chromosomes. *Cold Spring Harbor Symp. Quant. Biol.* **21**, 217–32.
Beermann, W. (1967*). La signification du 'puffing' dans les chromosomes géants des diptères. In *De l'Embryologie Expérimentale à la Biologie Moléculaire*, ed. E. Wolff, pp. 43–65. Dunod, Paris.
Belousov, L. V. (1963). The formation of excess interstitial cells as an index of the intensity of cell division in the morphogenesis of *Campanularia integra* and *Campanularia lacerata*. *Dohl. Biol. Sci. Sec.* **150**, 686–9.
Benazzi, M. (1966). Considerations on the neoblasts of planarians on the basis of certain karyological evidence. *Chromosoma* **19**, 14–27.
Berg, W. E. & Kato, Y. (1959). Localization of polynucleotide in the egg of *Ilyanassa*. *Acta Embryol. Morphol Exp.* **2**, 227–33.

Bergquist, P. R. & Green, C. R. (1977). An ultrastructural study of settlement and metamorphosis in sponge larvae. *Cahiers de Biol. mar.* **18**, 289–302.

Berrill, N. J. (1949a). Growth and form in gymnoblastic hydroids. I. Polymorphic development in *Bougainvillia* and *Aselomaris*. *J. Morphol.* **84**, 1–30.

Berrill, N. J. (1949b*). Developmental analysis of Scyphomedusae. *Biol. Rev.* **24**, 393–410.

Berrill, N. J. (1952a). Growth and form in gymnoblastic hydroids. II. Sexual and asexual reproduction in *Rathkea*. III. Hydranth and gonophore development in *Pennaria* and *Acaulis*. IV. Relative growth in *Eudendrium*. *J. Morphol.* **90**, 1–32.

Berrill, N. J. (1952b). Growth and form in gymnoblastic hydroids. V. Growth cycle in *Tubularia*. *J. Morphol.* **90**, 583–601.

Berrill, N. J. (1952c*). Regeneration and budding in worms. *Biol. Rev.* **27**, 401–38.

Berrill, N. J. (1961*). *Growth, Development and Pattern*, 555 pp. Freeman, San Francisco.

Berrill, N. J. (1975*). Chordata: Tunicata. In *Reproduction of Marine Invertebrates*, eds A. C. Giese & J. S. Pearse, vol. 2, pp. 241–82. Academic Press, London & New York.

Berrill, N. J. & Liu, C. K. (1948*). Germ plasm, Weismann and Hydrozoa. *Quart. Rev. Biol.* **23**, 124–32.

Berry, A. J. (1977*). Gastropoda: Pulmonata. In *Reproduction of Marine Invertebrates*, eds A. C. Giese & J. S. Pearse, vol. 4, pp. 181–226. Academic Press, London & New York.

Berry, R. O. (1941). Chromosome behavior in the germ cells and development of the gonads in *Sciara ocellaris*. *J. Morphol.* **68**, 547–83.

Bertzbach, R. (1960). Experimentelle Untersuchungen über den Einfluss von Röntgenstrahlen auf die Embryonalentwicklung der Honigbiene. *Wilhelm Roux' Arch. Entwicklungsmech. Organismen* **152**, 524–51.

Best, J. B., Hand, S. & Rosenvold, R. (1968). Mitosis in normal and regenerating planarians. *J. Morphol.* **168**, 157–67.

Best, J. B., Rosenvold, R., Sonders, J. & Wade, C. (1965). Studies on the incorporation of isotopically labelled nucleotides and amino acids in *Planaria*. *J. Exp. Zool.* **159**, 397–404.

Betchaku, T. (1967). Isolation of planarian neoblasts and their behaviour *in vitro* with some aspects of the mechanism of the formation of regeneration blastema. *J. Exp. Zool.* **164**, 407–34.

Bezem, J. J. & Raven, C. P. (1975). Computer simulation of early embryonic development. *J. Theor. Biol.* **54**, 47–61.

Bier, K. (1952). Beziehungen zwischen Nährzellkerngrösze und Ausbildung ribonukleinsäurehaltiger Strukturen in den Oocyten von *Formica rufa rufa-pratensis minor* Gösswald. *Verhandl. Zool. Gesellschaft. Freiburg* pp. 369–74.

Bier, K. (1954). Über den Saisondimorphismus der Oogenese von *Formica rufa rufa-pratensis minor* Gössw. und dessen Bedeutung für die Kastendetermination. *Biol. Zentralblatt* **73**, 170–90.

Bock, E. (1939). Bildung und Differenzierung der Keimblätter bei *Chrysopa perla* (L). *Z. Morphol. Ökol. Tiere* **35**, 615–702.

Bock, E. (1942). Wechselbeziehungen zwischen den Keimblättern bei der Organbildung von *Chrysopa perla* (L). I. Die Entwicklung des Ektoderms in mesodermdefekten Keimteilen. *Wilhelm Roux' Arch. Entwicklungsmech. Organismen* **141**, 157–247.

Bode, H., Berking, S., David, C. N., Gierer, A., Schaller, H. & Trenkner, E. (1973). Quantitative analysis of cell types during growth and morphogenesis in *Hydra*. *Wilhelm Roux' Arch. Entwicklungsmech. Organismen* **171**, 269–85.

Bode, H. R. & David C. N. (1978). Regulation of a multipotent stem cell, the interstitial cell of *Hydra*. *Prog. Biophys. Mol. Biol.* **33**, 189–206.
Bodine, M. W. (1970) The segmental origin of the appendages of the head and anterior body segments of a spiroboloid millipede *Narceus annularis*. *J. Morphol.* **132**, 47–67.
Bodo, F. & Bouillon, J. (1968). Etude histologique du développement embryonnaire de quelques hydroméduses de Roscoff: *Phialidium hemispaericum* (L.), *Obelia* sp. (Péron et Lesueur), *Sarsia eximia* (Allman), *Podocoryne carnea* (Sars), *Gonionemus vertens* (Agassiz) *Cahiers de Biol. mar.* **9**, 69–104.
Boelsterli, U. (1975). Notes on oogenesis in *Tubularia crocea* Agassiz (Athecata, Hydrozoa). *Pubbl. Staz. Zool. Napoli* **39**, suppl. 53–66.
Boelsterli, U. (1977). An electron microscopic study of early developmental stages, myogenesis, oogenesis and cnidogenesis in the anthomedusa, *Podocoryne carnea* M. Sars. *J. Morphol.* **154**, 259–89.
Boilly, B. (1968). Origine des cellules de régénération chez *Aricia foetida* Clap. (Annélide, polychéte). *Arch. Anat. Microsc. Morph. Exp.* **57**, 297–308.
Boilly, B. (1969*a*). Origine des cellules régénératrices chez *Nereis diversicolor* O. F. Müller (Annélide polychéte). *Wilhelm Roux' Arch. Entwicklungsmech. Organismen* **162**, 286–305.
Boilly, B. (1969*b*). Sur l'origine des cellules régénératrices chez les annélides polychétes. *Arch. Zool. Expér. Générale* **110**, 127–43.
Boilly, B. (1969*c*). Etude expérimentale de la location, par rapport au plan d'amputation, de la source de cellules de régénération mésodermiques chez une Annélide polychéte (*Syllis amica* Quatrefages). *J. Embryol. Exp. Morphol.* **21**, 193–206.
Boletzky, S. V. (1978*a*). Gut development in cephalopods: a correction. *Rev. Suisse Zool.* **85**, 379–80.
Boletzky, S. V. (1978*b*). Nos connaissances actuelles sur le développement des octopodes. *Vie Milieu* **28/29**, 85–120.
Bolla, R. I. & Roberts, L. S. (1971). Developmental physiology of cestodes. IX. Cytological characteristics of the germinative region of *Hymenolepis diminuta*. *J. Parasitol.* **57**, 267–77.
Borradaile, L. A. & Potts, F. A. (1963*). The phylum Annelida. In *The Invertebrata, a Manual for the Use of Students*, 4th edn, pp. 266–316. Cambridge University Press.
Bouillon, J. (1957). Etude monographique du genre *Limnocnida* (Limnomeduse). *Ann. Soc. Roy. Zool. Belg.* **87**, 253–500.
Bounoure, L. (1939*). *L'Origine des Cellules Reproductrices et le Problème de la Lignée Germinale*, 271 pp. Gauthiers-Villars, Paris.
Boveri, Th. (1887*a*). Über Differenzierung der Zellkerne während der Furchung des Eies von *Ascaris megalocephala*. *Anat. Anz.* **2**, 288–693.
Boveri, Th. (1887*b*). Die Bildung der Richtungskörper bei *Ascaris megalocephala* and *Ascaris lumbricoides*. *Zell Stud. Jena* **1**, 1–93.
Boveri, Th. (1888). Die Befruchtung und Teilung des Eies von *Ascaris megalocephala*. *Zell Stud. Jena* **2**, 1–198.
Boveri, Th. (1891). Befruchtung. *Ergeb. Anat. Entwicklungsgesch.* **1**, 386–485.
Boveri, Th. (1893). Über die Entstehung des Gegensetzes zwischen den Geschlechtszellen und den somatischen Zellen bei *Ascaris megalocephala*, nebst Bemerkungen über die Entwicklungsgeschichte der Nematoden. *Sitzungsber. Ges. Morphol. Physiol.* **8**, 114–25.
Boveri, Th. (1899). Die Entwicklung von *Ascaris megalocephala* mit besonderer Rücksicht auf die Kernverhältnisse. *Festschr. C. v. Kupffer Jena* pp. 383–430.

Boveri, Th. (1904). Protoplasma Differenzierung als auslösender Faktor für Kernverschiedenheit. *Sitzungsber. phys. med. Ges. Würzburg*, 16pp.

Boveri, Th. (1909*a*). Über 'Geschlechtschromosomen' bei Nematoden. *Arch. Zellforsch.* **4**, 132–41.

Boveri, Th. (1909*b*). Die Blastomerenkerne von *Ascaris megalocephala* und die Theorie der Chromosomenindividualität. *Arch. Zellforsch.* **3**, 181–268.

Boveri, Th. (1910*a*) Über die Teilung centrifugierter Eier von *Ascaris megalocephala*. *Arch. Entwicklungsmech.* **30**, 101–25.

Boveri, Th. (1910*b*). Die Potenzen der *Ascaris* Blastomeren bei abgeanderter Furchung. Zugleich ein Beitrag zur Frage qualitativ ungleicher Chromosomenteilung. *Festschr. 60 Geburtstag R. Hertwigs* **3**, 131–214.

Braem, F. (1897). Die geschlechtliche Entwicklung von *Plumatella fungosa*. *Zoologica* **10**, 1–94.

Bregman, A. A. (1975). Q-, C- and G-banding patterns in the germline and somatic chromosomes of *Miastor* sp. (Diptera: Cecidomyidae). *Chromosoma* **53**, 119–30.

Brien, P. (1932). Contribution à l'étude de la régénération naturelle chez les Spongillidae *Spongilla lacustris* (L.); *Ephydatia fluviatilis* (L.). *Arch. Zool. Exp. Gén.* **74**, 461–506.

Brien, P. (1953*). La pérennité somatique. *Biol. Rev.* **28**, 308–49.

Brien, P. (1954). Origine et localisation des cellules germinales chez les Hydroïdes gymnoblastiques. *Mém. 3ᵉ Congr. Nat. Sci.* **6**, 3pp.

Brien, P. (1956). Le bourgeonnement des endoproctes et leur phylogenèse; à propos du bourgeonnement chez *Pedicellina cernua* (Pallos). *Ann. Soc. Roy. Zool. Belg.* **87**, 26–43.

Brien, P. (1960*). Classe des Bryozoaires. In *Traité de Zoologie, Anatomie, Systématique, Biologie*, vol. 5, pp. 1053–335.

Brien, P. (1961). Etude d'*Hydra pirardi* (nov. spec.). Origine et repartition des nematocystes, gamétogenèse. Involution postgamétique. Evolution réversible des cellules interstitielles. *Bull. Biol. France Belg.* **95**, 301–63.

Brien, P. (1962). Contribution à l'étude de la biologie sexuelle. Induction gamétique et sexuelle chez les hydres d'eau douce par les greffes en parabiose. *Bull. Acad. Belg. Sci.* **48**, 825–47.

Brien, P. (1963). Contribution à l'étude de la biologie sexuelle chez les hydres d'eau douce. Induction gamétique et sexuelle par la méthode des greffes en parabiose. *Bull. Biol. France Belg.* **97**, 213–83.

Brien, P. (1964*). Blastogenèse et gamétogenèse. In *L'Origine de la Lignée Germinale chez les Vertébrés et chez quelques Groupes d'Invertébrés*, ed. E. Wolff, pp. 21–76. Hermann, Paris.

Brien, P. (1965*). Considérations à propos de la reproduction sexuélle des invertébrés. *Ann. Biol.* **4**, 329–65.

Brien, P. (1966*). Origine des cellules sexuelles chez les métazoaires; l'état gamétique. In *Biologie de la Reproduction Animale; Blastogenèse, Gamétogenèse, Sexualisation*, ed. P. Brien, 39–76. Masson & Cie, Paris.

Brien, P. & Huysmans, G. (1937). La croissance et le bourgeonnement du stolon chez les Stolonifera (*Bowerbankia* (Farre)). Evolution du stolon et de la zoecie chez les Bryozoaires. *Ann. Soc. Zool. Belg.* **68**, 13–40.

Brien, P. & Reniers-Decoen, M. (1949). La croissance, la blastogénèse, l'ovogénèse chez *Hydra fusca* (Pallas). *Bull. Biol. France Belg.* **83**, 295–386.

Brien, P. & Reniers-Decoen, M. (1955). La signification des cellules interstitielles des hydras d'eau douce et la problème de la réserve embryonnaire. *Bull. Zool. France Belg.* **89**, 258–325.

Brisson, P. (1973). Observation ultrastructurale des cellules germinales chez l'em-

bryon d'*Acroloxus lacustris* (L.) (Gastéropode Pulmoné Basommatophore). *C. R. Acad. Sci., Paris* **277**, 2205–8.
Brisson, P. & Besse, C. (1975). Etude ultrastructurale de l' ébauche gonadique chez l'embryon de *Lymnaea stagnalis* L. *Bull. Soc. Zool. France* **100**, 345–9.
Brisson, P. & Regondaud, J. (1971). Observation relatives à l'origine dualiste de l'appareil génital chez quelques gastéropodes pulmonés basommatophores. *C. R. Acad. Sci., Paris* **273**, 2339–41.
Brisson, P. & Regondaud, J. (1977). Origine et structure de l'ébauche de la gonade chez les gastéropodes pulmonés basommatophores. *Malacologia* **16**, 457–66.
Bronskill, J. F. (1959). Embryology of *Pimpla turionellae* (L) (Hymenoptera: Ichneumonidae). *Can. J. Zool.* **37**, 655–88.
Bronskill, J. F. (1964). Embryogenesis of *Mesoleius tenthredinis* Morl. (Hymenoptera: Ichneumonidae). *Can. J. Zool.* **42**, 439–53.
Brøndstedt, A. & Brøndstedt, H. V. (1961a). Influence of temperature on rate of regeneration in time-graded regeneration field in planarians. *J. Embryol. Exp. Morphol.* **9**, 159–66.
Brøndstedt, A. & Brøndstedt, H. V. (1961b). Number of neoblasts in the intact body of *Euplanaria torva* and *Dendrocoelum lacteum*. *J. Embryol. Exp. Morphol.* **9**, 167–72.
Brøndstedt, H. V. (1969*). *Planarian Regeneration*. Pergamon Press, Oxford.
Brooks, F. G. (1930). Studies on the germ cell cycle of trematodes. *Amer. J. Hyg.* **12**, 299–340.
Brown, S. W. & Bennett, F. D. (1957). On sex determination in the diaspine scale, *Pseudaulacaspis pentagona* (Targ.) (Coccoïdea). *Genetics* **42**, 510–23.
Brown, S. W. & De Lotto, G. (1959). Cytology and sex ratios of an African species of armoured scale insect (Coccoïdea: Diaspididae). *Amer. Nat.* **93**, 369–79.
Bruslé, J. (1968a). Aspects ultrastructuraux de la différenciation ovogénétique chez un hermaphrodite fonctionnel, *Asterina gibbosa* Pennant. *Ann. Sci. Nat. Zool. Biol. Anim.* **10**, 545–62.
Bruslé, J. (1968b). Aspects ultrastructuraux de la différenciation spermatogénique chez un hermaphrodite fonctionnel, *Asterina gibbosa* Pennant. Comparaison des deux lignées gamétogénétiques. *Ann. Sci. Nat. Zool. Biol. Anim.* **10**, 563–72.
Bruslé, J. (1970). Radiosensibilité des cellules germinales et somatiques de la gonade hermaphrodite d' *Asterina gibbosa* Penn. après irradiation X. Etude ultrastructural des radiolésions différentielles. *Arch. Biol.* **80**, 451–70.
Bruslé, J. & Delavault, R. (1968). Recherches sur la cytodifférenciation des gamètes chez un hermaphrodite fonctionnel: *Asterina gibbosa*. Ultrastructure des ovogonies et des ovocytes en prémeiose. *C. R. Acad. Sci., Paris* **266**, 21–3.
Buchner, P. (1957). Endosymbiosestudien an Schildläusen. VI. Die nicht in Symbiose lebende Gattung *Apiomorpha* und ihre ungewöhnliche embryonalentwicklung. *Z. Morphol. Ökol. Tiere* **46**, 481–528.
Burnett, A. L. (1961). The growth process in *Hydra*. *J. Exp. Zool.* **146**, 21–84.
Burnett, A. L., Davis, L. E. & Ruffing, F. E. (1966). A histological and ultrastructural study of germinal differentiation of interstitial cells arising from gland cells in *Hydra viridis*. *J. Morphol.* **120**, 1–8.
Burnett, A. L. & Diehl, N. A. (1964). The nervous system of *Hydra*. III. The initiation of sexuality with special reference to the nervous system. *J. Exp. Zool.* **157**, 237–50.
Cable, R. M. (1931). Studies on the germ cell cycle of *Cryptocotyle lingua* Creplin. I. Gametogenesis in the adult. *Quart. J. Microsc. Sci.* **74**, 564–89.
Cable, R. M. (1934). Studies on the germ cell cycle of *Cryptocotyle lingua*. II. Germinal development in the larval stages. *Quart. J. Microsc. Sci.* **76**, 573–614.

Cambar, R. (1956*). Les problèmes scientifiques et philosophiques de la lignée germinale chez les animaux: aspects et conceptions modernes. *Bull. Nat.* **43**, 78–103.

Camenzind, R. (1966). Die Zytologie der bisexuellen und parthenogenetischen Fortpflanzung von *Heteropeza pygmaea* Winnertz, einer Gallmücke mit pädogenetischer Vermehrung. *Chromosoma* **18**, 123–52.

Campbell, R. D. (1967a). Tissue dynamics of steady growth in *Hydra littoralis*. I. Patterns of cell division. *Develop. Biol.* **15**, 487–502.

Campbell, R. D. (1967b). Tissue dynamics of steady growth in *Hydra littoralis*. II. Patterns of tissue movements. *J. Morphol.* **121**, 19–28.

Campbell, R. D. (1967c). Tissue dynamics of steady growth in *Hydra littoralis*. III. Behaviour of specific cell types during tissue movements. *J. Exp. Zool.* **164**, 379–92.

Campbell, R. D. (1973). Vital staining of single cells in developing tissues: India ink injection to trace tissue movements in *Hydra*. *J. Cell Sci.* **13**, 651–61.

Campbell, R. D. (1974*). Cnidaria. In *Reproduction of Marine Invertebrates*, eds A. C. Giese & J. S. Pearse, vol. 1, pp. 133–99. Academic Press, London & New York.

Campbell, R. D. & David, C. N. (1974). Cell cycle kinetics and development of *Hydra attenuata*. II. Interstitial cells. *J. Cell Sci.* **16**, 349–58.

Cantwell, G. E., Nappi, A. J. & Stoffolano, J. G. (1976). Embryonic and postembryonic development of the house fly (*Musca domestica* L.). *Techn. Bull. US Dept. Agric.* No. 1519, 69 pp.

Cather, J. N. (1967). Cellular interactions in the development of the shell gland of the gastropod *Ilyanassa*. *J. Exp. Zool.* **166**, 205–24.

Cather, J. N. (1971*). Cellular interactions in the regulation of development in annelids and molluscs. *Adv. Morphogen.* **9**, 67–125.

Cather, J. N. & Verdonk, N. H. (1974). The development of *Bithynia tentaculata* (Prosobranchia, Gastropoda) after removal of the polar lobe. *J. Embryol. Exp. Morphol.* **31**, 415–22.

Cather, J. N. & Verdonk, N. H. (1979). Development of *Dentalium* following removal of D quadrant blastomeres at successive cleavage stages. *Wilhelm Roux' Arch. Develop. Biol.* **187**, 355–66.

Cather, J. N., Verdonk, N. H. & Dohmen, M. R. (1976). Role of the vegetal body in the regulation of development in *Bithynia tentaculata* (Prosobranchia, Gastropoda). *Amer. Zool.* **16**, 455–68.

Cavallin, M. (1971). La 'polyembryonie substitutive' et le problème de l'origine de la lignée germinale chez la phasme *Carausius morosus* Br. *C. R. Acad. Sci., Paris* **272**, 462–5.

Cavallin, M. (1976). La ségrégation de la lignée germinale chez le phasme *Carausius morosus* Br. *Bull. Soc. Zool. France* **101**, suppl. 4, 15–21.

Chandebois, R. (1960). Sur les sources de l'histogénèse régénératrice chez les planaires. *C. R. Acad. Sci., Paris* **251**, 146.

Chandebois, R. (1962). Role des éléments fixes et libres du parenchyme dans la régénération de *Planaria subtentaculata* Drap. *Bull. Biol. France Belg.* **96**, 203–27.

Chandebois, R. (1965*). Cell transformation systems in planarians. In *NATO Symposium on Regeneration in Animals and Related Problems*, eds V. Kiortsis & H. A. L. Trampusch, pp. 131–142, North-Holland, Amsterdam.

Chandebois, R. (1968). The respective roles of mitotic activity and of cell differentiation in planarian regeneration. *J. Embryol. Exp. Morphol.* **20**, 178–88.

Chandebois, R. (1976*). *Histogenesis and Morphogenesis in Planarian Regeneration*, ed. A. Wolsky, 182 pp. Karger, Basel & New York.

Chapron, C. & Relexans, J. C. (1971). Ultrastructure des gonocytes primordiaux et des gonies chez l'hermaphrodite *Eisenia foetida* (oligochète lombricidé). *C. R. Acad. Sci., Paris* **272**, 2916–19.
Charniaux-Cotton, H. (1964*). La lignée germinale chez les arthropodes. In *L'Origine de la Lignée Germinale chez les Vertebrés et chez quelques Groups d'Invertébrés*, ed. E. Wolff, pp. 139–74. Hermann, Paris.
Chen, P. S. (1971*). Biochemical aspects of insect development. *Monographs Develop. Biol.* **3**, 230 pp. Karger, Basel & New York.
Cheng, T. C. & James, H. A. (1960). Studies on the germ cell cycle, morphogenesis and development of the cercarial stage of *Crepidostomum cornutum* (Osborn, 1903) (Trematoda: Allocreadiidae). *Trans. Amer. Microsc. Soc.* **79**, 75–85.
Chia, F. S. & Crawford, B. J. (1973). Some observations on gametogenesis, larval development and substratum selection of the sea pen *Ptilosarcus guerneyi*. *Mar. Biol.* **23**, 73–82.
Chitwood, B. G. (1974*). Nemic embryology. In *Introduction to Nematology*, eds B. G. Chitwood & M. B. Chitwood, pp. 202–12. University Park Press, Baltimore.
Chrétien, M. (1957). Histologie et développement de l'ovaire chez *Alcyonidium gelatinosum* (L) (Bryozoaire cténostome) *Bull. Lab. marit. Dinard* **43**, 25–51.
Christophers, S. R. (1960*). *Aëdes aegypti (L.) The Yellow Fever Mosquito, its Life History, Bionomics and Structure*, pp. 131–57 and 177–93. Cambridge University Press.
Chuang, S. H. (1962). The embryonic and post-embryonic development of *Rhabditis teres* (A. Schneider). *Nematologica* **7**, 317–30.
Ciordia, H. (1956). Cytological studies of the germ cell cycle of the trematode family Bucephalidae. *Trans. Amer. Microsc. Soc.* **75**, 103–16.
Clark, M. E. & Clark R. B. (1962). Growth and regeneration in *Nephthys*. *Zool. Jahrb. (Physiol.)* **70**, 24–90.
Clark, R. B. & Evans, S. M. (1961). The effect of delayed brain extirpation and replacement on caudal regeneration in *Nereis diversicolor*. *J. Embryol. Exp. Morphol.* **9**, 97–105.
Clement, A. C. (1952). Experimental studies on germinal localisation in *Ilyanassa*. I. The role of the polar lobe in determination of the cleavage pattern and its influence in later development. *J. Exp. Zool.* **121**, 593–626.
Clement, A. C. (1956). Experimental studies on germinal localisation in *Ilyanassa*. II. The development of isolated blastomeres, *J. Exp. Zool.* **132**, 427–46.
Clement, A. C. (1960). Development of the *Ilyanassa* embryo after removal of the mesentoblast cell. *Biol. Bull.* **119**, 310.
Clement, A. C. (1962). Development of *Ilyanassa* following removal of the D macromere at successive cleavage stages. *J. Exp. Zool.* **149**, 193–216.
Clement, A. C. (1963). Effects of micromere deletion on development in *Ilyanassa*. *Biol. Bull.* **125**, 375.
Clement, A. C. (1967). The embryonic value of the micromeres in *Ilyanassa obsoleta*, as determined by deletion experiments. I. The first quartet cells. *J. Exp. Zool.* **166**, 77–88.
Clement, A. C. (1968). Development of the vegetal half of the *Ilyanassa* egg after removal of most of the yolk by centrifugal force, compared with the development of animal halves of similar composition. *Develop. Biol.* **17**, 165–86.
Clement, A. C. (1976). Cell determination and organogenesis in molluscan development: a reappraisal based on deletion experiments in *Ilyanassa*. *Amer. Zool.* **16**, 447–53.
Collier, J. R. (1960a). The localization of ribonucleic acid in the egg of *Ilyanassa obsoleta*. *Exp. Cell Res.* **21**, 126–36.

Collier, J. R. (1960b). The localisation of some phosphorus compounds in the egg of *Ilyanassa obsoleta. Exp. Cell Res.* **21**, 548–55.

Collier, J. R. (1961). The effect of removing the polar lobe on the protein synthesis of the embryo of *Ilyanassa obsoleta. Acta Embryol. Morphol. Exp.* **4**, 70–6.

Collier, J. R. (1965*). Morphogenetic significance of biochemical patterns in mosaic embryos. In *Biochemistry of Animal Development*, vol. 1, ed. R. Weber, pp. 203–44. Academic Press, London & New York.

Collier, J. R. (1975). Nucleic acid synthesis in the normal and lobeless embryo of *Ilyanassa obsoleta. Exp. Cell Res.* **95**, 254–62.

Connes, R. (1977). Contribution à l'étude de la gemmulogenèse chez la désmosponge marine *Suberites domuncula* (Olivi Nardo). *Arch. Zool. Exp. Gén.* **118**, 391–407.

Conrad, G. W. (1973*). Control of polar lobe formation in fertilised eggs of *Ilyanassa obsoleta* Stimpson. *Amer. Zool.* **13**, 961–80.

Conrad, G. W. & Williams, D. C. (1974a). Polar lobe formation and cytokinesis in fertilised eggs of *Ilyanassa obsoleta*. I. Ultrastructure and effects of cytochalasin B and colchicine. *Develop. Biol.* **36**, 363–78.

Conrad, G. W. & Williams, D. C. (1974b). Polar lobe formation and cytokinesis in fertilised eggs of *Ilyanassa obsoleta*. II. Large bleb formation caused by high concentrations of exogenous calcium ions. *Develop. Biol.* **37**, 280–94.

Cornman, I. (1943). Suspension of pole cell formation in *Drosophila* eggs by hydrostatic pressure. *J. Cell Comp. Physiol.* **22**, 197–8.

Cort, W. W. (1944*). The germ cell cycle in the digenetic trematodes. *Quart. Rev. Biol.* **19**, 275–84.

Cort, W. W., Ameel, D. J. & Van der Woude, A. (1954). Germinal development in the sporocysts and rediae of the digenetic trematodes. *Exp. Parasitol.* **3**, 185–216.

Cort, W. W. & Olivier, L. (1941). Early developmental stages of strigeid trematodes in the first intermediate host. *J. Parasitol.* **27**, 493–504.

Cort, W. W. & Olivier, L. (1943). The development of the sporocysts of schistosome, *Cercaria stagnicolae* Talbot, 1936. *J. Parasitol.* **29**, 164–76.

Counce, S. J. (1959). Comparative ontogeny of the polar granules in *Drosophila. Anat. Rec.* **134**, 546–7.

Counce, S. J. (1961*). The analysis of insect embryogenesis. *Ann. Rev. Entomol.* **6**, 295–512.

Counce, S. J. (1963). Developmental morphology of polar granules in *Drosophila* including observations on pole cell behavior and distribution during embryogenesis. *J. Morphol.* **112**, 129–45.

Counce, S. J. (1973*). The causal analysis of insect embryogenesis. In *Developmental Systems: Insects*, eds S. J. Counce & C. H. Waddington, vol. 2, pp. 1–156. Academic Press, London & New York.

Counce, S. J. & Ede, D. A. (1957). The effect on embryogenesis of a sex-linked female-sterility factor in *Drosophila melanogaster. J. Embryol. Exp. Morphol.* **5**, 404–21.

Counce, S. J. & Selman, G. G. (1955). The effects of ultrasonic treatment on embryonic development of *Drosophila melanogaster. J. Embryol. Exp. Morphol.* **3**, 121–41.

Coward, S. J. (1969*). Regeneration in planarians: some unresolved problems and questions. *J. Biol. Psychol.* **11**, 15–19.

Coward, S. J., Bennett, C. E. & Hazlehurst, B. L. (1974). Lysosomes and lysosomal enzyme activity in the regenerating planarian; evidence in support of dedifferentiation. *J. Exp. Zool.* **189**, 133–46.

Coward, S. J., Hirsh, F. M. & Taylor, J. H. (1970). Thymidine kinase activity during regeneration in the planarian *Dugesia dorotocephala*. *J. Exp. Zool.* **173**, 269–77.
Cowden, R. R. (1964). A cytochemical study of gonophore and oocyte development in *Pennaria tiarella*. *Acta Embryol. Morphol Exp.* **7**, 167–79.
Cowden, R. R. (1965). Cytochemical studies of embryonic development to metamorphosis in the gymnoblastic hydroid, *Pennaria tiarella*. *Acta Embryol. Morphol. Exp.* **8**, 221–31.
Crawford, B. J. & Chia, F. S. (1978). Coelomic pouch formation in the starfish *Pisaster ochraceus* (Echinodermata: Asteroidea). *J. Morphol.* **157**, 99–120.
Crofts, D. R. (1937). The development of *Haliotis tuberculata*, with special reference to the organogenesis during torsion. *Phil. Trans. Roy. Soc. ser.* B **228**, 219–68.
Crowell, J. (1964). The fine structure of the polar lobe of *Ilyanassa obsoleta*. *Acta Embryol. Morphol. Exp.* **7**, 225–34.
Curtis, S. K. & Cowden, R. R. (1971). Normal and experimentally modified buds in *Cassiopea* (phylum Coelenterata; class Scyphozoa). *Acta Embryol. Exp.* Suppl. 239–59.
Curtis, S. K. & Cowden, R. R. (1972). Regenerative capacities of the scyphistoma of *Cassiopea* (phylum, Coelenterata; class, Scyphozoa). *Acta Embryol. Exp.* suppl. 429–54.
Daniels, S. & Ehrman, L. (1974). Embryonic pole cells and mycoplasmalike symbionts in *Drosophila paulistorum*. *J. Invert. Pathol.* **24**, 14–19.
Dan-Sohkawa, M. & Satoh, N. (1978). Studies on dwarf larvae developed from isolated blastomeres of the starfish, *Asterina pectinifera*. *J. Embryol. Exp. Morphol.* **46**, 171–85.
Darden, W. H. (1973*). Hormonal control of sexuality in algae. In *Humoral Control of Growth and Differentiation*, eds J. Lobue & A. S. Gordon, vol. 2, pp. 101–19. Academic Press, London & New York.
D'Asaro, C. N. (1966). The egg capsules, embryogenesis, and early organogenesis of a common oyster predator, *Thais haemastoma floridana* (Gastropoda: Prosobranchia). *Bull. Mar. Sci.* **16**, 884–914.
D'Asaro, C. N. (1969). The comparative embryogenesis and early organogenesis of *Bursa corrugata* Perry and *Distorsio clathrata* Lamarck (Gastropoda, Prosobranchia). *Malacologia* **9**, 349–89.
Dautert, E. (1929). Die Bildung der Keimbläter bei *Paludina vivipara*. *Zool. Jahrb. Abt. Anat. Ontog. Tiere* **50**, 433–96.
Davenport, R. (1979*). The sea-urchin embryo. In *An Outline of Animal Development*, ed. R. Davenport, pp. 259–93. Addison-Wesley, Reading, Mass.
David, C. N. (1975). Stem cell differentiation in *Hydra*. *Microbiol.* 434–41.
David, C. N. & Challoner, D. (1974). Distribution of interstitial cells and differentiating nematocytes in nests in *Hydra attenuata*. *Amer. Zool.* **14**, 537–42.
David, C. N. & Murphy, S. (1977). Characterisation of interstitial stem cells in *Hydra* by cloning. *Develop. Biol.* **58**, 372–83.
Davidson, E. H. (1976*). *Gene Activity in Early Development*, 2nd edn, 375 pp. Academic Press, London & New York.
Davis, C. W. C. (1967). A comparative study of larval embryogenesis in the mosquito *Culex fatigans* Wiedemann (Diptera: Culicidae) and the sheep-fly *Lucilia sericata* Meigen (Diptera: Calliphoridae). I. Description of embryonic development. *Austr. J. Zool.* **15**, 547–79.
Davis, C. W. C. (1970). A comparative study of larval embryogenesis in the mosquito *Culex fatigans* Wiedemann (Diptera: Culicidae) and the sheep-fly *Lucilia sericata* Meigen (Diptera: Calliphoridae). II. Causal interactions in embryonic development. *Austr. J. Zool.* **18**, 125–54.

Davis, C. W. C., Krause, J. & Krause G. (1968). Morphogenetic movements and segregation of posterior egg fragments *in vitro*. *Wilhelm Roux' Arch. Entwicklungsmech. Organismen* **161**, 209–40.

Davis, L. E., Burnett, A. L., Haynes, J. F. & Mumaw, V. R. (1966). A histological and ultrastructural study of dedifferentiation and redifferentiation of digestive and gland cells in *Hydra viridis*. *Develop. Biol.* **14**, 307–29.

Dawydoff, C. (1928). Sur la réversibilité des processus du développement. Les phases extrêmes de la réduction des némertes. *C. R. Acad. Sci., Paris* **186**, 911–13.

Dawydoff, C. (1959). Classe des Echiuriens. In *Traité de Zoologie*, ed. P. Grassé, vol. 5, pp. 855–907. Masson et Cie, Paris.

Delavault, R. (1961). Observations complémentaires sur le cycle sexuel d'*Asterina gibbosa*, de Dinard. *Ext. Lab. maritime Dinard* **46**, 36–8.

Delavault, R. & Bruslé, J. (1968). Recherches sur la cytodifférenciation des gamètes chez un hermaphrodite fonctionnel: *Asterina gibbosa*. Ultrastructure des cellules de la lignée spermatogénétique et comparaison spermatogonies–ovogonies. *C. R. Acad. Sci., Paris* **266**, 710–12.

De Leo, G. (1972). Data on the cell-lineage and the early stages of development of *Sepiola rondeletti*. *Acta embryol. Exp.* **1**, 25–44.

De Vos, L. (1971). Etude ultrastructurale de la gemmulogenèse chez *Ephydatia fluviatilis*. I. Le vitellus-formation-teneur en ARN et glycogène. *J. Microsc.* **10**, 283–304.

Devriès, J. (1968). Les premières étapes de la segmentation (formation de la jeune blastula) chez le lombricien *Eisenia foetida*. *Bull. Soc. Zool. France* **93**, 87–97.

Devriès, J. (1969). Le développement des embryos d'*Eisenia foetida* après la destruction unilatérale des mésotéloblastes. *Bull. Soc. Zool. France* **94**, 663–71.

Devriès, J. (1970). Sur le rôle du mésoderme dans la différenciation de l'endoderme chez l'embryon d'*Eisenia foetida* (Lombricien). *Bull. Soc. Zool. France* **95**, 169–72.

Devriès, J. (1971). Origine de la lignée germinale chez le lombricien *Eisenia foetida*. *Ann. Embryol. Morphogen.* **4**, 37–43.

Devriès, J. (1973). Détermination précoce du développement embryonnaire chez le lombricien *Eisenia foetida*. *Bull. Soc. Zool. France* **98**, 405–17.

Devriès, J. (1974*a*). Le mésoderme, feuillet directeur de l'embryogenèse chez le lombricien *Eisenia foetida*. I. La détermination de la métamérie interne. *Acta embryol. Exp.* **2**, 105–22.

Devriès, J. (1974*b*). Le mésoderme, feuillet directeur de l'embryogenèse chez le lombricien *Eisenia foetida*. II. La différenciation du tube digestif et des dérivés ectodermiques. *Acta embryol. Exp.* **2**, 157–80.

Devriès, J. (1974*c*). Le mésoderme, feuillet directeur de l'embryogenèse chez le lombricien *Eisenia foetida*. III. La détermination des ectotéloblastes. *Acta embryol. Exp.* **2**, 181–90.

Diaz, J. P. (1974). De l'origine de certains endopinacocytes à partir de choanocytes chez la démosponge *Suberites massa* Nardo *Bull. Soc. Zool. France* **99**, 687–93.

Diaz, J. P., Connes, R. & Paris, J. (1973) Origine de la lignée germinale chez une démosponge de l'etang de Thau: *Suberites massa* Nardo. *C. R. Acad. Sci., Paris* **277**, 661–4.

Diaz, J. P., Connes, R. & Paris, J. (1975). Etude ultrastructurale de l'ovogenèse d'une démosponge: *Suberites massa* Nardo. *J. Microsc.* **24**, 105–16.

Diehl, F. A. (1969). Cellular differentiation and morphogenesis in *Cordylophora*. *Wilhelm Roux' Arch. Entwicklungsmech. Organismen* **162**, 309–35.

Diehl, F. A. (1973*). The developmental significance of interstitial cells during regeneration and budding. In *Biology of Hydra*, ed. A. L. Burnett, pp. 109–41. Academic Press, London & New York.

Diehl, F. A. & Burnett, A. L. (1965a). The role of interstitial cells in the maintenance of *Hydra*. II. Budding. *J. Exp. Zool.* **158**, 283–98.

Diehl, F. A. & Burnett, A. L. (1965b). The role of interstitial cells in the maintenance of *Hydra*. III. Regeneration of hypostome and tentacles. *J. Exp. Zool.* **158**, 299–318.

Dohle, W. (1964). Die Embryonalentwicklung von *Glomeris marginata* (Villers) im Vergleich zur Entwicklung anderer Diplopoden. *Zool. Jahrb. (Anat.)* **81**, 241–310.

Dohmen, M. R. & Lok, D. (1975). The ultrastructure of the polar lobe of *Crepidula fornicata* (Gastropoda, Prosobranchia). *J. Embryol. Exp. Morphol.* **34**, 419–38.

Dohmen, M. R. & Van de Mast, J. M. A. (1978). Electron microscopical study of RNA-containing cytoplasmic localisations and intercellular contacts in early cleavage stages of eggs of *Lymnaea stagnalis* (Gastropoda, Pulmonata). *Proc. Kon. Ned. Akad. Wetensch., ser. C* **81**, 403–14.

Dohmen, M. R. & Van der Mey, J. C. A. (1977). Local surface differentiations at the vegetal pole of the eggs of *Nassarius reticulatus, Buccinum undatum*, and *Crepidula fornicata* (Gastropoda, Prosobranchia). *Develop. Biol.* **61**, 104–13.

Dohmen, M. R. & Verdonk, N. H. (1974). The structure of a morphogenetic cytoplasm, present in the polar lobe of *Bithynia tentaculata* (Gastropoda, Prosobranchia). *J. Embryol. Exp. Morphol.* **31**, 423–33.

Dohmen, M. R. & Verdonk, N. H. (1979*). Cytoplasmic localisations in mosaic eggs. In *Maternal Effects in Development*, eds D. R. Newth & M. Balls, pp. 127–45. Cambridge University Press.

Dorsett, D. A. (1961). The reproduction and maintenance of *Polydora ciliata* (Johnst.) at Whitstable. *J. Mar. Biol. Ass.* **41**, 383–96.

Du Bois, A. M. (1932). A contribution to the embryology of *Sciara* (Diptera). *J. Morphol.* **54** 161–95.

Du Bois, A. M. (1933). Chromosome behaviour during cleavage in the eggs of *Sciara coprophila* (Diptera) in the relation to the problem of sex determination. *Z. Zellforsch. mikr. Anat.* **19**, 595–614.

Dubois, F. (1949). Contribution à l'étude de la migration des cellules de régénération chez les planaires dulcidoles. *Bull. Biol.* **83**, 213-83.

Duboscq, O. & Tuzet, O. (1937a). L'ovogénèse, la fécondation et les premiers stades du développement des Eponges calcaires. *Arch. Zool. Exp. Gén.* **79**, 157–316.

Duboscq, O. & Tuzet, O. (1937b). Fusome et cellules en croix des Eponges calcaires. *C. R. Acad. Sci., Paris* **204**, 1888–91.

Duboscq, O. & Tuzet, O. (1939). Les diverses formes des choanocytes des Eponges calcaires hétérocoeles et leur signification. *Arch. Zool. Exp. Gén.* **80**, 353–88.

Duboscq, O. & Tuzet, O. (1942a). Sur les cellules en croix des *Sycon (Sycon ciliatum* Fabr, *Sycon coronatum* Elles et Sol., *Sycon elegans* Bower) et leur signification. *Arch. Zool. Exp. Gén.* **82**, 151–63.

Duboscq, O. & Tuzet, O. (1942b). Recherches complémentaires sur l'ovogénèse, la fécondation et les premiers stades du développement des éponges calcaires. *Arch. Zool. Exp. Gén.* **81**, 395–466.

Dupont, M. (1942). Origine des cellules germinales et formation du gonophore chez les Tubulaires. *Arch. Biol., Liège* **53**, 493–511.

Du Praw, E. J. (1967). The honeybee embryo. In *Methods in Developmental Biology*. eds F. H. Wilt & N. H. Wessels, pp. 183–217. Crowell, New York.

Dunn, M. C. (1959). Studies on the germ cell cycle of *Neorenifer wardi* (Byrd, 1936). *Trans. Amer. Microsc. Soc.* **78**, 385–408.

Eddy, E. M. (1975*). Germ plasm and the differentiation of the germ cell line. *Int. Rev. Cytol.* **43**, 229–80.

Ede, D. A. (1964). An inherited abnormality affecting the development of the yolk plasmodium and endoderm in *Dermestes maculatus* (Coleoptera). *J. Embryol. Exp. Morphol.* **12**, 551–62.

Ede, D. A. & Counce, S. J. (1956). A cinematographic study of the embryology of *Drosophila melanogaster*. *Wilhelm Roux' Arch. Entwicklungsmech. Organismen* **148**, 402–15.

Ehrman, L. & Daniels, S. (1975). Pole cells of *Drosophila paulistorum*: Embryologic differentiation with symbionts. *Austr. J. Biol. Sci.* **28**, 133–44.

Elbers, P. F. (1959). Over de beginoorzaak van het Li-effekt in de morphogenese. Een electronen-microscopisch onderzoek aan eieren van *Limnaea stagnalis* en *Paracentrotus lividis*. PhD Thesis, State University of Utrecht, 59pp.

Elbers, P. F. (1969). The primary action of lithium chloride on morphogenesis in *Limnaea stagnalis*. *J. Embryol. Exp. Morphol.* **22**, 449–63.

Euzet, L. & Mokhtar-Maamouri, F. (1975). Développement embryonnaire de trois cestodes du genre *Acanthobothrium* (Tetraphyllidea, Onchobothriidae). *Ann. Parasitol. (Paris)* **50**, 675–90.

Euzet, L. & Mokhtar-Maamouri, F. (1976). Développement embryonnaire de deux Phyllobothriidae (Cestoda: Tetraphyllidea). *Ann. Parasitol. (Paris)* **51**, 309–27.

Falk, R., Orevi, N. & Menzl, B. (1973). A fate map of larval organs of *Drosophila* and preblastoderm determination. *Nature, New Biol.* **246**, 19–20.

Faulkner, C. H. (1930). The anatomy and the histology of bud formation in the serpulid *Filograna implexa*, together with some cytological observations on the nuclei of the neoblasts. *J. Linn. Soc. London, Zool.* **37**, 109–90.

Fedecka-Bruner, B. (1964). Radiodestruction des testicules suivis de régénération chez la planaire *Dugesia lugubris*. *C. R. Acad. Sci., Paris* **258**, 3353–6.

Fennhoff, F. J. (1978). Die Gonophorenbildung und Embryonalent wicklung der Hydroide *Tubularia crocea* Agassiz. *Zool. Jahrb. (Anat.)* **100**, 433–55.

Fielding, C. J. (1967). Developmental genetics of the mutant '*grandchildless*' of *Drosophila subobscura*. *J. Embryol. Exp. Morphol.* **17**, 375–84.

Fincher, J. A. (1940). The origin of the germ cells in *Stylotella heliophila* Wilson (Tetraxonida). *J. Morphol.* **67**, 175–97.

Fioroni, P. (1966). Zur Morphologie und Embryogenese des Darmtraktes und der transitorische Organe bei Prosobranchiern (Mollusca, Gastropoda). *Rev. Suisse Zool.* **73**, 621–876.

Fioroni, P. (1967). Molluskenembryologie und allgemeine Entwicklungsgeschichte. *Verh. Naturforsch. Ges. Basel* **78**, 283–307.

Fioroni, P. (1969). Zum embryonalen und postembyonalen Dotterabbau des Flusskrebses (*Astacus*; Crustacea, Malacostraca, Decapoda). *Rev. Suisse Zool.* **76**, 919–46.

Fioroni, P. (1971). Die Entwicklungstypen der Mollusken, eine vergleichend-embryologische Studie. *Z. wiss. Zool., Leipzig* **182**, 263–394.

Fioroni, P. (1974). Die Sonderstellung der Tintenfische. *Naturwiss. Rundschau* **27**, 133–43.

Fioroni, P. (1977*). Probleme und Ergebnisse der Entwicklungsbiologie der Mollusken. *Verh. deutsch. Zool. Ges.* **70**, 216–28.

Flinkinger, R. A. (1964). Isotopic evidence for a local origin of blastema cells in regenerating *Planaria*. *Exp. Cell Res.* **34**, 403–6.

Formigoni, A. (1954). Studio sullo sviluppo embrionale di *Musca domestica* L. *Boll. Zool. Agrar. Bachiocoltur.* **20**, 111–54.
Föyn, B. (1927). Studien über Geschlecht und Geschlechtszellen bei Hydroiden. II. Auspressungsversuche an *Clava squamata* (Müller) mit mischung von Zellen aus Polypen desselben oder verschiedenen Geschlechts. *Wilhelm Roux' Arch. Entwicklungsmech. Organismen* **110**, 89–148.
Föyn, B. (1929). Studien über Geschlecht und Geschlechtszellen bei Hydroiden. III. Bemerkungen über die Entstehung der Keimzellen und die Entwicklung der Gonophoren bei *Clava squamata* (Müller). *Wilhelm Roux' Arch. Entwicklungsmech. Organismen* **114**, 501–11.
Franzén, Å. (1977*). Gametogenesis of bryozoans. In *Biology of Bryozoans*, eds R. M. Woolacott & R. L. Zimmer, pp. 1–22. Academic Press, London & New York.
Freeman, G. (1977). The establishment of the oral-aboral axis in the ctenophore embryo. *J. Embryol. Exp. Morphol.* **42**, 237–60.
Fretter, V. & Graham, A. (1964*). Reproduction. In *Physiology of Mollusca*, eds R. M. Wilbur & C. M. Yorge, vol. 1, pp. 127–64. Academic Press, London & New York.
Fullilove, S. L. & Jacobson, A. G. (1971). Nuclear elongation and cytokinesis in *Drosophila montana*. *Develop. Biol.* **26**, 560–77.
Fux, T. (1975). Positional effects of the polar plasm on the karyotype differentiation in *Heteropeza pygmaea* (Cecidomyidae, Diptera). *Rev. Suisse Zool.* **82**, 679–80.
Gabriel, A. & Le Moigne, A. (1971). Action de l'actinomycine D sur la différenciation cellulaire au cours de la régénération des planaires qui viennent d'éclore. I. Etudes morphologiques, histologiques et ultrastructurales du pouvoir de régénération en présence de l'antibiotique. *Z. Zellforsch.* **115**, 426–41.
Garaudy-Tamarelle, M. (1969). Quelques observations sur le développement embryonnaire de l'ébauche génitale chez le Collembole *Anurida maritima* Guérin. *C. R. Acad. Sci.*, Paris **268**, 945–7.
Garaudy-Tamarelle, M. (1970). Observations sur la régénération de la lignée germinale chez le Collembole *Anurida maritima* Guérin. Explication de son caractère intravitellin. *C. R. Acad Sci.*, Paris **270**, 1149–52.
Garcia-Bellido, A., Lawrence, P. A. & Morata, G. (1979*). Compartments in animal development. *Sci. Amer.* **241**, 90–8.
Gates, G. E. (1943). Some further notes on regeneration in *Perionyx excavatus*. *Proc. Nat. Acad. Sci. India B* **13**, 168–79.
Gates, G. E. (1951). Regeneration in an Indian earth worm, *Perionyx millardi*. *Proc. Nat. Acad. Sci. India* **34**, 115–47.
Gehring, W. J. (1973*). Genetic control of determination in the *Drosophila* embryo. In *Genetic Mechanisms of Development*, ed. F. H. Ruddle, pp. 103–28. Academic Press, London & New York.
Gehring, W. J. (1976a*). Developmental genetics of *Drosophila*. *Ann. Rev. Gen.* **10**, 209–52.
Gehring, W. J. (1976b*). Determination. *Dahlem Workshop on Organisation and Expression of Chromosomes*, pp. 97–113.
Gehring, W. J., Wieschaus, E. & Holliger, M. (1976). The use of 'normal' and 'transformed' gynandromorphs in mapping the primordial germ cells and the gonadal mesoderm in *Drosophila*. *J. Embryol. Exp. Morphol.* **35**, 607–16.
Geigy, R. (1931). Action de l'ultra-violet sur le pôle germinal dans l'oeuf de *Drosophila melanogaster* (castration et mutabilité). *Rev. Suisse Zool.* **38**, 187–288.
Geigy, R. & Aboïm, A. N. (1944). Gonadenentwicklung bei *Drosophila* nach

frühembryonaler Ausschaltung der Geschlechtszellen. *Rev. Suisse Zool.* **51**, 410–17.

Geilenkirchen, W. L. M., Verdonk, N. H. & Timmermans, L. P. M. (1970). Experimental studies on morphogenetic factors localised in the first and the second polar lobe of *Dentalium* eggs. *J. Embryol, Exp. Morphol.* **23**, 237–43.

Geyer-Duszynska, I. (1959). Experimental research on chromosome elimination in Cecidomyidae (Diptera). *J. Exp. Zool.* **141**, 391–448.

Geyer-Duszynska, I. (1961). Spindle disappearance and chromosome behaviour after partial-embryo irradiation in Cecidomyidae (Diptera). *Chromosome* **12**, 233–47.

Geyer-Duszynska, I. (1966). Genetic factors in oögenesis and spermatogenesis in Cecidomyidae. *Chromosomes Today* **1**, 174–8.

Geyer-Duszynska, I. (1967). Experiments on nuclear transplantation in *Drosophila melanogaster*. Preliminary report. *Rev. Suisse Zool.* **74**, 614–15.

Ghirardelli, E. (1953). Sul determinante germinale in '*Spadella cephaloptera*' Busch. *Rend, Accad, Naz, Lincei, Ser. 8* **14**, 150–3.

Ghirardelli, E. (1965*). Differentiation of the germ cells and regeneration of the gonads in planarians. In *Proc. Regeneration in Animals and Related Problems*, eds V. Kiortsis & H. A. L. Trampusch, pp. 177–84. North-Holland, Amsterdam.

Ghirardelli, E. (1966a). Prime immagini electroniche del determinante germinale della uova di *Spadella cephaloptera* Busch (Chaetognatha). *Acta Med, Rom,* **4**, 68–72.

Ghirardelli, E. (1966b). Il determinante germinale nell'uovo e nella gastrula di *Spadella cephaloptera* Busch (Chaetognatha). Osservazioni al microscopio elettronico. *Arch. Zool. Ital.* **51**, 841–54.

Ghose, K. C. (1962). The cleavage, gastrulation and germ layer formation in the giant land snail, *Achatina fulica*. *Proc. Zool. Soc.* **15**, 47–54.

Ghose, K. C. (1963). Morphogenesis of the pericardium and heart, kidney and ureter, and gonad and gonoduct in the giant land snail, *Achatina fulica* Bowdich. *Proc. Zool. Soc.* **16**, 201–14.

Giudice, G. (1973*). *Developmental Biology of the Sea-Urchin Embryo*, 469 pp. Academic Press, London & New York.

Glätzer, K. H. (1971). Die Ei- und Embryonal-entwicklung von *Corydendrium parasiticum* mit besonderer Berücksichtigung der Oocyten-Feinstruktur während der Vitellogenese. *Helogoländer wiss. Meeresunters.* **22**, 213–80.

Goldschimdt, R. B. & Lin, T. P. (1947). Note on heterochromatic nature of diminution chromatin. *Science* **105**, 619.

Gontcharoff, M. (1960). Le développement post-embryonnaire et la croissance chez *Lineus ruber* et *Lineus viridis* (Nemertes, Lineidae). *Ann. Sci. Nat. Zool.* ser. **12**, 225–79.

Goss, R. J. (1952). The early embryology of the book louse, *Liposcelis divergens* Badonnel (Psocoptera, Liposcellidae). *J. Morphol.* **91**, 135–67.

Goss, R. J. (1953). The advanced embryology of the book louse, *Liposcelis divergens* Badonnel (Psocoptera, Liposcellidae). *J. Morphol.* **92**, 157–91.

Gould-Somero, M. (1975*). Annelids and echiurans. In *Reproduction of Marine Invertebrates*, eds A. C. Giese & J. S. Pearse, vol. 3, pp. 277–311. Academic Press, London & New York.

Grassé, P. P. (1961*). Classe des Dicyémides. *Traité Zool, Anat. Syst. Biol.* **4**, 707–29.

Graziosi, G. & Micali, F. (1974). Differential responses to ultraviolet irradiation of the polar cytoplasm of *Drosphila* eggs. *Wilhelm Roux' Arch. Entwicklungsmech. Organismen* **175**, 1–11.

Gremigni, V. (1974*). The origin and cytodifferentiation of germ cells in the planarians. *Boll. Zool.* **41**, 359–77.
Gremigni, V. & Puccinelli, I. (1977). A contribution to the problem of the origin of the blastema cells in planarians: a karyological and ultrastructural investigation. *J. Exp. Zool.* **199**, 57–71.
Gremigni, V., Miceli, C. & Picano, E. (1980). On the role of germ cells in planarian regeneration. II. Cytophotometric analysis of the nuclear Feulgen-DNA content in cells of regenerated somatic tissues. *J. Embryol. Exp. Morphol.* **55**, 65–76.
Gremigni, V., Miceli, C. & Puccinelli, I. (1980). On the role of germ cells in planarian regeneration. I. A karyological investigation. *J. Embryol. Exp. Morphol.* **55**, 53–63.
Grifford, B. (1977). Individualisation et organogenèse de la gonade embryonnaire de *Viviparus viviparus* L. (Mollusque gastéropode prosobranche à sexes séparés). *Wilhelm Roux'Arch. Develop. Biol.* **183**, 131–47.
Grifford, B. (1978). Sexualisation de la gonade de *Viviparus viviparus* L. (Mollusque gastéropode prosobranche à sexes séparés). *Wilhelm Roux' Arch. Develop. Biol.* **184**, 213–31.
Guerrier, P. (1967). Les facteurs de polarisation dans les premiers stades du développement chez *Parascaris equorum*. *J. Embryol Exp. Morphol.* **18**, 121–42.
Guerrier, P. (1968). Origine et stabilité de la polarité animale–végétative chez quelques *Spiralia*. *Ann. Embryol. Morphogén.* **1**, 119–39.
Guerrier, P. (1970a). Les caractères de la segmentation et la détermination de la polarité dorso-ventrale dans le développement de quelques Spiralia. I. Les formes à premier clivage égal. *J. Embryol. Exp. Morphol.* **23**, 611–37.
Guerrier, P. (1970b). Les caractères de la segmentation et la détermination de la polarité dorso-ventrale dans le developpement de quelques Spiralia. II. *Sabellaria alveolata* (Annélide polychète). *J. Embryol. Exp. Morphol.* **23**, 639–65.
Guerrier, P. (1970c). Les caractères de la segmentation et la détermination de la polarité dorso-ventrale dans le développement de quelques Spiralia. III. *Pholas dactylus* et *Spicula subtruncata* (Mollusques lamellibranches). *J. Embryol. Exp. Morphol.* **23**, 667–92.
Guerrier, P. (1970d). Nouvelles données expérimentales sur la segmentation et l'organogenèse chez *Limax maximus* (Gastéropode pulmoné). *Ann. Embryol. Morphogén.* **3**, 283–94.
Guerrier, P. (1971*). La polarisation cellulaire et les caractères de la segmentation au cours de la morphogenèse spirale (Gastéropodes pulmonés, Lamellibranches, Annélides polychètes). *Ann Biol.* **10**, 151–92.
Guerrier, P., Van den Biggelaar, J. A. M., Van Dongen, C. A. M. & Verdonk, N. H. (1978). Significance of the polar lobe for the determination of dorso-ventral polarity in *Dentalium vulgare* (da Costa). *Develop. Biol.* **63**, 233–42.
Guerrier, P. & Van den Biggelaar, J. A. M. (1979). Intercellular activation and cell interactions in so-called mosaic embryos. In *Cell Lineage, Stem Cells and Cell Determination*, ed. N. le Douarin, pp. 29–36. North-Holland, Amsterdam.
Guilford, H. G. (1958). Observations on the development of the miracidium and the germ cell cycle in *Heronimus chelydrae* MacCallum (Trematoda). *J. Parasitol.* **44**, 64–74.
Günther, J. (1971). Entwicklungsfähigkeit, Geschlechtsverhältnis und Fertilität von *Pimpla turionellae* L (Hymenoptera, Ichneumonidae) nach Röntgenbestrahlung oder Abschnürung des Eihinterpols. *Zool. Jahrb. (Anat.)* **88**, 1–46.
Gurdon, J. B. (1974a*). The genome in specialised cells, as revealed by nuclear transplantation in Amphibia. In *The Cell Nucleus*, ed. H. Busch, vol. 1, pp. 471–89. Academic Press, London & New York.

Gurdon, J. B. (1974b*). *The Control of Gene Expression in Animal Development*, 160 pp. Harvard University Press, Cambridge, Mass., USA.
Gurdon, J. B. & Woodland, H. R. (1968*). The cytoplasmic control of nuclear activity in animal development. *Biol. Rev.* **43**, 233–67.
Gustafsson, M. K. S. (1976). Studies on cytodifferentiation in neck region of *Diphyllobothrium dendriticum* Nitzch, 1824 (Cestoda, Pseudophyllidea). *Z. Parasitenk.* **50**, 323–9.
Gustafsson, M. K. S. (1977). Aspects of the cytology and histogenesis in cestodes with special reference to the genus *Diphyllobothrium*. PhD Thesis, Åbo Akademi, Finland.
Gustafsson, T. (1963). Cellular mechanisms in the morphogenesis of the sea-urchin. Cell contacts within the ectoderm and between mesenchyme and ectoderm cells. *Exp. Cell Res.* **32**, 570–89.
Guyard, A. (1970). Aspects ultrastructuraux de la différenciation gonocytaire au début de l'organogenèse de la glande hermaphrodite d'*Helix aspersa* Müll. (Gastéropode pulmoné). *Bull. Soc. Zool.* **95**, 471–4.
Guyomarc'h-Cousin, C. (1976). Organogenèse descriptive de l'appareil génital chez *Littorina saxatilis* (Olivi), Gastéropode prosobranche. *Bull. Soc. Zool. France* **101**, 465–76.
Hagan, H. R. (1951*). *Embryology of the Viviparous Insects*, 472 pp. Ronald Press.
Haget, A. (1950). Mise en évidence d'une induction exercée par l'ectoderme sur le mésoderme dans la morphogenèse embryonnaire de *Leptinotarsa* (Insecte, Col.). *C. R. Acad. Sci., Paris* **230**, 1788–90.
Haget, A. (1953a). Analyse expérimentale des conditions d'édification d'une gonade embryonnaire chez le Coléoptère *Leptinotarsa*. *C. R. Soc. Biol. (Bordeaux)* **147**, 673–5.
Haget, A. (1953b). Analyse expérimentale des facteurs de la morphogenèse chez le Coléoptère, *Leptinotarsa*. *Bull. Biol. France Belg.* **97**, 123–217.
Haget, A. (1969). Séparation expérimentale du soma et du germen de la gonade, par la transplantation des initiales germinales en position ectopique, chez les embryons du Coléoptère *Leptinotarsa*. *C. R. Acad. Sci., Paris* **269**, 2226–9.
Haget, A. (1970). Premières données infrastructurales sur la surface et le périplasme de l'oeuf du doryphore *Leptinotarsa decemlineata* Say. *C. R. Acad. Sci., Paris* **271**, 1303–6.
Haget, A. (1972). Caractéristiques ultrastructurales du pôle postérieur et de l'oosome, dans l'oeuf jeune du coléoptère *Leptinotarsa decemlineata* Say. *C. R. Acad. Sci., Paris* **275**, 2737–40.
Hämmerling, J. (1924). Die ungeschlechtliche Fortpflanzung und Regeneration bei *Aeolosoma Hemphrichii*. Histologische und experimentelle Untersuchungen. *Zool. Jahrb. allg. Zool. Physiol.* **41**, 581–652.
Hathaway, D. S. & Selman, G. G. (1961). Certain aspects of cell-lineage and morphogenesis studied in embryos of *Drosophila melanogaster* with an ultraviolet micro-beam. *J. Embryol. Exp. Morphol.* **9**, 310–25.
Hauschteck, E. (1962). Die Cytologie der Pädogenese und der Geschlechtsbestimmung einer heterogonen Gallmücke. *Chromosoma* **13**, 163–82.
Haven, N. (1977*). Cephalopoda: Nautiloidea. In *Reproduction of Marine Invertebrates*, eds A. C. Giese & J. S. Pearse, vol. 4, pp. 227–41. Academic Press, London & New York.
Hay, E. D. & Coward, S. J. (1975). Fine structure studies on the planarian, *Dugesia*. I. Nature of the 'neoblast' and other cell types in non-injured worms. *J. Ultrastruct. Res.* **50**, 1–21.

Haynes, J. & Burnett, A. L. (1963). Dedifferentiation and redifferentiation of cells in *Hydra viridis*. *Science* **142**, 1481–3.
Hegner, R. W. (1908). Effects of removing the germ-cell determinants from the eggs of some chrysomelid beetles. Preliminary report. *Biol. Bull.* **16**, 19–26.
Hegner, R. W. (1911). Experiments with chrysomelid beetles. III. The effects of killing parts of the egg of *Leptinotarsa decemlineata*. *Biol. Bull.* **20**, 237–51.
Hegner, R. W. (1914*). Studies on germ cells. I. The history of the germ cells in insects with special reference to the 'Keimbahn' determinants. II. The origin and significance of the 'Keimbahn' determinants in animals. *J. Morphol.* **25**, 375–509.
Heinig, S. (1967). Die Abänderung embryonaler Differenzierungsprozesse durch totale Röntgenbestrahlung im Ei von *Gryllus domestica*. *Zool. Jahrb. (Anat.)* **84**, 425–92.
Heming, B. S. (1979). Origin and fate of germ cells in male and female embryos of *Haplothrips verbasci* (Osborn) (Insecta, Thysanoptera, Phlaeothripidae). *J. Morphol.* **160**, 323–44.
Herlant-Meewis, H. (1946*b*). Contribution à l'étude de la régénération chez les Oligochètes. Reconstitution du germen chez *Lumbricillus lineatus* (Enchytraeidés). Première partie: éléments régénérateurs. *Ann. Soc. Zool. Belg.* **77**, 5–47.
Herlant-Meewis, H. (1946*b*). Contribution à l'étude de la régénération chez les Oligochètes. Reconstitution du germen chez *Lumbricillus lineatus* (Enchytraeidés), (deuxième partie). *Arch. Biol.* **57**, 197–306.
Herlant-Meewis, H. (1954). Etude histologique des Aeolosomatidae au cours de la reproduction asexuée. *Arch. Biol.* **65**, 73–134.
Herlant-Meewis, H. (1964*). Regeneration in annelids. *Adv. Morphogen.* **4**, 155–215.
Hermann, F. (1978/9*). Nemertini. In *Morphogenese der Tiere*, ed. F. Seidel, section 3, D5-1, 136 pp. Fischer-Verlag, Jena.
Hertwig, W. (1977*). *The Biological Problem of To-day. Preformation or Epigenesis? The Basis of a Theory of Organic Development.* 148 pp. Dabor Science, New York. (Translated by P. C. Mitchel from *Präformation oder Epigenese?* Reprint of the 1900 edition published by Macmillan, New York.)
Hess, O. (1956*a*). Die Entwicklung von Halbkeimen bei dem Süsswasser-Prosobranchier, *Bithynia tentaculata* L. *Wilhelm Roux' Arch. Entwicklungsmech. Organismen* **148**, 336–61.
Hess, O. (1956*b*). Die Entwicklung von Exogastrula-keimen bei dem Süsswasser-Prosobranchier *Bithynia tentaculata* L. *Wilhelm Roux' Arch. Entwicklungsmech. Organismen* **148**, 474–88.
Hess, O. (1957). Die Entwicklung von Halbkeimen bei dem Süsswasser-Pulmonaten, *Limnaea stagnalis* L. *Wilhelm Roux' Arch. Entwicklungsmech. Organismen* **150**, 124–45.
Hess, O. (1962*). Entwicklungsphysiologie der Mollusken. *Fortschr. Zool.* **14**, 130–63.
Heymons, R. (1897). Entwicklungsgeschichtliche Untersuchungen an *Lepisma saccharina* L. *Z. wiss. Zool.* **62**, 583–631.
Heymons, R. (1901). *Die Entwicklungsgeschichte der Scolopender. Zoologica, Stuttgart* **13**, 1–224.
Hill, S. D. (1970). Origin of the regeneration blastema in polychaete annelids. *Amer. Zool.* **10**, 101–12.
Hill, S. D. (1972). Caudal regeneration in the absence of a brain in two species of sedentary polychaetes. *J. Embryol. Exp. Morphol.* **28**, 667–80.
Hogg, N. A. S. & Wijdenes, J. (1979). A study of gonadal organogenesis, and the

factors influencing regeneration following surgical castration in *Deroceras reticulatum* (Pulmonata: Limnaeidae). *Cell Tissue Res.* **198**, 295–307.

Holland, N. D. (1976). The fine structure of the embryo during the gastrula stage of *Comanthus japonica* (Echinodermata: Crinoidea). *Tissue & Cell* **8**, 491–510.

Hope, W. D. (1974*). Nematoda. In *Reproduction of Marine Invertebrates*, eds A. C. Giese & J. S. Pearse, vol. 1, pp. 391–460. Academic Press, London & New York.

Hörstadius, S. (1927). Studien über die Determination bei *Paracentrotus lividus*. *Wilhelm Roux' Arch. Entwicklungsmech. Organismen* **112**, 239–46.

Hörstadius, S. (1928). Über die Determination des Keimes bei Echinodermen. *Acta Zool. Stockholm* **9**, 1–191.

Hörstadius, S. (1935). Über die Determination im Verlaufe der Eiachse bei Seeigeln. *Pubbl. Staz. Zool. Napoli* **14**, 251–429.

Hörstadius, S. (1936a). Über die zeitliche Determination im Keim von *Paracentrotus lividus* Lk. *Wilhelm Roux' Arch. Entwicklungsmech. Organismen* **135**, 1–39.

Hörstadius, S. (1936b). Weitere Studien über die Determination im Verlaufe der Eiachse bei Seeigeln. *Wilhelm Roux' Arch. Entwicklungsmech. Organismen* **135**, 40–68.

Hörstadius, S. (1937). Experiments on determination in the early development of *Cerebratulus lacteus*. *Biol. Bull.* **73**, 317–42.

Hörstadius, S. (1939*). The mechanics of sea-urchin development studied by operative methods. *Biol. Rev.* **14**, 132–79.

Hörstadius, S. (1973*). *Experimental Embryology of Echinoderms*, 192 pp. Clarendon Press, Oxford.

Hörstadius, S. & Josefsson, L. (1972). Morphogenetic substances from sea-urchin eggs. Isolation of animalising substances from developing eggs of *Paracentrotus lividus*. *Acta Embryol. Morphol. Exp.* **14**, 7–23.

Howland, R. B. (1941). Structure and development of centrifuged eggs and early embryos of *Drosophila melanogaster*. *Proc. Amer. Phil. Soc.* **84**, 605–16.

Huet, M. (1965). Action des rayons X sur la lignée germinale et la régénération de l'appareil génital d'*Asterina gibbosa* Penn. (Echinoderme). *C. R. Acad. Sci., Paris* **260**, 707–9.

Huet, M. (1966). Etude de la régénération chez *Asterina gibbosa* Penn. (Echinoderme Astéride) par la méthode de l'irradiation aux rayons X. *C. R. Séan. Soc. Biol.* **160**, 466–96.

Huet, M. (1967). Etude expérimentale du role du système nerveux dans la régénération du bras de l'étoile de mer *Asterina gibbosa* Penn. (Echinoderme–Astéride). *Bull. Soc. Zool. France* **92**, 641–5.

Huet, M. (1972). Etude expérimentale de la régénération du bras et de l'appareil génital d'*Asterina gibbosa* Penn. (Echinoderme–Astéride). *Bull. Soc. Zool. France* **97**, 609–18.

Huet, M. (1974). La lignée germinale chez les Echinodermes: origine et évolution des cellules germinales au cours de la régénération de la gonade chez l'étoile de mer *Asterina gibbosa* Penn. *J. Embryol. Exp. Morphol.* **31**, 787–806.

Huet, M. (1975). Le rôle du système nerveux au cours de la régénération du bras chez une étoile de mer: *Asterina gibbosa* Penn. (Echinoderme, Astéride). *J. Embryol. Exp. Morphol.* **33**, 535–52.

Hutchins, R. & Brandhorst, B. P. (1979). Commitment to vegetalized development in sea-urchin embryos. Failure to detect changes in patterns of protein synthesis. *Wilhelm Roux' Arch. Develop. Biol.* **186**, 95–102.

Hyman, L. H. (1940*). *The Invertebrates*, vol. 1, *Protozoa through Ctenophora*, 726 pp. McGraw-Hill, New York & London.

Hyman, L. H. (1951a*). *The Invertebrates*, vol. 2, *Platyhelminthes and Rhynchocoela. The Acoelomate Bilateralia*, 550 pp. McGraw-Hill, New York & London.
Hyman, L. H. (1951b*). *The Invertebrates*, vol. 3, *Acanthocephala, Aschelminthes and Entoprocta. The Pseudocoelomate Bilateralia*, 572 pp. McGraw-Hill, New York & London.
Hyman, L. H. (1955*). *The Invertebrates*, vol. 4, *Echinodermata. The Coelomate Bilateralia*, 763 pp. McGraw-Hill, New York & London.
Hyman, L. H. (1959*). *The Invertebrates*, vol. 5, *Smaller Coelomate Groups: Chaetognatha, Hemichordata, Pogonophora, Phoronida, Ectoprocta, Brachiopoda, Sipunculida. The Coelomate Bilateralia*, 783 pp. McGraw-Hill, New York & London.
Hyman, L. H. (1967*). *The Invertebrates*, vol. 6, *Mollusca I. Aplacophora, Polyplacophora, Monoplacophora, Gastropoda. The Coelomate Bilateralia*, 792 pp. McGraw-Hill, New York & London.
Idris, B. E. M. (1960a). Die Entwicklung im normalen Ei von *Culex pipiens* L. (Diptera). *Z. Morphol. Ökol. Tiere* **49**, 387–429.
Idris, B. E. M. (1960b). Die Entwicklung im geschnürten Ei von *Culex pipiens* L. (Diptera). *Wilhelm Roux' Arch. Entwicklungsmech. Organismen* **152**, 230–62.
Ihle, J. E. & Nierstrasz, H. F. (1928*). *Leerboek der bijzondere Dierkunde*, phylum 7, *Annelida*, pp. 172–208. Oosthoek, Utrecht.
Illmensee, K. (1968). Transplantation of embryonic nuclei into unfertilised eggs of *Drosophila melanogaster*. *Nature, Lond.* **219**, 1268–9.
Illmensee, K. (1972). Developmental potencies of nuclei from cleavage, preblastoderm, and syncytial blastoderm transplanted into unfertilised eggs of *Drosophila melanogaster*. *Wilhelm Roux' Arch. Entwicklungsmech. Organismen* **170**, 267–98.
Illmensee, K. (1973). The potentialities of transplanted early gastrula nuclei of *Drosophila melanogaster*. Production of their imago descendants by germ-line transplantation. *Wilhelm Roux' Entwicklungsmech. Organismen* **171**, 331–43.
Illmensee, K. (1976*). Nuclear and cytoplasmic transplantation in *Drosophila*. In *Insect Development, 8th Sym. Roy. Entom. Soc. London*, ed. P. A. Lawrence, pp. 76–96. Blackwell Scientific Publications, Oxford.
Illmensee, K. (1978*). *Drosophila* chimaeras and the problem of determination. In *Results and Problems in Cell Differentiation: Genetic Mosaics and Cell Differentiation*, ed. W. J. Gehring, **9**, 51–69. Springer Verlag, Berlin & Heidelberg.
Illmensee, K. & Mahowald, A. P. (1974). Transplantation of posterior polar plasm in *Drosophila*. Induction of germ cells at the anterior pole of the egg. *Proc. Nat. Acad. Sci. USA* **7**, 1016–20.
Illmensee, K. & Mahowald, A. P. (1976). The autonomous function of germ plasm in a somatic region of the *Drosophila* egg. *Exp. Cell. Res.* **97**, 127–40.
Illmensee, K., Mahowald, A. P. & Loomis, M. R. (1976). The ontogeny of germ plasm during oogenesis in *Drosophila*. *Develop. Biol.* **49**, 40–68.
Ivanova-Kasas, O. M. (1958). Biology and embryonic development of *Eurytoma aciculata* Ratz. (Hymenoptera, Eurytomidae). *Rev. Entomol. USSR* **37**, 5–23.
Ivanova-Kasas, O. M. (1972*). Polyembryony in insects. In *Developmental Systems: Insects*, eds S. J. Counce & C. H. Waddington, vol. 1, pp. 242–71. Academic Press, London & New York.
Ivanova-Kasas, O. M. (1977a). Analysis of the early development of Crustacea. I. The principal variants of early development. *Biol. Morya* **1**, 3–19.
Ivanova-Kasas, O. M. (1977b). Analysis of early development of Crustacea. II. Problem of evolution of cleavage processes. *Biol. Morya* **3**, 3–13.
Ivanova-Kasas, O. M. (1979*). *Comparative Embryology of the Invertebrates*.

Arthropoda, (*Onychophora, Tardigrada, Pentastomida, Pantopoda, Trilobita, Chelicerata and Crustacea*), 223 pp. Publishing House Nauka, Moscow, (in Russian).
Iwanoff, P. P. (1928). Die Entwicklung der Larvalsegmente bei den Anneliden. *Z. Morphol. Okol. Tiere* **10**, 62–161.
Izumi, S. (1953). Biological studies on *Ascaris* eggs. IV. *Ascaris* ontogenesis observed by nucleic acid stainings, and distribution of fat or glycogen following the development of *Ascaris* eggs. *Jap. J. Med. Sci. Biol.* **6**, 371–83.
Jägersten, G. (1964). On the morphology and reproduction of entoproct larvae. *Zool. Bidr. Uppsala* **36**, 296–314.
James, B. L. & Bowers, E. A. (1967). Reproduction in the daughter sporocyst of *Cercaria bucephalopsis haimeana* (Lacaze-Duthiers, 1854) (Bucephalidae) and *Cercaria dichotoma* Lebour, 1911 (non Müller) (Gymnophallidae). *Parasitol.* **57**, 607–25.
James, B. L., Bowers, E. A. & Richards, J. G. (1966). The ultra-structure of the daughter sporocyst of *Cercaria bucephalopsis haimeana* Lacaze-Duthiers, 1854 (Digenea: Bucephalidae) from the edible cockle, *Cardium edule* L. *Parasitol.* **56**, 753–62.
Janet, C. (1923*). *Le Volvox*, 179 pp. Mém. 3, Paris.
Janning, W. (1974). Entwicklungsgenetische Untersuchungen an Gynandern von *Drosophila melanogaster*. II. Der morphogenetische Anlageplan. *Wilhelm Roux' Arch. Entwicklungsmech. Organismen* **174**, 349–59.
Janning, W. (1978*). Gynandromorph fate maps in *Drosophila*. In *Results and Problems in Cell Differentiation*, ed. W. J. Gehring, vol. 9, pp. 1–28. Springer Verlag, Berlin & Heidelberg.
Janning, W., Pfreundt, J. & Tiemann, R. (1979*). The distribution of anlagen in the early embryo of *Drosophila*. In *Cell Lineage, Stem Cells and Cell Determination*, ed. N. le Douarin, pp. 83–100. North-Holland, Amsterdam.
Jazdowska-Zagrodziňská, B. (1966). Experimental studies on the role of 'polar granules' in the segregation of pole cells in *Drosophila melanogaster*. *J. Embryol Exp. Morphol.* **16**, 391–9.
Jazdowska-Zagrodziňská, B. & Matuszewski, B. (1978). Nuclear lamellae in the germ-line of gall midges (Cecidomyiidae, Diptera). *Experientia* **34**, 777–8.
Jennison, B. L. (1979). Gametogenesis and reproductive cycles in the sea anemone *Anthopleura elegantissima* (Brandt, 1835). *Can. J. Zool.* **57**, 403–11.
Josefsson, L. & Hörstadius, S. (1969). Morphogenetic substances from sea-urchin eggs. Isolation of animalising and vegetalising substances from unfertilised eggs of *Paracentrotus lividus*. *Develop. Biol.* **20**, 481–500.
Jung, E. (1966*a*). Untersuchungen am Ei des Speisebohnenkäfers *Bruchidius obtectus* Say (Coleoptera). I. Entwicklungsgeschichtliche Ergebnisse zur Kennzeichung des Eitypus. *Z. Morphol. Ökol. Tiere* **56**, 444–80.
Jung, E. (1966*b*). Untersuchungen am Ei des Speisebohnenkäfers *Bruchidius obtectus* Say (Coleoptera). II. Entwicklungsphysiologische Ergebnisse der Schnürungsexperimente. *Wilhelm Roux' Arch. Entwicklungsmech. Organismen* **157**, 320–92.
Jung, E. & Krause, G. (1967). Experimente mit Verlagerung polnahen Eimaterials zur Analyse der Bedingungen für die metamere Gliederung des Embryos von *Bruchidius* (Coleoptera). *Wilhelm Roux. Arch. Entwicklungsmech. Organismen* **159**, 89–126.
Jura, C. (1964*a*). Cytological and experimental observations on the origin and fate of the pole cells in *Drosophila virilus* Sturt. Part I. Cytological analysis. *Acta Biol. Cracov., ser. Zool.* **7**, 59–73.
Jura, C. (1964*b*). Cytological and experimental observations on the origin and fate of

the pole cells in *Drosophila virilis* Sturt. Part II. Experimental analysis. *Acta Biol. Cracov., ser. Zool.* **7**, 89–103.

Jura, C. (1965). Embryonic development of *Tetrodontophora bielanensis* (Waga) (Collembola) from oviposition till germ band formation stage. *Acta Biol. Cracov., ser. Zool.* **8**, 141–57.

Jura, C. (1966). Origin of the endoderm and embryogenesis of the alimentary system in *Tetrodontophora bielanensis* (Waga) (Collembola). *Acta Biol. Cracov., ser. Zool.* **9**, 93–102.

Jura, C. (1967a). Origin of germ cells and gonads formation in embryogenesis of *Tetrodontophora bielanensis* (Waga) (Collembola). *Acta Biol. Cracov., ser. Zool.* **10**, 97–103.

Jura, C. (1967b). The significance and function of the primary dorsal organ in embryonic development of *Tetrodontophora bielanensis* (Waga) (Collembola). *Acta Biol. Cracov., ser. Zool.* **10**, 301–10.

Jura, C. (1972*). Development of apterygote insects. In *Developmental Systems: Insects*, eds S. J. Counce & C. H. Waddington, vol. 1, pp. 49–94. Academic Press, London & New York.

Kalthoff, K. & Sander, K. (1968). Der Entwicklungsgang der Miszbildung 'Doppelabdomen' im partiell UV-bestrahlten Ei von *Smittia parthenogenetica* (Dipt., Chironomidae). *Wilhelm Roux' Arch. Entwicklungsmech. Organismen* **161**, 129–46.

Kanellis, A. (1952). Anlagenplan und Regulationserscheinungen in der Keimanlage des Eies von *Gryllus domesticus. Wilhelm Roux' Arch. Entwicklungsmech. Organismen* **145**, 417–61.

Karstner, A. (1965/7*). *Lehrbuch der Speziellen Zoologie. Bd. I. Wirbellose. Teil 1. & 2.*, 1242 pp. Fischer Verlag, Stuttgart.

Kato, K. (1968*). Platyhelminthes. In *Invertebrate Embryology*, eds M. Kumé & K. Dan (translated by J. C. Dan), pp. 125–43. Nolit, Belgrade.

Kautzch, G. (1910). Über die Entwicklung von *Agelena labyrinthica* Clerck. *Zool. Jahrb. (Anat. Ont.)* **30**, 535–602.

Kenk, R. (1941). Induction of sexuality in the asexual form of *Dugesia tigrina* (Girard). *J. Exp. Zool.* **87**, 55–69.

Kessel, E. L. (1939). The embryology of fleas. *Smithson. Miscel. Coll.* **98**, 1–78.

Kille, F. R. (1942). Regeneration of the reproductive system following binary fission in the sea-cucumber, *Holothuria parvula* (Selenka). *Biol Bull.* **83**, 55–66.

Kimble, J. & Hirsh, D. (1979). The postembryonic cell lineages of the hermaphrodite and male gonads in *Coenorhabditis elegans. Develop. Biol.* **70**, 396–417.

King, R. C. (1970*). *Ovarian Development in* Drosophila melanogaster. 225 pp. Academic Press, London & New York.

King, R. L. & Beams, H. W. (1937/8). An experimental study of chromatin diminution in *Ascaris. J. Exp. Zool.* **77**, 425–38.

Klag, J. (1977). Differentiation of primordial germ cells in the embryonic development of *Thermobia domestica*, Pack. (Thysanura): an ultrastructural study. *J. Embryol. Exp. Morphol.* **38**, 93–114.

Klima, J. (1962). Elektronenmikroskopische Studien über die Feinstrukturen der Tricladen (Turbellaria). *Protoplasma* **54**, 101–56.

Kochert, G. (1975*). Developmental mechanisms in *Volvox* reproduction. In *The Developmental Biology of Reproduction*, eds C. L. Markert & J. Papaconstantinou, pp. 55–90. Academic Press, London & New York.

Kochert, G. (1977*). Sexual hormones and cell differentiation in *Volvox carteri*. In *Eucaryotic Microbes as Model Developmental Systems*, eds D. H. O'Day & P. A. Horgen, pp. 235–51. Marcel Dekker, New York.

Krause, G. (1938a). Eizelbeobachtungen und typischer Gesamtbilder der Entwicklung von Blastoderm und Keimanlage im Ei der Gewächshausschrecke *Tachycines asynomorus* Adelung. *Z. Morphol. Ökol. Tiere* **34**, 1–78.

Krause, G. (1938b). Die Ausbildung der Körpergrundgestalt im Ei der Gewächshausschrecke *Tachycines asynomorus. Z. Morphol. Ökol. Tiere* **34**, 499–564.

Krause, G. (1939*). Die Eitypen der Insekten. *Biol. Zentralbl.* **59**, 495–536.

Krause, G. (1953). Die Aktionsfolge zur Gestaltung des Keimstreifs von *Tachycines* (Saltatoria), insbesondere das morphogenetische Konstruktionsbild bei duplicitas parallela. *Wilhelm Roux' Arch. Entwicklungsmech. Organismen* **146**, 275–370.

Krause, G. (1957*). Neue Beiträge zur Entwicklungsphysiologie des Insektenkeimes. *Verh. deutsch. Zool. Ges. Graz* **51**, 396–424.

Krause, G. (1958*). Induktionssysteme in der Embryonalentwicklung von Insekten. *Ergebn. Biol.* **20**, 159–98.

Krause, G. (1961*). Preformed ooplasmic reaction systems in insect eggs. In *Symposium on Germ Cells and Development*, ed. S. Ranzi, pp. 302–37. I.I.E. & Fond. Baselli, Milan.

Krause, G. & Krause, J. (1957). Die Regulation der Embryonalanlage von *Tachycines* (Saltatoria) im Schnittversuch. *Zool. Jahrb.* **75**, 481–500.

Krause, G. & Sander, K. (1962*). Ooplasmic reaction systems in insect embryogenesis. *Adv. Morphogen.* **2**, 259–303.

Krichinskaya, E. V. & Efimova, G. V. (1978). The restoration of the whole worm from a small fragment of the body in *Dugesia tigrina* following the repeated removal of regenerates. *Ontogenesis* **9**, 510–14.

Kunz, W. (1970). Genetische Aktivität der Keimbahnchromosomen während des Eiwachstums von Gallmücken (Cecidomyiidae). *Verh. deutsch. Zool. Ges.* **64**, 42–6.

Kunz, W. & Eckhardt, R. A. (1974). The chromosomal distribution of satellite DNA in the germ-line and somatic tissues of the gall midge, *Heteropeza pygmaea. Chromosoma* **47**, 1–19.

Kuwana, J. & Takami, T. (1968*). Insecta. In *Invertebrate Embryology*, eds M. Kumé & K. Dan (translated by J. C. Dan), pp. 405–84. Nolit, Belgrade.

Lallier, R. (1978). Recherches sur la polarité de l'oeuf de l'oursin *Paracentrotus lividus. Acta Embryol. Exp.* **1**, 47–57.

Lameere, A. (1919). Contribution à la connaissance des Dicyémides. *Bull. Biol. France Belg.* **53**, 234–75.

Lattaud, C. (1973). Autodifférenciation ovarieene chez l'Annélide Oligochète *Eisenia foetida f. typica* Sev. démonstrée au moyen de culture organotypique. *C. R. Acad. Sci., Paris* **276**, 1737–40.

Lattaud, C. (1974). Etude en culture organotypique du contrôle du sexe des gamétogenèses chez l'Annélide Oligochète. *C. R. Acad. Sci., Paris* **279**, 935–8.

Laugé, G. & Prudhommeau, C. (1971). Irradiation UV des cellules polaires de l'oeuf chez *Drosophila melanogaster* Meig. I. Structure histologique des testicules agamétiques chez les adultes. *Bull. Soc. Zool. France* **96**, 247–63.

Lawrence, P. A. (1971*). The organization of the insect segment. In *Control Mechanisms of Growth and Differentiation*, eds D. D. Davies & M. Balls, pp. 379–90. Cambridge University Press.

Lawrence, P. A. & Morata, G. (1976*). The compartment hypothesis. In *Insect Development*, ed. P. A. Lawrence, pp. 132–49. Blackwell Scientific Publications, Oxford.

Lemaire, J. (1970). Table de développement embryonnaire de *Sepia officinalis* L. (Mollusque Céphalopode). *Bull. Soc. Zool. France* **95**, 773–82.

Lemaire, J. (1972a). Différenciation sexuelle de la gonade embryonnaire de *Sepia officinalis* L. cultivée *in vitro*. *C. R. Acad. Sci., Paris* **275**, 475–8.
Lemaire, J. (1972b). Origine et évolution du système coelomique et de l'appreil génital de *Sepia officinalis* L. (Mollusque Céphalopode). *Ann. Embryol. Morphogén.* **5**, 43–59.
Lemaire, J. & Richard, A. (1970). Evolution embryonnaire de l'appareil génital: différenciation du sexe chez *Sepia officinalis* L. *Bull. Soc. Zool. France* **95**, 475–8.
Le Moigne, A. (1963). Etude du développement embryonnaire de *Polycelis nigra* (Turbellarié, Triclade). *Bull. Soc. Zool. France* **88**, 403–23.
Le Moigne, A. (1966a). Etude du développement embryonnaire et recherches sur les cellules de régénération chez l'embryon de la planaire *Polycelis nigra* (Turbellarié, Triclade). *J. Embryol. Exp. Morphol.* **15**, 39–60.
Le Moigne, A. (1966b). Etude au microscope électronique des cellules d'embryons de *Polycelis* (Turbellarié, Triclade), au début de leur développement. *C. R. Acad. Sci., Paris* **263**, 550–3.
Le Moigne, A. (1967a). Mise en évidence au microscope électronique de la persistance de cellules indifférenciées au cours du développement embryonnaire de la planaire *Polycelis nigra*. *C. R. Acad. Sci., Paris* **265**, 242–4.
Le Moigne, A. (1967b). Etude au microscope électronique de le différenciation des principaux types cellulaires chez l'embryon de la planaire *Polycelis nigra*. *Bull. Soc. Zool. France* **92**, 617–28.
Le Moigne, A. (1969). Etude du développement et de la régénération embryonnaires de *Polycelis nigra* (Ehr.) et *Polycelis tenuis* (Iyima) (Turbellariés Triclades). *Ann. Embryol. Morphogén.* **2**, 51–69.
Le Moigne, A. & Gabriel, A. (1971). Action de l'actinomycine D. sur la différenciation au cours de la régénération de Planaires qui viennent d'éclore. II. Etudes autoradiographiques, histologiques et ultrastructurales de l'action de l'antibiotique sur les synthèses d'ARN. *Z. Zellforsch.* **115**, 442–60.
Le Moigne, A., Sauzin, M. J., Lender, T. & Delavault, R. (1965). Quelques aspects des ultrastructures du blastème de régénération et des tissues voisins chez *Dugesia gonocephala* Turbellarié, Triclade). *C. R. Séan. Soc. Biol.* **159**, 530–4.
Le Moigne, A., Sauzin, M. J. & Lender, T. (1966). Comparaison de l'ultrastructure du néoblaste et de la cellule embryonnaire des planaires d'eau douce. *C. R. Acad. Sci., Paris* **263**, 627–9.
Lender, T. (1962*). Factors in morphogenesis of regenerating freshwater planarians. *Adv. Morphogen.* **2**, 305–31.
Lender, T. (1965*). La régénération des gonades d'*Asterina gibbosa* (Echinoderme, Astéride). In *Regeneration in Animals and Related Problems*, eds V. Kiortsis & H. A. L. Trampusch, pp. 278–82. North-Holland, Amsterdam.
Lender, T. & Delavault, R. (1964*). La lignée germinale chez les échinodermes. In *L'Origine de la Lignée Germinale chez les Vertébrés et chez quelques Groupes d'Invertébrés*, ed. E. Wolff, pp. 177–89. Hermann, Paris.
Lender, T. & Gabriel, A. (1960). Etude histochimique des néoblastes de *Dugesia lugubris* (Turbellarié, Triclade) avant et pendant la régénération. *Bull. Soc. Zool. France* **85**, 100–10.
Lentz, T. L. (1965). The fine structure of differentiating interstitial cells in *Hydra*. *Z. Zellforsch.* **67**, 547–60.
Lentz, T. L. (1966). *The Cell Biology of Hydra*. Chapter 6: Fine structure of differentiating interstitial cells, pp. 81–105. North-Holland, Amsterdam.
Levin, V. L., Romashkina, T. B., Shvarzman, P. Y. & Shelomova, L. F. (1974). Some regularities in occurrence of the morphose 'agametic gonads' as a result of UV irradiation of *Drosophila* embryos. *Tsitologiya* **16**, 211–16.

Lin, T. P. (1954). The chromosomal cycle in *Parascaris equorum* (*Ascaris megalocephala*): oogenesis and diminution. *Chromosoma* **6**, 175–98.
Liu, C. K. & Berrill, N. J. (1948). Gonophore formation and germ cell origin in *Tubularia*. *J. Morphol.* **83**, 39–59.
Lohs-Schardin, M., Lender, K., Cremer, C., Cremer, T. & Zorn, C. (1979). Localized ultraviolet laser microbeam irradiation of early *Drosophila* embryos: fate maps based on location and frequency of adult defects. *Develop. Biol.* **68**, 533–45.
Loosli, M. (1935). Über die Entwicklung und den Bau der indifferenten und männlichen Larven von *Bonellia viridis* Rol. *Pubbl. Staz. Zool. Napoli* **15**, 16–59.
Lubet, P., Herlin-Houtteville, P. & Mathieu, M. (1976*). La lignée germinale des mollusques pelecypodes; origine et évolution. *Bull. Soc. Zool. France* **101**, suppl. 4, 22–7.
Luchtel, D. (1972*a*). Gonadal development and sex determination in pulmonate molluscs. I. *Arion circumscriptus*. *Z. Zellforsch.* **130**, 279–301.
Luchtel, D. (1972*b*). Gonadal development and sex determination in pulmonate molluscs. II. *Arion ater rufus* and *Deroceras reticulatum*. *Z. Zellforsch.* **130**, 302–11.
McConnaughey, B. H. (1951*). The life cycle of the dicyemid mesozoa. *Univ. Calif. Publ. Zool.* **55**, 295–336.
Mackiewicz, J. S. (1968). Vitellogenesis and egg shell formation in *Caryophyllaeus laticeps* (Pallas) and *Caryophyllaeides fennica* (Schneider) (Cestoidea: Caryophyllaeidea). *Z. Parasitenk.* **30**, 18–32.
McLaren, D. J. (1973). Oogenesis and fertilisation in *Dipetalonema vitaeae* (Nematoda, Filaroidea). *Parasitol.* **66**, 465–72.
Mahowald, A. P. (1962). Fine structure of pole cells and polar granules in *Drosophila melanogaster*. *J. Exp. Zool.* **151**, 201–16.
Mahowald, A. P. (1963*a*). Electron microscopy of the formation of the cellular blastoderm in *Drosophila melanogaster*. *Exp. Cell. Res.* **32**, 457–68.
Mahowald, A. P. (1963*b*). Ultrastructural differentiations during formation of the blastoderm in the *Drosophila melanogaster* embryo. *Develop. Biol.* **8**, 186–204.
Mahowald, A. P. (1968*a*) Polar granules of *Drosophila*. II. Ultrastructural changes during early embryogenesis. *J. Exp. Zool.* **167**, 237–62.
Mahowald, A. P. (1968*b*). Fine structure of polar granules in *Miastor*. *J. Cell Biol.* **39**, 84*a*.
Mahowald, A. P. (1971*a**). Origin and continuity of polar granules. In *Results and Problems in Cell Differentiation*, eds J. Rienert & H. Ursprung, vol. 2, pp. 158–69. Springer Verlag, Berlin & Heidelberg.
Mahowald, A. P. (1971*b*). Polar granules of *Drosophila*. III. The continuity of polar granules during the life cycle of *Drosophila*. *J. Exp. Zool.* **176**, 329–44.
Mahowald, A. P. (1971*c*). Polar granules of *Drosophila*. IV. Cytochemical studies showing loss of RNA from polar granules during early stages of embryogenesis. *J. Exp. Zool.* **176**, 345–52.
Mahowald, A. P. (1972*). Oogenesis. In *Developmental Systems: Insects I*, eds S. J. Counce & C. H. Waddington, pp. 1–47. Academic Press, London & New York.
Mahowald, A. P. (1975). Ultrastructural changes in the germ plasm during the life cycle of *Miastor* (Cecidomyidae, Diptera). *Wilhelm Roux' Arch. Entwicklungsmech. Organismen* **176**, 223–40.
Mahowald, A. P. (1977). The germ plasm of *Drosophila*: an experimental system for the analysis of determination. *Amer. Zool.* **17**, 551–63.

Mahowald, A. P. (1979*). Genetic control of oogenesis in *Drosophila*. In *Mechanisms of Cell Change*, eds J. D. Ebert & T. S. Okada, pp. 101–17. Wiley, New York.

Mahowald, A. P., Allis, C. D., Karrer, R. M., Underwood, E. M. & Waring, G. L. (1979*). Germ plasm and pole cells of *Drosophila*. In *Determinants of Spatial Organisation*, eds S. Subtelny & I. R. Konigsberg, pp. 127–46. Academic Press, London & New York.

Mahowald, A. P., Caulton, J. H. & Gehring, W. J. (1979). Ultrastructural studies of oocytes and embryos derived from female flies carrying the *grandchildless* mutation in *Drosophila subobscura*. *Develop. Biol.* **69**, 118–32.

Mahowald, A. P., Illmensee, K. & Turner, F. R. (1976). Interspecific transplantation of polar plasm between *Drosophila* embryos, *J. Cell Biol.* **70**, 358–73.

Mahr, E. (1957). Bewegungsvorgänge im Dotter- Entoplasmasystem des Insekteneies. *Naturwiss.* **7**, 226–7.

Mahr, E. (1960*a*). Struktur und Entwicklungsfunktion des Dotter-Entoplasmasystems im Ei des Heimchens (*Gryllus domesticus*). *Wilhelm Roux' Arch. Entwicklungsmech. Organismen* **152**, 263–302.

Mahr, E. (1960*b*). Normale Entwicklung, Pseudofurchung und die Bedeutung des Furchungszentrums im Ei des Heimchens (*Gryllus domesticus*). *Z. Morphol. Ökol. Tiere* **49**, 263–311.

Mahr, E. (1961). Bewegungssysteme in der Embryonalentwicklung von *Gryllus domesticus*. *Wilhelm Roux' Arch. Entwicklungsmech. Organismen* **152**, 662–724.

Malakhov, V. V. & Cherdantzev, V. G. (1976). Embryogenesis of a free-living marine nematode *Pontonema vulgare*. *Zool Zh.* **54**, 165–74.

Malaquin, A. (1925). La ségrégation, au cours de l'ontogenèse, de deux cellules sexuelles primordiales, souches de la lignée germinale, chez *Salmacina dysteri* (Huxley). *C. R. Acad. Sci., Paris* **180**, 324–7.

Malaquin, A. (1934). Nouvelles observations sur la lignée germinale de l'annélide *Salmacina dysteri* (Huxley). *C. R. Acad. Sci., Paris*. **198**, 1804–6.

Mangold-Wirz, K. & Fioroni, P. (1970). Die Sonderstellung der Cephalopoden. *Zool. Jahrb. (Syst.)* **97**, 522–631.

Manton, S. M. (1928). On the embryology of a Mysid crustacean, *Hemimysis lamornae*. *Phil. Trans. Roy. Soc., ser. B* **216**, 363–463.

Manton, S. M. (1934). On the embryology of the crustacean *Nebalia bipes*. *Phil. Trans. Roy. Soc., ser. B* **223**, 163–238.

Manton, S. M. (1949). Studies on the Onychophora. VII. The early embryonic stages of *Peripatopsis*, and some general considerations concerning the morphology and phylogeny of the Arthropoda. *Phil. Trans. Roy. Soc., ser. B* **233**, 483–580.

Manton, S. M. (1964). Mandibular mechanisms and the evolution of the arthropods. *Phil. Trans. Roy. Soc., ser. B* **247**, 1–183.

Manton, S. M. (1970*). Arthropods: introduction. *Chem. Zool.* **5**, 1–34.

Manton, S. M. (1972). The evolution of arthropod locomotory mechanisms. Part 10. *J. Linn. Soc. Zool.* **51**, 203–400.

Marcum, B. A. & Campbell, R. D. (1978). Development of *Hydra* lacking nerve and interstitial cells. *J. Cell. Sci.* **29**, 17–33.

Maresquelle, H. J. (1978). La notion d'épigenèse, essai de définition. *C. R. Acad. Sci., Paris* **286**, 481–3.

Mariscal, R. N. (1965). The adult and larval morphology and life history of the entoproct *Barentsia gracilis* (M. Sars, 1835). *J. Morphol.* **116**, 311–38.

Mariscal, R. N. (1975*). Entoprocta. In *Reproduction of Marine Invertebrates*, eds A. C. Giese & J. S. Pearse, vol. 2, pp. 1–41. Academic Press, London & New York.

Marthy, H. J. (1968). Die Organogenese des Coelomsystems von *Octopus vulgaris* Lam. PhD Thesis, University of Basel.

Marthy, H. J. (1972). Sur la localisation et la stabilité du plan d'ébauches d'organes chez l'embryon de *Loligo vulgaris* (Mollusque, céphalopode). *C. R. Acad. Sci., Paris* **275**, 1291–3.

Marthy, H. J. (1973). An experimental study of eye development in the cephalopod *Loligo vulgaris*: determination and regulation during formation of the primary optic vesicle. *J. Embryol. Exp. Morphol.* **29**, 347–61.

Marthy, H. J. (1975). Organogenesis in Cephalopoda: further evidence of blastodisc-bound developmental information. *J. Embryol. Exp. Morphol.* **33**, 75–83.

Marthy, H. J. (1976). Les déterminismes dans la morphogenèse. Contribution à l'embryologie expérimentale des Céphalopodes. DSc Thesis, University of Pierre & Marie Curie, Paris Vi, 55 pp.

Marthy, H. J. (1977). Sur le rôle du més-entoblaste dans la morphogenèse des Céphalopodes. *Ann. Soc. Franc. Biol. Dévelop. Caen*, p. 23. Symp. Interactions cellulaires et tissulaires dans la morphogenèse.

Marthy, H. J. (1978). Recherches sur le rôle morphogénétique du més-endoderme dans l'embryogenèse de *Loligo vulgaris* (Céphalopode). *C. R. Acad. Sci., Paris* **287**, 1345–8.

Martin, V. J. & Thomas, M. B. (1977). A fine-structural study of embryonic and larval development in the gymnoblastic hydroid *Pennaria tiarella*. *Biol. Bull.* **153**, 198–218.

Maul, V. (1967). Dynamik und Erbverhaltung plasmatischer Eibereiche der Honigbiene. *Zool. Jahrb. (Anat.)* **84**, 63–166.

Meewis, H. (1934). Eléments régénérateurs dans le bourgeonnement pygidial et dans la zone de scissiparité chez *Chaetogaster diaphanus* Gruith. *Ann. Soc. Zool. Belg.* **65**, 9–39.

Meewis, H. (1937). Etude de l'organogénèse lors de la reproduction asexuée chez *Chaetogaster diaphanus* Gruith. *Ann. Soc. Zool. Belg.* **68**, 147–94.

Melander, Y. (1963). Cytogenetic aspects of embryogenesis in *Paludicola*, Tricladida. *Hereditas* **49**, 119–66.

Meng, C. (1968). Strukturwandel und histochemische Befunde insbesondere am Oosom während der Oogenese und nach der Ablage des Eies von *Pimpla turionellae* L. (Hymenoptera, Ichneumonidae). *Wilhelm Roux' Arch. Entwicklungsmech. Organismen* **161**, 162–208.

Meng, C. (1970). Autoradiographische Untersuchungen am Oosom im der Oocyte von *Pimpla turionellae* L. (Hymenoptera). *Wilhelm Roux' Arch. Entwicklungsmech. Organismen* **165**, 35–52.

Mergner, H. (1957). Die Ei- und Embryonalentwicklung von *Eudendrium racemosum* Cavolini. *Zool. Jahrb. Anat. Ontog. Tiere* **76**, 63–164.

Metschnikoff, E. (1866). Embryologische Studien an Insekten. *Z. wiss. Zool.* **16**, 389–500.

Metz, C. W. (1938). Chromosome behaviour, inheritance and sex determination in *Sciara*. *Amer. Nat.* **72**, 485–520.

Metz, C. W. (1957). Interactions between chromosomes and cytoplasm during early embryonic development in *Sciara* (Diptera). *Biol. Bull.* **113**, 323 (abstract).

Meyer, A. (1931). Cytologische Studien über die Gonoblasten und andere ähnliche Zellen in der Entwicklung von *Tubifex*. *Z. Morphol. Okol. Tiere* **22**, 269–86.

Micali, F., Marzari, R., Cristini, F. de & Graziozi, G. (1978). Differential responses to ultraviolet irradiation of the polar cytoplasm of *Drosophila* egg: III. Temperature effect. *Acta Embryol. Exp.* **2**, 247–55.

Miya, K. (1955). Studies on the development of the gonad in the silkworm *Bombyx*

mori L. 5. On the differentiation of the germ cells of the 'new additional crescent'. *J. Fac. Agric. Iwate Univ.* **2**, 239–44.
Miya, K. (1957). Studies on the development of the gonad in the silkworm, *Bombyx mori* L. 7. Differentiation of germ cells in eggs cauterized at the cleavage stage. *Jap. J. Genet.* **32**, 153–7.
Miya, K. (1958). Studies on the embryonic development of the gonad in the silkworm *Bombyx mori* L. 1. Differentiation of germ cells. *J. Fac. Agric. Iwate Univ.* **3**, 436–67.
Mokhtar-Maamouri, F. & Swiderski, Z. (1975). Etude en microscopie électronique de la spermatogénèse de deux cestodes *Acanthobothrium filicalle benedenii* Loenberg, 1889 et *Onchobothrium uncinatum* (Rud., 1819) (Tetraphyllidea, Onchobothriidae). *Z. Parasitenk.* **47**, 269–81.
Moor, B. (1977). Zur Embryologie von *Bradybaena* (*Eulota*) *fruticum* Müller (Gastropoda, Pulmonata Stylommatophora). *Zool. Jahrb. (Anat.)* **97**, 323–99.
Morita, M. (1967). Observations on the fine structure of the neoblast and its cell division in the regenerating planaria. *Sci. Rep. Tôhoku Univ. ser. 4 (Biol.)* **33**, 399–406.
Morita, M., Best, J. B. & Noel, J. (1969). Electron microscopic studies of planarian regeneration. I. Fine structure of neoblasts in *Dugesia dorotocephala*. *J. Ultrastruct. Res.* **27**, 7–23.
Moritz, K. B. (1967*a*). Die Blastomerendifferenzierung für Soma und Keimbahn bei *Parascaris equorum*. I. Cytochemische und photometrische Untersuchungen. *Wilhelm Roux' Arch. Entwicklungsmech. Organismen* **159**, 31–88.
Moritz, K. B. (1967*b*). Die Blastomerendifferenzierung für Soma und Keimbahn bei *Parascaris equorum*. II. Untersuchungen mittels UV-Bestrahlung und Zentrifugierung. *Wilhelm Roux' Arch. Entwicklungsmech. Organismen* **159**, 203–66.
Moritz, M. (1957). Zur Embryonalentwicklung der Phalangiiden (Opiliones, Palpatores) unter besonderer Berücksichtigung der ausseren Morphologie, der Bildung des Mitteldarmes und der Genitalanlage. *Zool. Jahrb. (Anat. Ont.)* **76**, 331–70.
Morrill, J. B. (1963). Development of centrifuged *Limnaea stagnalis* eggs with giant polar bodies. *Exp. Cell Res.* **31**, 490–8.
Morrill, J. B., Blair, C. A. & Larsen, W. J. (1973). Regenerative development in the pulmonate gastropod, *Lymnaea palustris* as determined by blastomere deletion experiments. *J. Exp. Zool.* **183**, 47–56.
Morrill, J. B. & Gottesman, D. M. (1960). Development of isolated blastomeres of *Limnaea palustris*. *Anat. Rec.* **137**, 383.
Moser, J. G., Bode, H. J., Nünemann, H., Collatz, S., Feldhege, A. & Herzfeld, A. (1970). Differenzierung des Aktomyosinsystems während der Morphogenese der Hausgrille, *Acheta domesticus* L. *Verh. deutsch. Zool. Ges.* **64**, J., 56–60.
Mukai, H. & Makioka, T. (1978). Studies on the regeneration of an entoproct, *Barentsia discreta*. *J. Exp. Zool.* **205**, 261–76.
Müller, M. (1957/8). Entwicklung und Bedeutung der Vitellophagen in der Embryonalentwicklung der Honigbiene. Ein Beitrag zur Frage nach der Bedeutung des Dottersystems. *Zool. Jahrb. Allg. Zool. Physiol. Tiere* **67**, 111–50.
Müller, W. A. (1967). Differenzierungspotenzen und Geschlechtsstabilität der I-Zellen von *Hydractinia echinata*. *Wilhelm Roux' Arch. Entwicklungsmech. Organismen* **159**, 412–32.
Nachtwey, R. (1925). Untersuchungen über die Keimbahn, Organogenese und Anatomie von *Asplanchna priodonta* Gosse. *Z. wiss. Zool.* **126**, 239–492.
Naef, A. (1928). Die Cephalopoden (Embryologie). *Fauna Flora del Golgi di Napoli*, 35 monogr., p. 148. R. Fiedländer & Sohn, Berlin.

Nagao, Z. (1965). Studies on the development of *Tubularia radiata* and *Tubularia venusta* (Hydrozoa). *Publ. Akkeshi Mar. Biol. Stat.* **15**, 9–35.

Nelson, O. E. (1934*). The segregation of the germ cells in the grasshopper, *Melanoplus differentialis* (Acrillidae, Orthoptera). *J. Morphol.* **55**, 545–75.

Neumann, R. (1977). Polyp morphogenesis in a scyphozoan: evidence for a head inhibitor from the presumptive foot end in vegetative buds of *Cassiopeia andromeda*. *Wilhelm Roux' Arch. Develop. Biol.* **183**, 79–83.

Nicklas, R. B. (1959). An experimental and descriptive study of chromosome elimination on *Miastor* sp. (Cecidomyidae; Diptera). *Chromosoma* **10**, 301–36.

Nicklas, R. B. (1960). The chromosome cycle of a primitive Cecidomyid – *Mycophila speyeri*. *Chromosoma* **11**, 402–18.

Nielson, C. (1966). On the life cycle of some Loxosomatidae (Entoprocta). *Ophelia* **3**, 221–47.

Nielson, C. (1967). Metamorphosis of the larva of *Loxosomella murmanica* (Nilus) (Entoprocta). *Ophelia* **4**, 85–9.

Nieuwkoop, P. D. & Sutasurya, L. A. (1979*). *Primordial Germ Cells in the Chordates*, 187 pp. Cambridge University Press.

Nigon, V. (1965*). Développement et reproduction des Nématodes. In *Traité de Zoologie*, ed. P. P. Grassé, vol. 4, pp. 218–386. Masson et Cie, Paris.

Nigon, V., Guerrier, P. & Monin, H. (1960). L'architecture polaire de l'oeuf et les mouvements des constituants cellulaires au cours des premières étapes du développement chez quelques nématodes. *Bull. Biol. France Belg.* **93**, 131–202.

Nissani, M. (1977). Cell lineage analysis of germ cells of *Drosophila*. *Nature, Lond.* **265**, 729–30.

Noda, K. & Kanai, C. (1977). An ultrastructural observation on *Pelmatohydra robusta* at sexual and asexual stages, with a special reference to 'germinal plasm'. *J. Ultrastruct. Res.* **61**, 284–94.

Nouvel, H. (1947). Les Dicyémides. 1re partie: systematique, generations vermiformes, infusorigène et sexualité. *Arch. Biol. (Paris)* **58**, 59–230.

Nouvel, H. (1948). Les Dicyémides. 2e partie: infusoriforme, teratologie, specificité du parasitisme, affinités. *Arch. Biol. (Paris)* **59**, 147–223.

Nyholm, K. G. (1943). Zur Entwicklung und Entwicklungsbiologie der Ceriantharien und Actinien. *Zool. Bidr. Uppsala* **22**, 87–248.

Nyholm, K. G. (1959). On the development of the primitive Actinia *Protanthea simplex*, Carlgren. *Zool. Bidr. Uppsala* **33**, 69–77.

Oelhafen, F. (1961). Zur Embryogenese von *Culex pipiens*: Markierungen und Exstirpationen mit UV. Strahlenstich. *Wilhelm Roux' Arch. Entwicklungsmech. Organismen* **153**, 120–57.

Ogren, R. E. (1962). Continuity of morphology from oncosphere to early cysticercoid in the development of *Hymenolepis diminuta* (Cestoda: Cyclophillidea). *Exp. Parasitol.* **12**, 1–6.

Okada, K. (1968*). Annelida. In *Invertebrate Embryology*, eds M. Kumé & K. Dan, (translated by J. C. Dan), pp. 192–239. Nolit, Belgrade.

Okada, M., Kleinman, I. A. & Scheiderman, H. A. (1974*a*). Restoration of fertility in sterilised *Drosophila* eggs by transplantation of polar cytoplasm. *Develop. Biol.* **37**, 43–54.

Okada, M., Kleinman, I. A. & Schneiderman, H. A. (1974*b*). Chimeric *Drosophila* adults produced by transplantation of nuclei into specific regions of fertilised egg. *Develop. Biol.* **39**, 286–94.

Okugawa, K. I. (1957). An experimental study of sexual induction in the asexual form of Japanese fresh-water planaria, *Dugesia gonocephala* (Dugès). *Bull. Kyoto Gakugei Univ., ser. B* **11**, 8–27.

Ortolani, G. (1963a). Ricerche sui territori organo formativi nell' uovo dei Ctenofori. *Boll. Zool.* **30**, 25–31.
Ortolani, G. (1963b). Sulla origine del mesoderma nei Ctenofori. *Rend. Accad. Naz. Lincei, ser. 8* **34**, 434–5.
Ortolani, G. (1964). Orgine dell'organo apicale e di derivati mesodermici nello sviluppo embrionale di Ctenofori. *Acta Embryol. Morphol. Exp.* **7**, 191–200.
Otto, J. J. & Campbell, R. D. (1977). Budding in *Hydra attenuata*: bud stages and fate map. *J. Exp. Zool.* **200**, 417–28.
Pai, S. (1928). Die Phasen des Lebenscyclus der *Anguillula aceti* Ehrbg. und ihre experimentell-morphologische Beeinflussung. *Z. wiss. Zool.* **131**, 293–344.
Painter, T. S. (1966). The role of the E-chromosomes in Cecidomyidae. *Proc. Nat. Acad. Sci. USA* **56**, 853–5.
Panelius, S. (1968). Germ line and oogenesis during paedogenetic reproduction in *Heteropeza pygmaea* Winnertz. (Diptera: Cecidomyidae). *Chromosoma* **23**, 333–45.
Panijel, J. & Pasteels, J. (1951). Analyse cytochimique de certains phénomènes de recharge en ribonucléoprotéines: le cas de l'oeuf de '*Parascaris equorum*' lors de la fécondation. *Arch, Biol.* **62**, 353–70.
Parisi, E., Filosa, S., De Petrocellis, B. & Monroy, A. (1978). The pattern of cell division in the early development of the sea-urchin, *Paracentrotus lividus*. *Develop. Biol.* **65**, 38–49.
Pasteels, J. (1948). Recherches sur le cycle germinal chez l'*Ascaris*. Etude cytochimique des acides nucléiques dans l'oogenèse, la spermatogenèse et le développement chez *Parascaris equorum* Goerze. *Arch. Biol.* **59**, 405–46.
Pease, D. C. (1940). The influence of centrifugal force on the bilateral determination and the polar axis of *Cumingia* and *Chaetopterus* eggs. *J. Exp. Zool.* **84**, 387–411.
Pedersen, K. J. (1959). Cytological studies on the planarian neoblast. *Z. Zellforsch.* **50**, 799–817.
Pedersen, K. J. (1972). Studies on regeneration blastemas of the planarian *Dugesia tigrina* with special reference to differentiation of the muscle–connective tissue filament system. *Wilhelm Roux' Arch. Entwicklungsmech. Organismen* **169**, 134–69.
Peltrera, A. (1940). La capacità regolative dell'uovo de *Aplysia limacina* L. studiate con la centrifugazione e con la reazioni vitali. *Publ. Staz. Zool. Napoli* **18**, 20–49.
Penners, A. (1934). Die Herkunft der Urkeimzellen bei *Tubifex* (Experimentelle Prüfung der Frage; zugleich ein Nachweis von Funktionswechselpotenzen einselner Segmente). *Z. wiss. Zool.* **145**, 388–98.
Penners, A. (1936). Regulation am Keim von *Tubifex rivulorum* Lam. nach Ausschaltung des ektodermalen Keimstreifs. *Z. wiss. Zool. (A)* **149**, 86–130.
Penners, A. (1938). Abhängigkeit der Formbildung vom Mesoderm im *Tubifex*-Embryo. *Z. wiss. Zool.* **150**, 305–57.
Penners, A. & Stäblein, A. (1930). Über die Urkeimzellen bei Tubificiden (*Tubifex rivulorum* Lam. and *Limnodrilus udekemianus* Claparède). *Z. wiss. Zool.* **137**, 606–26.
Philiptschenko, J. (1912a). Beiträge zur Kenntnis der Apterygoten. III. Die Embryonalentwicklung von *Isotoma cinerea* Nic. *Z. wiss. Zool.* **103**, 519–660.
Philiptschenko, J. (1912a). Zur Kenntnis der Apterygotenembryologie. *Zool. Anz.* **39**, 43–9.
Pianka, H. D. (1974*). Ctenophora. In *Reproduction of Marine Invertebrates*, eds A. C. Giese & J. S. Pearse, vol. 1, pp. 201–65. Academic Press, London & New York.

Pieper, S. M. B. (1953). The life history and germ cell cycle of *Spirorchis artericola* (Ward., 1921). *J. Parasitol.* **39**, 310–25.
Pocock, M.A. (1933). *Volvox* and associated algae from Kimberley, South Africa. *S. Afr. Mus. Ann.* **16**, 473–545.
Porchet, M., Dhainaut, A. & Porchet-Hennere, E. (1979). Evidence of coelomic substances inducing genital maturation in *Perinereis cultrifera* (Annelida Polychaeta). *Wilhelm Roux' Arch. Develop. Biol.* **186**, 129–37.
Potswald, H. E. (1969). Cytological observations on the so-called neoblasts in the serpulid *Spirorbis*. *J. Morphol.* **128**, 241–60.
Poulson, D. F. (1945). Chromosomal control of embryogenesis in *Drosophila*. *Amer. Nat.* **79**, 340–63.
Poulson, D. F. (1947). The pole cells of Diptera, their fate and significance. *Proc. Nat. Sci. USA* **33**, 182–4.
Poulson, D. F. (1950). Histogenesis, organogenesis and differentiation in the embryo of *Drosophila melanogaster* Meigen. In *Biology of* Drosophila, ed. M. Demerec, pp. 168–274. Wiley, New York.
Poulson, D. F. & Waterhouse, D. F. (1960). Experimental studies on pole cells and midgut differentiation in Diptera. *Austr. J. Biol. Sci.* **13**, 541–67.
Price, R. D. (1958). Observations on a unique monster embryo of *Wyeomyia smithii* (Coquillett) (Diptera: Culicidae). *Ann. Entomol. Soc. Amer.* **51**, 600–4.
Prudhommeau, C. & Laugé, G. (1972). Irradiation UV des cellules polaires de l'oeuf chez *Drosophila melanogaster*. II. Estimation du nombre de cellules polaires à l'origine de la lignée germinale. *Mutation Res.* **14**, 43–52.
Rabinowitz, M. (1941). Studies on the cytology and early embryology of the egg of *Drosophila melanogaster*. *J. Morphol.* **69**, 1–49.
Rattenbury, J. C. & Berg, W. E. (1954). Embryonic segregation during early development of *Mytilus edulis*. *J. Morphol.* **95**, 393–413.
Raven, C. P. (1946). The development of the egg of *Limnaea Stagnalis* L. from the first cleavage till the trochophore stage, with special reference to its 'chemical embryology'. *Arch. Néerl. Zool.* **7**, 353–434.
Raven, C. P. (1948*). The chemical and experimental embryology of *Limnaea*. *Biol Rev.* **23**, 333–69.
Raven, C. P. (1952). Morphogenesis in *Limnaea stagnalis* and its disturbance by lithium. *J. Exp. Zool.* **121**, 1–78.
Raven, C. P. (1958). Information versus preformation in embryonic development. *Arch. Néerl. Zool.* **13**, suppl. 185–93.
Raven C. P. (1963*a*). The nature and origin of the cortical morphogenetic field in *Limnaea*. *Develop. Biol.* **7**, 130–43.
Raven, C. P. (1963*b*). Differentiation in molluscan eggs. *Symp. Soc. Exp. Biol. No. 17, Cell Differentiation*, pp. 274–84.
Raven, C. P. (1964*). Development. In *Physiology of Mollusca*, eds K. M. Wilbur & C. M. Yonge, vol. 1. pp. 165–95. Academic Press, London & New York.
Raven, C. P. (1966*). *Morphogenesis: The Analysis of Molluscan Development*, 2nd edn, 365pp. Pergamon Press, Oxford.
Raven, C. P. (1967). The distribution of special cytoplasmic differentiations of the egg during early cleavage in *Limnaea stagnalis*. *Develop. Biol.* **16**, 407–37.
Raven, C. P. (1970*). The cortical and subcortical cytoplasm of the *Lymnaea* egg. *Int. Rev. Cytol.* **28**, 1–44.
Raven, C. P. (1974). Further observations on the distribution of cytoplasmic substances among the cleavage cells in *Lymnaea stagnalis*. *J. Embryol. Exp. Morphol.* **31**, 37–59.

Raven, C. P. (1975*). Development. In *Pulmonates*, eds V. Fretter & J. Peake, vol. 1, pp. 367–400. Academic Press, London & New York.

Raven C. P. (1976). Morphogenetic analysis of spiralian development. *Amer. Zool.* **16**, 395–403.

Raven, C. P. & Bezem, J. J. (1971). Computer simulation of embryonic development. I and II. Radialised development of the *Lymnaea* egg. *Proc. Kon. Ned. Akad. Wetensch.* **74**, 209–33.

Raven C. P. & Bezem, J. J. (1972). Computer simulation of embryonic development. III. Differentiation in the radialised embryo. *Proc. Kon. Ned. Akad. Wetensch.* **75**, 20–33.

Raven, C. P. & Bezem, J. J. (1973). Computer simulation of embryonic development. IV. Normal development of the *Lymnaea* egg. *Proc. Kon. Ned. Akad. Wetensch.* **76**, 23–35.

Raven, C. P., Bezem, J. J. & Baretta-Bekker, J. G. (1973). Computer simulation of embryonic development. V. Localisation of the induction centre in the normal development of the *Lymnaea* egg. *Proc. Kon. Ned. Akad. Wetensch.* **76**, 319–40.

Raven, C. P. & Van der Wal U. P. (1964). Analysis of the formation of the animal pole plasm in the eggs of *Lymnaea stagnalis*. *J. Embryol. Exp. Morphol.* **12**, 123–39.

Rees, G. (1940). Studies on the germ cell cycle of the digenetic trematode *Parorchis acanthus* Nicoll. Part II. Structure of the miracidium and germinal development in the larval stages. *Parasitology* **32**, 372–91.

Reggiani, M. P. (1956). Origine delle cellule germinali e maturazione dei gameti in *Spirorbis pagenstecheri*. *Boll. Zool.* **23**, 581–7.

Reggiani, M. P. (1957). Osservazioni sull'origine delle cellule germinali e il differenziamento dei gameti nel policheti ermafrodita *Spirorbis pagenstecheri*. *Quatr. Rev. Biol.* **49**, 245–61.

Reinhardt, E. (1960). Kernverhältnisse, Eisystem und Entwicklungsweise von Drohnen- und Arbeiterinneneiern der Honigbiene (*Apis mellifera*). *Zool. Jahrb. (Anat.)* **78**, 167–234.

Reitberger, A. (1939/40). Die Cytologie des pädogenetischen Entwicklungszyklus der Gallmücke *Oligarces paradoxus* Mein. *Chromosoma*, **1**, 391–473.

Reith, F. (1931). Versuche über die Determination der Keimesanlage bei *Camponotus ligniperda*. *Z. wiss. Zool.* **139**, 664–734.

Relexans, J. C. (1970*a*). Contribution à l'étude expérimentale de la différenciation sexuelle chez un hermaphrodite simultané *Eisenia foetida* Lav. PhD Thesis, University of Bordeaux, 81 pp.

Relexans, J. C. (1970*b*). Mise en évidence expérimentale, chez le lombricien *Eisenia foetida*, d'une ségrégation précoce de la lignée germinale par application d'une température élevée au cours du développement embryonnaire. *C. R. Acad. Sci., Paris* **270**, 977–80.

Relexans, J. C. (1973). Transplantations de gonades indifférenciées chez l'hermaphrodite simultané *Eisenia foetida* (Oligochète, Lumbricidae). Mise en évidence du facteurs locaux (inducteurs ?) de la différenciation sexuelle. *J. Embryol. Exp. Morphol.* **30**, 143–61.

Relexans, J. C. (1975). Factors of primary sexual differentiation in the simultaneous hermaphrodite *Eisenia foetida* (Oligochaeta: Lumbricidae). In *Intersexuality in the Animal Kingdom*, ed. R. Reinboth, pp. 72–83. Springer Verlag, Berlin & Heidelberg.

Rice, T. B. & Garen, A. (1975). Localised defects of blastoderm formation in maternal effect mutants of *Drosophila*. *Develop. Biol.* **43**, 277–86.

Richard, A. & Lemaire, J. (1975). Détermination et différenciation sexuelles chez la seiche *Sepia officinalis* L. (Mollusque Céphalopode). *Publ. Staz. Zool. Napoli* **39**, 574–94.

Richard-Mercier, N. (1972). Embryogenèse et différenciation sexuelle de la gonade du doryphore, *Leptinotarsa decemlineata* Say. (Coléoptère, Chrysomelide). *Ann. Embryol. Morphogén.* **5**, 191–201.

Richard-Mercier, N. (1977). Organogenèse d'ovaries et de testicules stériles après cautérisation des cellules polaires de l'embryon du Doryphore (*Leptinotarsa decemlineata* Say). *Wilhelm Roux' Arch. Develop. Biol.* **183**, 171–6.

Richards, A. G. & Miller, A. (1937*). Insect development analysed by experimental methods: a review. Part I. Embryonic stages. *J. New York Entomol. Soc.* **45**, 1–60.

Rieffel, S. M. & Crouse, H. V. (1966). The elimination and differentiation of chromosomes in the germ line of *Sciara*. *Chromosoma* **19**, 231–76.

Royer, M. (1973). La formation des ébauches gonadiques dans l'embryon de l'insect hermaphrodite *Icerya purchasi*. *C. R. Acad. Sci., Paris* **276**, 1605–8.

Runnström, J. (1928). Zur experimentellen Analyse der Wirkung des Lithiums auf den Seeigelkeim. *Acta Zool.* **9**, 365–424.

Rybicka, K. (1964*a*). Embryonic development of *Moniezia expansa* (Rud., 1810) (Cyclophyllidea, Anoplocephalidae). *Acta Parasitol. Polonica* **12**, 313–26.

Rybicka, K. (1964*b*). Gametogenesis and embryonic development in *Dipylidium caninum*. *Exp. Parasitol.* **15**, 293–313.

Rybicka, K. (1966*a*). Embryogenesis in *Hymenolepis diminuta*. I. Morphogenesis. *Exp. Parasitol.* **19**, 366–79.

Rybicka, K. (1966*b**). Embryogenesis in Cestodes. *Adv. Parasitol.* **4**, 107–87.

Sabbadin, A. & Zaniolo, G. (1979). Sexual differentiation and germ cell transfer in the colonial ascidian, *Botryllus schloeseri*. *J. Exp. Zool.* **207**, 289–304.

Sacarrǎo, G. F. (1962). On the position of the ontogeny of cephalopods in relation to the development of the other molluscs. *Rev. Fac. Cien. Lisboa*, C **10**, 5–54.

Sacks, M. (1955). Observations on the embryology of an aquatic gastrotrich, *Lepidodermella squammata* Dujarden 1841. *J. Morphol.* **96**, 473–96.

Sander, K. (1956). The early embryology of *Pyrilla perpusilla* Walker (Homoptera), including some observations on the later development. In *On Indian Insect Types*, ed. M. B. Mirza, pp. 1–61. Aligarh Muslim Univ. Publ. Zool. Ser. IV.

Sander, K. (1959/60*a*). Analyse des ooplasmatischen Reaktions-systems von *Euscelis plebejus* Fall. (Cicadina) durch Isolieren und Kombinieren von Keimteilen. I. Die Differenzierungsleistungen vorderer und hinterer Eiteile. *Wilhelm Roux' Arch. Entwicklungsmech. Organismen* **151**, 430–97.

Sander, K. (1959/60*b*). Analyse des ooplasmatischen Reaktions-systems von *Euscelis plebejus* Fall (Cicadina) durch Isolieren und Kombinieren von Keimteilen. II. Die Differenzierungsleistungen nach Verlagern von Hinterpolmaterial. *Wilhelm Roux' Arch. Entwicklungsmech. Organismen* **151**, 660–707.

Sander, K. (1967). Mechanismen der Keimeseinrollung (Anatrepsis) im Insecten-Ei. *Verh. deutsch. Zool. Ges. Heidelberg* pp. 81–9.

Sander, K. (1968). Entwicklungsphysiologische Untersuchungen am embryonalen Mycetom von *Euscelis plebejus* F. (Homoptera, Cicadina). *Develop. Biol.* **17**, 16–38.

Sander, K. (1976*a**). Specification of the basic body pattern in insect embryogenesis. *Adv. Insect Physiol.* **12**, 125–238.

Sander, K. (1976*b**). Morphogenetic movements in insect embryogenesis. In *Insect Development*, ed. P. A. Lawrence, pp. 35–52. Blackwell Scientific Publications, Oxford.

Sander, K. & Vollmar, H. (1967). Vital staining of insect eggs by incorporation of trypan blue. *Nature, Lond.* **216**, 174–5.

Santamaria, P. (1975). Transplantation of nuclei between eggs of different species of *Drosophila*. *Wilhelm Roux' Arch. Entwicklungsmech. Organismen* **178**, 89–98.
Sarà, M. & Orsi, L. R. (1975). Sex differentiation in *Sycon* (Porifera Calcispongiae). *Pubbl. Staz. Zool. Napoli* **39**, 618–34.
Sauzin, M. J. (1966). Etude au microscope électronique du néoblaste de la planaire *Dugesia gonocephala* (Turbellarié, Triclade) et de ses changements ultrastructuraux au cours des premiers stades de la régénération. *C. R. Acad. Sci., Paris* **263**, 605–8.
Sauzin, M. J. (1967*a*). Etude ultrastructurale de la différenciation du néoblaste au cours de la régénération de la planaire *Dugesia gonocephala*. I. Différenciation en cellule nerveuse. *Bull. Soc. Zool. France* **92**, 313–18.
Sauzin, M. J. (1967*b*). Etude ultrastructurale de la différenciation au cours de la régénération de la planaire *Dugesia gonocephala*. II. Différenciation musculaire. *Bull. Soc. Zool. France* **92**, 613–16.
Sauzin, M. J. (1968). Présence d'emissions nucléaires dans les cellules différenciées et en différenciation de la planaire adulte *Dugesia gonocephala*. *C. R. Acad. Sci., Paris* **267**, 1146–8.
Schäller, G. (1960). Beitrag zum Problem der Keimzellenbildung bei Trematodenlarven. *Z. Parasitenk.* **20**, 146–51.
Schleip, W. (1924). Die Herkunft der Polarität des Eies von *Ascaris megalocephala*. *Arch. mikrosk. Anat. Entw. Mech.* **100**, 573–98.
Schleip, W. (1925). Die Furchung dispermer *Dentalium*-eier. *Wilhelm Roux' Arch. Entwicklungsmech. Organismen* **106**, 68–123.
Schmidt, G. A. (1964). Embryonic development of littoral nemertines *Lineus desori*. *Zool. Pol.* **14**, 75–122.
Schnetter, W. (1965). Experimente zur Analyse der morphogenetischen Funktion der Ooplasmabestandteile in der Embryonalentwicklung des Kartoffelkäfers (*Leptinotarsa decemlineata* Say). *Wilhelm Roux' Arch. Entwicklungsmech. Organismen* **155**, 637–92.
Schnetter, W. (1967). Transplantation von Furchungs- und Blastodermkernen in entkernte Eier bei *Leptinotarsa decemlineata* (Coleoptera). *Verh. deutsch. Zool. Ges. Göttingen*, suppl. **30**, 494–9.
Schubiger, M. & Schneiderman, H. A. (1971). Nuclear transplantation in *Drosophila melanogaster*. *Nature, Lond.* **230**, 185–6.
Schulze, F. E. (1875). Ueber den Bau und die Entwicklung von *Sycandra raphanus*. *Z. wiss. Zool.* **25**, suppl. 3, 247–80.
Schüpbach, T., Wieschaus, E. & Nöthiger, R. (1978). A study of the female germ line in mosaics of *Drosophila*. *Wilhelm Roux' Arch. Develop. Biol.* **184**, 41–56.
Schwalm, F. E. (1965). Zell- und Mitosenmuster der normalen und nach Röntgenbestrahlung regulierenden Keimanlage von *Gryllus domesticus*. *Z. Morphol. Ökol. Tiere* **55**, 915–1023.
Schwalm, F. E. (1974). Autonomous structural changes in polar granules of unfertilised eggs of *Coelopa frigida* (Diptera). *Wilhelm Roux' Arch. Entwicklungsmech. Organismen* **175**, 129–33.
Schwalm, F. E., Simpson, R. & Bender, H. A. (1971). Early development of the kelp fly, *Coelopa frigida* (Diptera). Ultrastructural changes within the polar granules during pole cell formation. *Wilhelm Roux' Arch. Entwicklungsmech. Organismen* **166**, 205–18.
Seck, P. (1938). Zur Entwicklungsmechanik des Essigälchens. *Wilhelm Roux' Arch. Entwicklungsmech. Organismen* **137**, 57–85.
Seidel, F. (1924). Die Geschlechtsorgane in der embryonalen Entwicklung von *Pyrrhocoris apterus* L. *Z. Morphol. Ökol. Tiere* **1**, 429–506.

Seidel, F. (1929). Untersuchungen über das Bildungsprinzip der Keimanlage im Ei der Libelle *Platycnemis pennipes* I–V. *Wilhelm Roux' Arch. Entwicklungsmech. Organismen* **119**, 322–440.

Seidel, F. (1934). Das Differenzierungszentrum im Libellenkeim. I. Die dynamischen Voraussetzungen der Determination und Regulation. *Wilhelm Roux' Arch. Entwicklungsmech. Organismen* **131**, 135–87.

Seidel, F. (1952*). Entwicklungsphysiologie der Wirbellosen. *Fortschr. der Zool. über 1945–50*.

Seidel, F. (1961*). Entwicklungsphysiologische Zentren im Eisystem der Insekten. *Verh. deutsch. Zool. Ges. Bonn/Rhein, 1960* pp. 121–42.

Seidel, F. (1966*). Das Eisystem der Insekten und die Dynamik seiner Aktivierung. *Verh. deutsch. Zool. Ges. Jena, 1965* pp. 166–87.

Seidel, F., Bock, E. & Krause, G. (1940*). Die Organisation des Insekteneies (Reaktionsablauf, Induktionsvorgänge, Eitypen). *Naturwiss.* **28**, 433–46.

Seiler, J. (1959). Untersuchungen über die Entstehung der Parthenogenese bei *Solenobia triquetrella* F. R. (Lepidoptera, Psychidae). I. Die Zytologie des bisexuellen *S. triquetrella*, ihr Verhalten und ihr Sexualverhältnis. *Chromosoma* **10**, 73–114.

Seiler, J. (1960). Untersuchungen über die Entstehung der Parthenogenese bei *Solenobia triquetrella* F. R. (Lepidoptera, Psychidae). II. Analyse der diploid parthenogenetischen *S. triquetrella*. Verhalten, Aufzuchtresultate und Zytologie. *Chromosoma* **11**, 29–102.

Seilern-Aspang, F. (1956). Frühentwicklung einer marinen Triclade (*Procerodes lobata*, O. Schmidt). *Wilhelm Roux' Arch. Entwicklungsmech. Organismen* **148**, 589–95.

Seilern-Aspang, F. (1958). Entwicklungsgeschichtliche Studien an Paludicolen Tricladen. *Wilhelm Roux' Arch. Entwicklungsmech. Organismen* **150**, 425–80.

Sharov, A. G. (1966*). *Basic Arthropod Stock*, 271 pp. Pergamon Press, Oxford.

Short, R. B. & Damian, R. T. (1967). Oogenesis, fertilisation and first cleavage of *Dicyema aegira* McConnaughey and Kritzer 1952 (Mesozoa: Dicyemidae). *J. Parasitol.* **53**, 186–95.

Siewing, R. (1977). Mesoderm bei Ctenophoren. *Z. Zool. Syst. Evolutionsforsch.* **15**, 1–8.

Silén, L. (1945). Main features of development of the ovum, and ooecium in the ooeciferous Bryozoa Gymnoloemata. *Ark. Zool.* **35**, 1–34.

Silén, L. (1966). On the fertilisation problem in the gymnoloematous Bryozoa. *Ophelia* **3**, 113–40.

Simpson, T. L. & Gilbert, J. J. (1973). Gemmulation, gemmule hatching, and sexual reproduction in freshwater sponges. I. The life cycle of *Spongilla lacustris* and *Tubella pennsylvanica*. *Trans. Amer. Microsc. Soc.* **92**, 422–33.

Smith, L. D. & Williams, M. A. (1975*). Germinal plasm and determination of the primordial germ cells. In *The Developmental Biology of Reproduction*, eds C. L. Markert & J. Papaconstantinou, pp. 3–24. Academic Press, London & New York.

Sonnenblick, B. P. (1941). Germ cell movements and sex differentiation of the gonads in the *Drosophila* embryo. *Proc. Nat. Acad. Sci. USA* **27**, 484–9.

Sonnenblick, B. P. (1950). The early embryology of *Drosophila melanogaster*. In *Biology of* Drosophila, ed. M. Demerec, pp. 62–167. Wiley, New York; Chapman, London.

Spek, J. (1926). Über gesetzmässige Substanzverteilungen bei der Furchung des Ctenophoreneies und ihre Beziehungen zu den Determinationsproblemen. *Wilhelm Roux' Arch. Entwicklungsmech. Organismen* **107**, 54–73.

Spiegelman, M. & Dudley, P. L. (1973). Morphological stages of regeneration in the planarian *Dugesia tigrina*: a light and electron microscopical study. *J. Morphol.* **139**, 155–84.
Stagni, A. (1959). Primi appunti ed osservazioni sulla ricostituzione degli elementi germinali durante la rigenerazione di *Spirorbis pagenstecheri*. *Rend. Accad. Naz. Lincei, ser. 8* **27**, 71–5.
Stagni, A. (1961*a*). Ancora sulla genesi e sessualizzazione degli elementi germinali en esemplari rigeneranti di *Spirorbis pagenstecheri*. *Rend. Accad. Naz. Lincei, ser. 8* **30**, 928–32.
Stagni, A. (1961*b*). Notazioni sulla morfologia vasale in *Spirorbis pagenstecheri* (Polichete Serpulide) e rapporti dei vasi con gli elementi germinali. *Ann. Univ. Ferrara, ser. 13 Anat. Fisiol. compar.* **1**, 99–108.
Starr, R. C. (1970*). Control of differentiation in *Volvox*. In *Develop. Biol.* suppl. **4**, *Changing Synthesis in Development*, ed. M. N. Runner, pp. 59–100. Academic Press, London & New York.
Stefani, R. (1959). I fenomeni cariologici nella segmentazione dell'uovo ed i loro rapporti con la partenogenesi rudimentale ed accidentale negli embiotteri. *Caryologia* **12**, 1–70.
Stefani, R. (1961). La formazione dei foglietti embryonali, l'origine dell'epithelio intestinale e la determinazione della linea germinale femminile nell' *Haploembia solieri* (Embioptera). *Caryologia* **14**, 1–30.
Steinberg, S. N. (1963). The regeneration of whole polyps from ectodermal fragments of scyphistoma larvae of *Aurelia aurita*. *Biol. Bull.* **124**, 337–43.
Stéphan-Dubois, F. (1961). Les cellules de régénération chez la planaire *Dendrocoelom lacteum*. *Bull. Soc. Zool. France* **86**, 172–85.
Stéphan-Dubois, F. (1964*). La lignée germinale des Turbellariés et des Annélides dans l'évolution normale et la régénération. In *L'Origine de la Lignée Germinale chez les Vertébrés et chez quelques Groups d'Invertébrés*, ed. E. Wolff, pp. 115–36, Hermann, Paris.
Stéphan-Dubois, F. (1978*a*). Absence ou présence de néoblastes dans la région blessée en cas de suppression de segments antérieurs, chez l'Annélide oligochète *Tubifex tubifex* Müller. *C. R. Acad. Sci., Paris* **287**, 33–5.
Stéphan-Dubois, F. (1978*b*). Régénération antérieure de l'annélide *Tubifex tubifex*. *103ᵉ Congrès Nat. Soc. savantes, Nancy* 89–98.
Stolte, H. A. (1936*). Die Herkunft des Zellmaterials bei regenerativen Vorgängen der wirbellosen Tiere. *Biol. Rev*, **11**, 1–49.
Stolte, H. A. (1938). Gestaltung, Zeichnung und Organabbau unter dem Einflusz normaler und 'alternder' Gonaden bei *Polyophthalmus pictus* Duf. (Polychaeta). *Z. wiss. Zool.* **150**, 107–54.
Stunkard, H. W. (1937). The physiology, life cycles and phylogeny of the parasitic flatworms. *Amer. Mus. Novit.* **908**, 1–27.
Sulgostowska, T. (1972). The development of organ systems in cestodes. I. A study of histology of *Hymenolepis diminuta* (Rudolphi, 1819) (Hymenolepididae). *Acta Parasitol. Pol.* **20**, 449–62.
Sulgostowska, T. (1974). The development of organ systems in cestodes. II. Histogenesis of the reproductive system in *Hymenolepis diminuta* (Rudolphi, 1819) (Hymenolepididae). *Acta Parasitol. Pol.* **22**, 179–90.
Sulgostowska, T. (1976). Histology and histogenesis of the reproductive system in hermaphroditic cestodes, in cestodes with a trend towards dioeciousness, and in dioecious ones. *Rozpraw. Nauk.* **84**, 1–54. (English summary.)
Summers, R. G. & Haynes, J. F. (1969). The ontogeny of interstitial cells in *Pennaria tiarella*. *J. Morphol.* **129**, 81–8.

Swiderski, Z. (1967). Embryonic development of the cestode *Drepanidotaenia lanceolata* (Bloch, 1782). *Acta Parasitol. Pol.* **14**, 409–18.

Swiderski, Z. (1968). An electron microscopic evidence of the degeneration of some micromeres during the embryonic development of the cestode; *Catenotaenia pusilla* (Goeze, 1782) (Cyclophyllidae Catenotaeniidae). *Zool. Pol.* **18**, 469–73.

Tadano, M. (1968). Nemathelminthes. In *Invertebrate Embryology*, eds M. Kumé & K. Dan, (translated by J. C. Dan), pp. 159–91. Nolit, Belgrade.

Tanaka, A. (1976). Stages in the embryonic development of the german cockroach, *Blattella germanica* L. (Blattaria, Blattellidae). *Kontyû (Entomol. Soc. Japan)* **44**, 512–25.

Tardent, P. (1954). Axiale Verteilungs-Gradienten der interstitiellen Zellen bei *Hydra* und *Tubularia* und ihre Bedeutung für die Regeneration. *Wilhelm Roux' Arch. Entwicklungsmech. Organismen* **146**, 593–649.

Tardent, P. (1968). Experiments about sex determination in *Hydra attenuata* Pall. *Develop. Biol.* **17**, 483–511.

Tardent, P. (1975*). Sex and sex determination in coelenterates. In *Intersexuality in the Animal Kingdom*, ed. R. Reinbolt, pp. 1–13. Springer Verlag, Berlin & Heidelberg.

Tardent, P. (1978*). Coelenterata, Cnidaria. In *Morphogenese der Tiere*, ed. F. Seidel, pp. 69–398. Fischer Verlag, Jena.

Tardent, P. & Morgenthaler, U. (1966). Autoradiographische Untersuchungen zum Problem der Zellwanderungen bei *H. attenuata* Pall. *Rev. Suisse Zool.* **73**, 468–80.

Tardy, J. (1967). Régénération de la gonade après castration chirurgicale chez quelques Aeolidiidae (Mollusques Nudibranches). *C. R. (Séanc.) Soc. Biol.* **161**, 2013.

Tardy, J. (1970*). Organogenèse de l'appariel génital chez les mollusques. *Bull. Soc. Zool. France* **95**, 407–27.

Tawfik, M. F. S. (1957). Alkaline phosphatase in the germ cell determinant of the egg of *Apanteles* (Hymenoptera). *J. ins. Physiol.* **1**, 286–91.

Taylor, A. R. (1960). The spermatogenesis and embryology of *Litomosoides carinii* and *Dirofiralia immitis*. *J. Helminthol.* **34**, 3–12.

Terpitowska, B. (1976). Germ-line cells in copepods against a background of this phenomenon in animal's world. *Przeglad Zool.* **20**, 33–48.

Thierry-Mieg, D. (1976). Study of a temperature-sensitive mutant *grandchildless-like* in *Drosophila melanogaster*. *J. Microsc. Biol. Cellul.* **25**, 1–6.

Thierry-Mieg, D., Masson, M. & Gans, M. (1972). Mutant de stérilité à effect retardé de *Drosophila melanogaster*. *C. R. Acad. Sci., Paris* **275**, 2751–4.

Tiegs, O. W. (1940). The embryology and affinities of the Symphylan, based on a study of *Hanseniella agilis*. *Quart. J. Microsc. Sci.* **82**, 1–225.

Tiegs, O. W. (1947). The development and affinities of the Pauropoda, based on a study of *Pauropus sylvaticus*. *Quart. J. Microsc. Sci* **88**, 165–267 and 275–336.

Tiegs, O. W. & Manton, S. M. (1958*). The evolution of the Arthropoda. *Biol. Rev.* **33**, 255–337.

Tobler, H. (1976*). Genetic differences between germ line and somatic DNA in *Ascaris lumbricoides*. In *Progress in Differentiation Research*, ed. N. Müller-Bérat, pp. 147–54. North-Holland, Amsterdam.

Truckenbrodt, W. (1964). Zytologische und entwicklungsphysiologische Untersuchungen am besamten und am parthenogenetischen Ei von *Kalothermes flavicollis* Fabr. Reifung, Furchungsablauf und Bildung der Keimanlage. *Zool. Jahrb. (Anat.)* **81**, 359–434.

Turner, F. R. & Mahowald, A. P. (1976). Scanning electron microscopy of *Drosophila* embryogenesis. I. The structure of the egg envelopes and the formation of the cellular blastoderm. *Develop. Biol.* **50**, 95–108.

Turner, F. R. & Mahowald, A. P. (1977). Scanning electron microscopy of *Drosophila melanogaster* embryogenesis. II. Gastrulation and segmentation. *Develop. Biol.* **57**, 403–16.

Turner, F. R. & Mahowald, A. P. (1979). Scanning electron microscopy of *Drosophila melanogaster* embryogenesis. III. Formation of the head and caudal segments. *Develop. Biol.* **68**, 96–109.

Tuzet, O. (1964*). L'origine de la lignée germinale et la gamétogenèse chez les Spongiaires. In *L'Origine de la Lignée Germinale chez les Vertébrés et chez quelques Groups d'Invertébrés*, ed. E. Wolff, pp. 79–111. Hermann, Paris.

Tuzet, O. (1970*). La polarité de l'oeuf et la symétrie de la larve des éponges calcaires. *Proc. Symp. Zool. Soc. London* **25**, 437–48.

Tuzet, O., Garrone, R. & Pavans de Ceccatty, M. (1970*a*). Origine choanocytaire de la lignée germinale mâle chez la démosponge *Aplysilla rosea* Schulze (Dendroceratides). *C. R. Acad. Sci., Paris* **270**, 955–7.

Tuzet, O., Garrone, R. & Pavans de Ceccatty, M. (1970*b*). Observations ultrastructurales sur la spermatogenèse chez la Démosponge *Aplysilla rosea* Schulze (Dendroceratides): une métaplasie exemplaire. *Ann. Sci. Nat. Zool. Biol. Anim.* **12**, 27–50.

Uchida, T. & Yamada, M. (1968*a**). Coelenterata. In *Invertebrate Embryology*, eds M. Kumé & K. Dan, (translated by J. C. Dan), pp. 86–116. Nolit, Belgrade.

Uchida, T. & Yamada, M. (1968*b**). Ctenophora. In *Invertebrate Embryology*, eds M. Kumé & K. Dan, (translated by J. C. Dan), pp. 117–24. Nolit, Belgrade.

Ullmann, S. L. (1964). The origin and structure of the mesoderm and the formation of the coelomic sacs in *Tenebrio molitor* L. (Insecta, Coleoptera). *Phil. Trans. Roy. Soc., ser. B* **248**, 245–77.

Ullmann, S. L. (1965). Epsilon granules in *Drosophila* pole cells and oöcytes. *J. Embryol. Exp. Morphol.* **13**, 73–81.

Underwood, E. M., Turner, F. R. & Mahowald, A. P. (1980). Analysis of cell movements and fate mapping during early embryogenesis in *Drosophila melanogaster*. *Develop. Biol.* **74**, 286–301.

Underwood, E. M., Caulton, J. H., Allis, C. D. & Mahowald, A. P. (1980). Developmental fate of the pole cells in *Drosophila melanogaster*. *Develop. Biol.* **77**, 303–14.

Van Beneden, E. (1876). Recherches sur les Dicyémides, survivants actuels d'un enbranchement des Mézozoaires. *Bull. Acad. Roy. Belg. ser. 2* **41/42**, 111 pp.

Van Dongen, C. A. M. (1976). The development of *Dentalium* with special reference to the significance of the polar lobe. V. Differentiation of the cell pattern in lobeless embryos of *Dentalium vulgare* (da Costa) during late larval development. *Proc. Kon. Ned. Akad. Wetensch. ser. C* **79**, 245–66.

Van Dongen, C. A. M. & Geilenkirchen, W. L. M. (1974). The development of *Dentalium* with special reference to the significance of the polar lobe. I–III. Division chronology and development of the cell pattern in *Dentalium dentale* (Scaphopoda). *Proc. Kon. Ned. Akad. Wetensch. ser. C* **77**, 57–100.

Van Dongen, C. A. M. & Geilenkirchen, W. L. M. (1975). The development of *Dentalium* with special reference to the significance of the polar lobe. IV. Division chronology and development of the cell pattern in *Dentalium dentale* after removal of the polar lobe at first cleavage. *Proc. Kon. Ned. Akad. Wetensch. ser. C* **78**, 358–75.

Van de Vyver, G. (1967). Etude du développement embryonnaire des hydraires athécates (gymnoblastiques) à gonophores. I. Formes à planula. *Arch. Biol. Liège* **78**, 451–518.

Van de Vyver, G. (1968*a*). Etude du développement embryonnaire des hydraires athécates (gymnoblastiques) à gonophores. II. Formes à actinules. *Arch. Biol. Liège* **79**, 327–63.

Van de Vyver, G. (1968*b*). Etude du développement embryonnaire des hydraires athécates (gymnoblastiques) à gonophores. III. Discussion et conclusions générales. *Arch. Biol. Liège* **79**, 365–79.

Van de Vyver, G. & Bouillon, J. (1969). Etude du développement embryonnaire et de l'histogenèse de *Eleutheria dichotoma* (de Quatrefages) (Anthoméduse Eleutheriidae). *Ann. Embryol. Morphogén.* **2**, 317–27.

Van den Biggelaar, J. A. M. (1976). Development of dorsoventral polarity preceding the formation of the mesentoblast in *Lymnaea stagnalis*. *Proc. Kon. Ned. Akad. Wetensch.* **79**, 112–26.

Van den Biggelaar, J. A. M. (1977). Development of dorsoventral polarity and mesentoblast determination in *Patella vulgata*. *J. Morphol.* **154**, 157–86.

Van den Biggelaar, J. A. M. & Guerrier, P. (1979). Dorsoventral polar and mesentoblast determination as concomitant results of cellular interactions in the mollusc *Patella vulgata*. *Develop. Biol.* **68**, 462–71.

Van der Woude, A. (1954). Germ cell cycle of *Megalodiscus temperatus* (Stafford, 1905) Harwood 1932 (Paramphistomidae: Trematoda). *Amer. Midl. Nat.* **51**, 172–202.

Van der Woude, A., Cort, W. W. & Ameel, D. J. (1953). The early development of the daughter sporocyst of the Strigeoidea (Trematoda). *J. Parasitol.* **39**, 38–44.

Vandel, A. (1921). Recherches expérimentales sur les modes de reproduction des planaires Triclades paludicoles. *Bull. Biol. France Belg.* **55**, 343–518.

Vannini, E. (1947). Neoblasti e regenerazione dei segmenti genetali nel Serpulide ermafrodita *Salmacina incrustans* Clop. Nota preliminare. *1st Veneto di S. L. ed. A* **105**, 50–6.

Vannini, E. (1963). Sul concetto di 'gradiente di sessualita' in alcuni animali ermafroditi. *Monit. Zool. Ital.* **70**, 543–66.

Vannini, E. & Stagni, A. (1959). Nuove indagini sull'origine ed il differenziamento delle cellule germinali nel Polichete ermarodite *Spirorbis pagenstecheri*. *Rend. Accad. Naz. Lincei ser. 8* **26**, 259–65.

Verdonk, N. H. (1965). Morphogenesis of the head region in *Limnaea stagnalis* L. PhD Thesis, University of Utrecht, 107 pp.

Verdonk, N. H. (1968*a*). The determination of bilateral symmetry in the head region of *Limnaea stagnalis*. *Acta Embryol. Morphol. Exp.* **10**, 211–27.

Verdonk, N. H. (1968*b*). The effect of removing the polar lobe in centrifuged eggs of *Dentalium*. *J. Embryol. Exp. Morphol.* **19**, 33–42.

Verdonk, N. H. (1968*c*). The relation of the two blastomeres to the polar lobe in *Dentalium*. *J. Embryol. Exp. Morphol.* **20**, 101–5.

Verdonk, N. H. & Cather, J. N. (1973). The development of isolated blastomeres in *Bithynia tentaculata* (Prosobranchia, Gastropoda). *J. Exp. Zool.* **186**, 47–62.

Verdonk, N. H., Geilenkirchen, W. L. M. & Timmermans, L. P. M. (1971). The localisation of morphogenetic factors in uncleaved eggs of *Dentalium*. *J. Embryol. Exp. Morphol.* **25**, 57–63.

Vignau, J. (1967). Le remplacement régulier des embryons abortifs par des ébauches embryonnaires nouvelles dans les oeufs du Phasme (*Carausius morosus* Br.). Observations histologiques sur cette 'polyembryonie substitutive'. *C. R. Acad. Sci., Paris* **265**, 1404–7.

Villa, L. (1976). An ultrastructural investigation of the polar plasm of the egg of *Sternaspis* (Annelida, Polychaeta). *Acta Embryol. Exp.* **2**, 153–65.
Vollmar, H. (1972). Die Einrollbewegung (Anatrepsis) des Keimstreifs im Ei von *Acheta domesticus* (Orthopteridea, Gryllidae). *Wilhelm Roux' Arch. Entwicklungsmech. Organismen* **170**, 135–51.
Von Borstel, R. C. (1955). Feulgen-negative nuclear division in *Habrobracon* eggs after exposure to X-rays or nitrogen mustard. *Nature, Lond.* **175**, 342.
Von Borstel, R. C. (1957*). Nucleocytoplasmic relations in early insect development. In *The Beginnings of Embryonic Development*, ed. A. Tyler, R. C. von Borstel & C. B. Metz, pp. 175–99. Amer. Ass. Advanc. Sci., Washington, D. C.
Von Ubisch, L. (1943). Über die Bedeutung der Diminution von *Ascaris megalocephala*. *Acta Biotheor.* **7**, 163–82.
Wada, S. K. (1968*). Mollusca, I. Amphineura, Gastropoda, Scaphopoda, Pelecypoda. In *Invertebrate Embryology*, eds K. Kumé & K. Dan, (translated by J. C. Dan), pp. 485–525. Nolit, Belgrade.
Waring, G. L., Allis, C. D. & Mahowald, A. P. (1978). Isolation of polar granules and the identification of polar granule-specific protein. *Develop. Biol.* **66**, 197–206.
Warn, R. (1972). Manipulation of the pole plasm of *Drosophila melanogaster*. *Acta Embryol. Exp.* suppl. 415–27.
Warn, R. (1975). Restoration of the capacity to form pole cells in UV-irradiated *Drosophila* embryos. *J. Embryol Exp. Morphol.* **33**, 1003–11.
Warn, R. M. (1979*). The pole plasm of *Drosophila*. In *Maternal Effects in Development*, eds D. R. Newth & M. Balls, pp. 199–219. Cambridge University Press.
Webber, H. H. (1977*). Gastropoda: Prosobranchia. In *Reproduction of Marine Invertebrates*, eds A. C. Giese & J. S. Pearse, vol. 4, pp. 1–97. Academic Press, London & New York.
Weber, R. (1958). Über die submikroskopische Organisation und die biochemische Kennzeichnung embryonaler Entwicklungsstadien von *Tubifex*. *Wilhelm Roux' Arch. Entwicklungsmech. Organismen* **150**, 542–80.
Weber, R. (1960). Progressive maturation of the organisation pattern on fertilised and cleaving eggs in Spiralia. In *Symposium on Germ Cells and Development*, ed. S. Ranzi, pp. 225–54. I.I.E. & Fond. Baselli, Milan.
Webster, G. (1971*). Morphogenesis and pattern formation in hydroids. *Biol. Rev.* **46**, 1–46.
Webster, G. & Hamilton, S. (1972). Budding in *Hydra*: the role of cell multiplication and cell movement in bud initiation. *J. Embryol. Exp. Morphol.* **27**, 301–16.
Weiler-Stolt, B. (1960). Über die Bedeutung der interstitiellen Zellen fur die Entwicklung und Fortpflanzung mariner Hydroiden. *Wilhelm Roux' Arch. Entwicklungsmech. Organismen* **152**, 398–454.
Weismann, A. (1883). Die Entstehung der Sexualzellen bei den Hydromedusen. *Biol. Centralbl.* **4**, 12–31.
Weismann, A. (1885*). *Die Continuität des Keimplasma's als Grundlage einer Theorie des Vererbung*, 122 pp. Fischer Verlag, Jena.
Weismann, A. (1892*). *Das Keimplasma. Eine Theorie der Vererbung*, 628 pp. Fischer Verlag, Jena.
Wells, M. J. & Wells, J. (1977*). Cephalopoda: Octopoda. In *Reproduction of Marine Invertebrates*, eds. A. C. Giese and J. S. Pearse, vol. 4, pp. 291–336. Academic Press, London & New York.
West, J. A., Cantwell, G. E. & Shortino, T. J. (1968). Embryology of the house fly, *Musca domestica* (Diptera: Muscidae), to the blastoderm stage. *Ann. Entomol. Soc. Amer.* **61**, 13–17.

Weygoldt, P. (1963*). Grundorganisation und Primitiventwicklung bei Articulaten. *Zool. Anz.* **171**, 363–76.
White, M. J. D. (1946). The cytology of the Cecidomyidae (Diptera). II. The chromosome cycle and anomalous spermatogenesis of *Miastor*. *J. Morphol.* **79**, 323–69.
White, M. J. D. (1950). Cytological studies on gall midges. *Univ. Texas Publ.* **5007**, 1–80.
White, M. J. D. (1973*). *Animal Cytology and Evolution*, 3rd edn Cambridge University Press, 969 pp.
Wieschaus, E. & Szabad, J. (1979). The development and formation of the female germ line in *Drosophila melanogaster*: a cell lineage study. *Develop. Biol.* **68**, 29–46.
Wilby, O. K. & Webster, G. (1970). Experimental studies on axial polarity in *Hydra*. *J. Embryol. Exp. Morphol.* **24**, 595–613.
Wildermuth, H. (1970). Determination and transdetermination in cells of the fruit fly. *Sci. Prog., Oxford* **58**, 329–58.
Wolf, R. (1967). Der Feinbau des Oosoms normaler und zentrifugierter Eier der Gallmücke *Wachtliella persicariae* L. (Diptera). *Wilhelm Roux' Arch. Entwicklungsmech. Organismen* **158**, 459–62.
Wolf, R. (1969). Kinematik und Feinstruktur plasmatischer Faktorenbereiche des Eies von *Wachtliella persicariae* L. (Diptera). *Wilhelm Roux' Arch. Entwicklungsmech. Organismen* **162**, 121–60.
Wolf, R. (1973). Kausalmechanismen der Kern-bewegung und -teilung während der frühen Furchung im Ei des Gallmücke *Wachtliella persicariae* L. *Wilhelm Roux' Arch. Entwicklungsmech. Organismen* **172**, 28–57.
Wolf, R. (1975). Ein neurartiger Migrationsmechanismus bei Furchungskernen auf der Basis des Astersystems. *Verh. deutsch. Zool. Ges.* 1974, 174–8.
Wolf, R. (1977). Embryonalentwicklung der Gallmücke *Wachtliella persicariae* (Diptera). *Publ. wiss. Filmen, Sektion Biol.* **10**, no. 34, 3–24.
Wolf, R. (1978). The cytaster, a colchicine-sensitive migration organelle of cleaving nuclei in an insect egg. *Develop. Biol.* **62**, 464–72.
Wolf, R. & Krause, G. (1971). Die Ooplasmabewegungen wärend der Furchung von *Pimpla turionellae* L. (Hymenoptera), eine Zeitrafferfilmanalyse. *Wilhelm Roux' Arch. Entwicklungsmech. Organismen* **167**, 266–87.
Wolff, E. & Dubois, F. (1948). Sur la migration des cellules de régénération chez les Planaires. *Rev. Suisse Zool.* **55**, 218–27.
Wolff, E. & Lender, T. (1962*). Les néoblastes et les phénomènes d'induction et d'inhibition dans la régénération des planaires. *Ann. Biol.* **1**, 499–529.
Wolff, E., Lender, T. & Ziller-Sengel, C. (1964*). Le rôle de facteurs auto-inhibiteurs dans la régénération des planaires. (Une interprétation nouvelle de la théorie des gradients physiologiques de Child.) *Rev. Suisse Zool.* **71**, 75–98.
Woodhead, A. E. (1950). Germ cell cycle in the trematode family Brachylaemidae. *J. Parasitol.* **36**, section 2, suppl. 28.
Woodhead, A. E. (1954). Bisexual reproduction in the mother sporocyst of *Paragonimus kellicotti* (Trematoda). *Trans. Amer. Microsc. Soc.* **73**, 16–28.
Woodhead, A. E. (1955). The germ cell cycle in the trematode family Brachylaemidae. *Trans. Amer. Microsc. Soc.* **74**, 28–33.
Woodhead, A. E. (1957). Germ cell development in the first and second generations of *Schistosomatium douthitti* (Cort, 1914) Price, 1931 (Trematoda: Schistosomatidae). *Trans. Amer. Microsc. Soc.* **76**, 173–6.
Woodland, J. T. (1957). A contribution to our knowledge of lepismatid development. *J. Morphol.* **101**, 523–77.

Woodruff, L. S. & Burnett, A. L. (1965). The origin of the blastemal cells in *Dugesia tigrina*. *Exp. Cell. Res.* **38**, 295–305.
Woods, F. H. (1931). History of the germ cells in *Sphaerium striatinum* (Lam.) *J. Morphol. Physiol.* **51**, 545–95.
Woods, F. H. (1932). Keimbahn determinants and continuity of the germ cells in *Sphaerium striatinum* (Lam.). *J. Morphol.* **53**, 345–65.
Yajima, H. (1960). Studies on embryonic determination of the Harlequin-fly, *Chironomus dorsalis*. I. Effects of centrifugation and of its combination with constriction and puncturing. *J. Embryol. Exp. Morphol.* **8**, 198–215.
Yajima, H. (1964). Studies on embryonic determination of the Harlequin-fly, *Chironomus dorsalis*. II. Effects of partial irradiation of the egg by ultraviolet light. *J. Embryol. Exp. Morphol.* **12**, 89–100.
Yang, C. M. (1977). The egg development of *Paracalanus crassirostris* Dahl, 1894 (Copepoda, Calanoida). *Crustaceana* **33**, 33–8.
Yao, T. (1949). Cytochemical studies on the embryonic development of *Drosophila melanogaster*. I. Protein sulphydryl groups and nucleic acids. *Quart. J. Microsc. Sci.* **90**, 401–9.
Yao, T. (1950). Cytochemical studies on the embryonic development of *Drosophila melanogaster*. II. Alkaline and acid phosphatases. *Quart. J. Microsc. Sci.* **91**, 79–88.
Zaffagnini, F. & Lucchi, M. L. (1970). Osservazioni ultrastrutturali sul determinante germinale dei Dafnidi. *Rend. Accad. Naz. Lincei, ser. 8* **49**, 141–6.
Zalokar, M. (1971). Transplantation of nuclei in *Drosophila melanogaster*. *Proc. Nat. Acad. Sci. USA* **68**, 1539–41.
Zalokar, M. (1973). Transplantation of nuclei into the polar plasm of *Drosophila* eggs. *Develop. Biol.* **32**, 189–93.
Zalokar, M. (1976). Autoradiographic study of protein and RNA formation during early development of *Drosophila* eggs. *Develop. Biol.* **49**, 425–37.
Zalokar, M., Erk, I. & Santamaria, P. (1980). Distribution of ring-X chromosomes in the blastoderm of gynandromorphic *D. melanogaster*. *Cell* **19**, 133–41.
Zilch, R. (1978). Embryologische Untersuchungen an der holoblastischen Ontogenese von *Penaeus trisulcatus* Leach. *Zoomorphol.* **90**, 67–100.
Zissler, D. & Sander, K. (1973). The cytoplasmic architecture of the egg cell of *Smittia* sp. (Diptera, Chironomidae). I. Anterior and posterior polar regions. *Wilhelm Roux' Arch. Entwicklungsmech. Organismen* **172**, 175–86.
Zissler, D. & Sander, K. (1977). The cytoplasmic architecture of the egg cell of *Smittia* sp. (Diptera, Chironomidae). II. Periplasm and yolk endoplasm. *Wilhelm Roux' Arch. Develop. Biol.* **183**, 233–48.
Zur Strassen, O. (1896). Embryonalentwicklung der *Ascaris megalocephala*. *Arch. Entwicklungsmech. Organismen* **3**, 26–105.
Zur Strassen, O. (1898). Über die Riesenbildung bei *Ascaris* Eiern. *Arch. Entwicklungsmech. Organismen* **7**, 642–76.
Zur Strassen, O. (1903/6). Die Geschichte der T-Riesen von *Ascaris megalocephala* als Grundlage zu einer Entwicklungsmechanik dieser Spezies. *Zoologica* **17**, 1–342.
Zur Strassen, O. (1959). Neue Beiträge zur Entwicklungsmechanik der Nematoden. *Zoologica* **38**, 1–142.
Zwilling, E. (1963). Formation of endoderm from ectoderm in *Cordylophora*. *Biol. Bull.* **124**, 368–78.

Author index

References to figures are in bold type.

Aboim, 135, 145
Achtelig, 138
Agrell, 125
Alléaume, 124, 145
Allis, 135 (Underwood et al.), 136
    (Mahowald et al.), 138 (Mahowald et
    al.), 139, 187
Ameel, 48, 49
Amy, 122
Anderson, 69, 70, **70**, 71, **72**, 73, 74, 89,
    104, 105, 106, **106**, 108, 109, **110**,
    111, 112, **114**, 115, 116, 120, **121**,
    122, 123, 127, 128, **134**, 135, 147,
    148, 149, **149**, 150, 151, **151**, 152,
    154, 155, **156**
Ando, 122
André, 75
Arnold, 98, **99**, 100, 101, 103
Astaurov, 144
Atkinson, 95
Austin, 163

Balbiani, 129
Banchetti, 40, 46
Bantock, 142, 143
Baretta-Bekker, 88
Barnes, 5
Bauer, 140, 148
Bautz, 38
Bayreuther, 140
Bazitov, 52
Beams, **56**, 59, 61, 135, 143, 152
Bednarz, 48, 49, 50
Beeman, 87, 102, 103, 122
Beermann, 2, 140, 143, 148, 152, **153**
Belousov, 25
Benazzi, 40, 45
Bender, 130, 133
Bennett, 38, 133, 147
Berg, 88, 92, 95
Bergquist, 16
Berrill, 10, **11**, 13, **23**, 29, 30, 34, 35, 42,
    67, 68, 162, 178
Berry, 87, 102, 103, 148
Bertzbach, 124

Besse, 102
Best, 38, 39, 42
Betchaku, 39, 41
Bezem, 69, 88
Bier, 147
Blair, 97
Bock, 122, 124, 125, 126
Bode, 26, 124 (Moser et al.)
Bodine, 110
Bodo, 27
Boelsterli, 30, 31
Boilly, 67
Boletzky, 97, 100
Bolla, 52
Borradaile, 66
Bouillon, 26, 27, 28, 29, **30**, 31
Bounoure, xi, 1, 7
Boveri, 55, **57**, 59, 60
Bowers, 50
Braem, 63
Brandhorst, 82
Bregman, 143
Brien, **15**, 16, 20, 23, 24, 25, 26, 27, 30, 31,
    31, 32, 40, **63**, 64, **65**, 145, 152, 166,
    166, 178, 181
Brisson, 102
Bruslé, 83, 84
Brooks, **48**, 49
Bronskill, 122
Brøndstedt, A., 38
Brøndstedt, H. V., 38, 39
Brown, 133, 147
Buchner, 120, 145
Burnett, **22**, 24, 26, 27, 32, 39, 40

Cable, 49
Cambar, 1
Camenzind, 142, 143
Campbell, 21, 23, 24, 26, 27, 34, 35
Cantwell, 122
Cather, 69, 71, 88, 89, 91, 92, **93**, 95, 96,
    98
Caulton, 135 (Underwood et al.), 138
Cavallin, 128
Chalboner, 26

Chandebois, 38, 40, 42
Chapron, 74
Charniaux-Cotton, 135, 140, 147, 152
Chen, 139
Cheng, 49
Cherdantzev, 55, 58, **58**
Chia, 35, 81
Chitwood, 55, 56, 58
Chrétien, 64
Chuang, 55
Ciordia, 48
Clark, M. E., 67
Clark, R. E., 67, 68
Clement, 88, 91, 93, 94, 95
Connes, 15, 18
Conrad, 91, 92
Collatz, 124 (Moser et al.)
Collier, 92, 95
Cornman, 136
Cort, 48, 49
Counce, 117, 123, 124, 125, 126, 131, 135, 136, 137, 139, 142, 147
Coward, 38, 39, 40, **41**, 42, 185
Cowden, 28, 30, 34
Crawford, 35, 81
Cremer, C., 138 (Lohs-Schardin et al.)
Cremer, T., 138 (Lohs-Schardin et al.)
Cristini, 137 (Micali et al.)
Christophers, 122
Crofts, 88
Crouse, 143, 148
Crowell, 91, 92
Curtis, 34

Damian, 163
Dan-Sohkawa, 81
Daniels, 133
Darden, 13
D'Asaro, 88
Dautert, 88
Davant, 75
Davenport, 79, 81, 83
David, 26
Davidson, 2, 92, 173
Davis, 26, 122, 125, 147
Dawydoff, 162, 168
Delavault, 39 (Le Moigne et al.), 83, 84
De Leo, 98
De Lotto, 147
De Petrocellis, 79 (Parisi et al.)
De Vos, 15
Devriès, 72, 74
Diaz, 14, 18
Diehl, 24, 26, 27, 32
Dohle, 110, 111
Dohmen, 89, 91, 92, **93**, **94**, 96, 103
Dorsett, 73

Dubois, 38, 140, 142, **142**, 148
Duboscq, 15, 16, 17, **17**, 18, 20
Dudley, 42, 185
Dunn, 48, 49
Dupont, 29, 30
Du Praw, 123

Eckhardt, 143
Eddy, 133, 169
Ede, 124, 125, 135
Ehrman, 133
Elbers, 95
Erk, 138
Euzet, 51, 52
Efimova, 40
Evans, 68

Falk, 122
Faulkner, 64, 67, 73
Fedecka-Bruner, 45
Feeman, 159
Feldhege, 124 (Moser et al.)
Fennhoff, 28, 30
Fielding, 136, 137
Filosa, 79 (Parisi et al.)
Fincher, 18
Fioroni, 87, 89, 97, 150
Flickinger, 38
Formigoni, 122
Föyn, 24, 29
Franzén, 64, 65
Fretter, 103
Fullilove, 125
Fux, 142

Gabriel, 40, 41
Gans, 137
Garaudy-Tamarelle, 115
Garcia-Bellido, 125
Garen, 133
Garrone, 18
Gates, 74
Gehring, 4, 136, 137, 138, 139, 146
Geigy, 136, 145
Geilenkirchen, 88, 92, 93, 95
Geyer-Duszynska, 142, 143
Ghirardelli, 46, 169, **171**
Ghose, 88, 102
Gilbert, 16
Giudice, 79, 81, 83
Glätzer, 28, 31
Goldschmidt, 59, 163
Gontcharoff, 161
Goss, 120
Gottesmann, 97
Gould-Somero, 168
Graham, 103

Grassé, 163, **164**
Graziozi, 137
Green, 16
Gremigni, 40, 41, 45, 46
Grifford, 102, 103
Guerrier, 56, 59, 60, 71, 88, 92, 93, 95, 96, 97
Guilford, 48, 49
Günther, 138
Gurdon, 2
Gustafsson, 53, **53**, 54, 79
Guyard, 102
Guyomarc'h-Cousin, 102

Hagan, 148
Haget, 122, 124, 130, 145, 146
Hamilton, 27
Hämmerling, 68
Hand, 39
Hathaway, 135, 137, 145
Hauschteck, 142, 144
Haven, 98
Hay, 40, **41**, 42, 185
Haynes, 26, 28
Hazlehurst, 38
Hegner, 4, 136
Heinig, 124
Heming, 135
Herlant-Meewis, 67, 68, 69, 73, 74
Herlin-Houtteville, 101
Hermann, 162
Hertwig, xii
Herzfeld, 124 (Moser et al.)
Hess, 96, 97
Heymans, **111**, 115
Hill, 67, 68
Hirsh, 39, 57
Hissani, 138
Hogg, 101
Holland, 79
Holliger, 138, 148
Hope, 55, 59
Hörstadius, 79, **80**, 81, 82, **82**, 83, 161, **161**
Howland, 125
Huet, 78, 84, **84**
Hutchins, 82
Huysman, 62
Hyman, 5, 10, 14, 21, 37, 47, 51, 55, 62, 77, 78, 81, 87, 158, 159, 163, 166, 169

Idris, 122
Ihle, 66
Illmensee, 2, 125, 126, 130, 136
Ivanova-Kasas, 108, 128, 129, 150
Iwanoff, 67, 73, 74
Izumi, 55

Jacobson, 125
Jägersten, 168
James, 49, 50
Janet, **11**
Janning, 138
Jazdowska-Zogrodzińska, 139, 143, 185
Jennison, 35
Josefsson, 82
Jung, 122, 124
Jura, 113, 115, 129, 135, 137, 145

Kalthoff, 122, 133, 184
Kanai, 32, **33**, 36, 42, 184
Kanellis, 120
Karrer, 135 (Mahowald et al.), 136 (Mahowald et al.), 138 (Mahowald et al.)
Karstner, 5
Kato, 44, 92
Kautzsch, 155
Kenk, 44
Kessel, 61, 122, 135, 143, 152
Kille, 84
Kimble, 57
King, **56**, 59, 60, 136
Klag, 115, **116**
Kleinman, 126, 137
Klima, 40
Kochert, 13
Krause, 118, 120, 121, 124, 125, 126, 129, 135, 138
Krause, J., 124
Krichinskaya, 40
Kunz, 143, 144
Kuwana, 122

Lallier, 83
Lameere, 163
Lapkalo, 52
Larsen, 97
Lattaud, 75
Laugé, 135, 145
Lawrence, 125
Lender, 38, 39, 40, 41, 42, 45, 46, 83, 84, 138 (Lohs-Schardin et al.)
Lentz, 24, **25**, 26
Lemaire, 100, 103
Le Moigne, 39, 40, 41, 42, **43**, 44, 45, 185
Levin, 137, 145
Lin, 59, 163
Liu, 29, 35
Lohs-Schardin, 138
Lok, 91, 92, **94**, 103
Loomis, 136
Loosli, 168
Lubet, 101

Lucchi, 152
Luchtel, 102, 103

Mackiewicz, 52
Mahowald, 122, 126, 130, 131, **131**, **132**, 133, 135, 136, 138, 139, 144, 187
Mahr, 120, 124
Malokhov, 55, 58, **58**
Malaquin, 73
Mangold-Wirz, 97
Manton, 105, 106, 108, 109, 152
Marcum, 24, 26, 27
Maresquelle, xii
Mariscal, 166, 167, 168
Marthier, 101
Marthy, 100, 101, 103
Martin, 28
Marzali, 137 (Micali *et al.*)
Masson, 137
Matuszewski, 143
Maul, 123
McConnaughey, 162, 163
McLaren, 55
Meewis, 68
Melander, 44
Meng, 130
Menzl, 122
Mergner, 28, 30
Metschnikoff, 129
Metz, 140, 142, 147
Meyer, 73
Micali, 137
Micelli, 46
Miller, 125, 126
Miya, 127, 138, 146
Mokhtar-Maamouri, 51, 52
Monin, 56, 59, 60
Monroy, 79 (Parisi *et al.*)
Moor, 88
Morata, 125
Morgenthaler, 24
Morrill, 97
Morita, 39, 40, 42
Moritz, 56, 59, 60, 155
Moser, 124
Müller, 27, 32
Mumaw, 26 (Davis *et al.*)
Murphy, 26

Nachtwey, 164
Naef, 100
Nagao, 28
Nappi, 122
Nelson, 127
Neumann, 34
Nicklas, 140, **141**, 142, 143, 144, 147
Nielson, 168

Nierstrasz, 66
Nieuwkoop, xi, 4, 7, 172, 176, 178, 184, 186
Nigon, 55, 56, 59, 60
Noda, 32, **33**, 36, 42, 184
Noel, 42
Norris, 122
Nöthiger, 138
Nouvel, 162, 163
Nünemann, 124 (Moser *et al.*)
Nyholm, 35

Oelhafen, 122
Ogren, 52
Okada, 67, 69, 71, 74, 126, 137
Okugawa, 44
Olivier, 48, 49
Orevi, 122
Orsi, 16, 18
Ortolani, 159
Ostriakova-Varshaver, 144
Otto, 27

Pai, 55, 56, 58
Painter, 143
Panelius, 144
Panijel, 55
Paris, 18
Parisi, 79
Pasteels, 55, 59, 61
Pavans de Ceccatty, 18
Pease, 95
Pederson, 40, 41
Penners, 72, 73
Pettrera, 88, 95
Pfreundt, 138
Philiptschenko, 115
Pianka, 159
Picano, 46
Pieper, 48, 49
Pocock, **12**
Porchet, 74
Porchet-Hennere, 74
Potts, 66
Poulsen, 135, 137, 145
Price, 184
Prudhommeau, 135, 145
Puccinelli, 40, 46

Rabinowitz, 122, 135
Rattenbury, 88, 95
Raven, xii, 69, 88, 89, 90, 91, 92, 93, 95, 96, 100
Rees, 48, 49
Reggiani, 73
Regondaud, 102
Reinhardt, 147

Reith, 122
Reitberger, 142, 143, 147
Relexans, 74, 75
Renniers-Decoen, 23, 25, 26, 27, 31
Rice, 133
Richard, 50, 103
Richard-Mercier, 145
Richards, 125, 126
Rieffel, 143, 148
Roberts, 52
Romashkina, 137 (Levin et al.), 145 (Levin et al.)
Rosenvold, 38 (Best et al.), 39
Royer, 133
Ruffing, 26
Runnström, 82
Rybicka, 52

Sabbadin, 178
Sacarrão, 97
Sacks, 165
Sander, **118**, 119, 120, 122, 123, 124, 125, 130, 133, 135, 184
Santamaria, 126, 138
Sarà, 16, 18
Satoh, 81
Sauzin, 39 (Le Moigne et al.), 40, 41, 42
Schäller, 48, 49
Schleip, 92
Schmidt, 161
Schneiderman, 126, 137
Schnetter, 123, 125, 126, 145
Schubiger, 126
Schulze, **17**
Schüpbach, 138
Schwalm, 124, 130, 133
Seck, 58
Seidel, 120, 124, 125, 126, 127
Seiler, 144
Seilern-Aspang, 44
Selman, 135, 136, 145, 147
Sharov, 115
Shelomova, 137 (Levin et al.), 145 (Levin et al.)
Short, 163
Shortino, 122
Shvarzman, 137 (Levin et al.), 145 (Levin et al.)
Siewing, 159, **161**
Silén, 64
Simpson, 16, 130, 132
Smith, 135
Sonders, 38 (Best et al.)
Sonnenblick, 129, 133, 135
Spiegelman, 42, 185
Stäblein, 73
Stagni, 68, 73, 74

Starr, 13
Stefani, 120, 127, 144
Steinberg, 34
Stéphan-Dubois, 39, 44, 45, 68, 73, 74
Stoffolano, 122
Stolte, 15, 16, 24, 37, 42, 63, 67, 68, 73, 74, 162
Stunkard, 162
Sulgostowska, 53
Summers, 28
Sutasurya, xi, 4, 7, 172, 176, 178, 184, 186
Swiderski, 52
Szabad, 138

Tadano, 55, 56
Takami, 122
Tanaka, 120
Tardent, 21, 22, 24, 26, 32, 34, 35
Tardy, 101
Tawfik, 129
Taylor, 39, 55, 56
Terpitowska, 152
Thiemann, 138
Thierry-Mieg, 137
Thomas, 28
Tiegs, 108, 110
Timmermans, 92, 95
Tobler, **57**, 61
Truckenbrodt, 120, 144
Turner, 122, 136, 138
Tuzet, 15, 16, 17, **17**, 18, **19**, 20

Uchida, 21, 34, 35, 159
Ullmann, 122, 130
Underwood, 135, 136 (Mahowald et al.), 138 (Mahowald et al.)

Van Beneden, 162
Van Dongen, 88, 93, 92 (Guerrier et al.)
Van de Vijver, 28
Van den Biggelaar, 88, 89, **90**, 92 (Guerrier et al.), 93, 96, 97
Van der Mast, 89, 96
Van der Mey, 91
Van der Wal, 95
Van der Woude, 48, 49
Vandel, 42, 45
Vannini, 32, 46, 73, 74
Verdonk, 88, 89, 90, 91, **91**, 92, **93**, 95, 96, 103
Vignau, 128
Villa, 70
Vollmar, 124, 125
Von Borstel, 122, 144
Von Ubisch, 61

Wada, 88

Author index

Wade, 38 (Best et al.)
Waring, 135 (Mahowald et al.), 136 (Mahowald et al.), 138 (Mahowald et al.), 139, 187
Warn, 138
Waterhouse, 135, 137, 145
Webber, 102, 103
Weber, 71
Webster, 22, 27
Weiler-Stolt, 28, 32
Weismann, 1, 29, 35
Wells, J., 98, 103
Wells, M. J., 98, 103
West, 122
Weygoldt, 69, 71, 105, 150
White, 143, 144
Wieschaus, 138, 146
Wijdenes, 101
Wilby, 22
Wildermuth, **146**
Williams, 91, 135
Williams-Arnold, 98, 100, 103

Wolf, 122, 123, 125, 129, 139, 140, 142, 185
Wolff, 38, 39, 42, 46
Woodhead, 48, 50
Woodland, 2, 114, 115
Woodruff, 39, 40
Woods, 101

Yajima, 184
Yamada, 21, 34, 35, 159
Yang, 150
Yao, 125

Zaffagnini, 152
Zalokar, 125, 126, 138
Zaniolo, 178
Zilch, 150
Ziller-Sengel, 42, 46
Zissler, 122, 130
Zorn, 138 (Lohs-Schardin et al.)
Zur Strassen, 55, 59
Zwilling, 24

Taxonomic index

Reference to figures are in bold type.

Acanthobothrium, 51
Acanthocephala, 5, 158
Acarina, 154, 155
Acaulis, 30
Achatina, 88
Achatina fulica, 102
Acheta, 124
Acoela, 37, 44
Acoelomata, 5
Acroloxus, 102
Aëdes, 122
Aeolidiidae, 101
Aeolosoma hemphrichii, 68
Alcyonidium, 64
Amphipoda, 151
Anaspides, 152
Anguillula aceti, 55
Annelida, 6, 66, 106, 108
Anopleura, 116, 148
Anostraca, 150
Anthomedusae, 27
Anthozoa, 21, 35
Anurida maritima, 115
Apanteles, 129
Apiomorpha, 120
Apis, 122, 124, 125
Apis mellifera, 147
Aplacophora, 86, 89
Aplysella rosea, 18
Aplysia, 88
Apterygota, 6, 112, 113
Arachnoidea, 152
Archaeogastropoda, 89
Archiannelida, 66
Arenicola, 70
Argonauta, 103
Arion, 102, 103
Artemia, 150
Arthropoda, 6, 104, 107, 108, 109, 148, 155, 157
Ascaris, 55, 56, 140, 152
Ascaris equorum, 55
Ascaris equorum univalens, 56
Ascaris megalocephala, 57
Aselomaris, 30

Aspidiotus simulans, 147
Aspidiotus destructor, 147
Asplanchna, 164
Asterina gibbosa, 78, 81, 83, 84
Asteroidea, 76, 77

Barentsia discreta, 167
Beroe ovata, **161**
Bilateralia, 5, 37, 49, 50, 55, 66, 158, 162, 165, 168
Bithynia, 88, 91, 95, 96, 97
Bithynia tentaculata, **93**
Blatella, 120
Blattodea, 148
Bombyx, 122
Bombyx mori, 138, 144, 146
Bonella viridis, 168
Bongainvillia, 30
Brachiopoda, 6, 158
Bradybaena, 88
Branchiomma nigromaculata, 68
Branchiopoda, 151
Branchiura, 137
Bruchidius, 122, 124
Bryozoa, 62, **63**
Buccinum, 91, 92, 103
Bursa, 88

Calcarea, 14, 15, 16, 18
Calliphora, 124, 125, 145
Camallanus, 55
Campanularia, 21, 22, 24, 25, 28
Camponotus, 122
Capitata, 28
Carausius morosus, 128
Cassiopeia, 34
Caulobothrium, 51
Cecidomyidae, 140, 143, 144, 174
Cephalopoda, 87, 97, 103
Cercaria bucephalopsis haimaena, 50
Cerebratulus, 161
Cerebratulus lacteus, **161**
Cestoda, 5, 50, 51, 178, 183
Cestodaria, 51
Chaetogaster diaphanus, 68

Taxonomic index

Chaetognatha, 6, 168, 169, 171
Chaetopterus variopedatus, 68
Cheleutoptera, 116, 118, 119
Chelicerata, 6, 108, 152, 154, 155, 157
Chilopoda, 109, 110, 111
Chironomidae, 140, 174
Chironomus, 140, 184
Ciliata, 9, 10
Cirripedia, 150
Chordata, 6, 178
Chrysopa, 122, 124
Cladonema, 24, 28
Cladocera, 150
Calva, 30
Clava squamata, 24, 29
Cnidaria, 5, 21
Coelenterata, 5, 21, 32, 35, 158
Coelopa frigida, 130, 133
Coenorhabditis elegans, 57
Coleoptera, 116, 120, 127, 128, 129, 136, 148, 157
Collembola, 112, 113, 156
Copepoda, 150, 151
Cordylophora, 24, 26
Corrodentia, 148
Corydendrium, 28
Corydendrium parasiticum, 31
Crepidostomum cornutum, 49
Crepidula, 91, 92, 103
Crepidula formicata, 94
Crinoidea, 76, 77
Crustacea, 6, 108, 148
Ctenolepisma, 115
Ctenophora, 5, 158, 171
Culex, 122, 147
Cyclops, 150, 151, 152, 157, 174
Cyclops strenuus, 152, **153**
C. divulsus, 152
C. furcifer, 152
Cyprideis, 150

Dacus, 121, 122
Dacus tryoni, **134**
Daphnia, 157
D. magna, 152
D. pulex, 152
Dendrocoelium lacteum, 38, 39
Dentalium, 88, 91, 92, 95
Dermaptera, 116, 117, 118, 127, 148
Dermestes, 124
Deroceras, 101, 102, 103
Desmospongia, 14, 18
Dictyoptera, 116, 117, 118, 119
Dicyema aegira, 163
Digenea, 47, 49
Dipetalonema, 55
Diphyllobothrium, 53

Diphyllobothrium dendriticum, 53
Diplopoda, 109, 110
Diplura, 112, 113
Diptera, 116, 120, 121, 122, 123, 125, 128, 129, 136, 140, 148, 157, 182, 184
Dirofilaria, 55
Distorsio, 88
Drosophila, 122, 125, 126, 130, 131, 133, 136, 137, 140, 145, 146, 147, 182, 184, 185
Drosophila hydei, **132**
Drosophila melanogaster, 126, 129, 130, 131, 135, 137, 138, **146**
Drosophila paulistorum, 133
Drosophila subobscura, 136
Drosophila virilis, 129, 137
Dugesia, 37, 39, 45, 46
Dugesia dorotocephala, 39, 41
Dugesia gonocephala, 40, 44
Dugesia lugubris, 38, 40, 46
Dugesia tigrina, 40
Dynamena, 26
Dytiscus, 140

Echinodermata, 6, 76
Echinoidea, 76, 77
Echiurida, 6, 168, 171
Ectoprocta, 6, 62, 65, 178
Eisenia, 72, 74, 75
Eisenia foetida, 75
Eleutheria, 24, 28, 32
Eleutherozoa, 76
Embioptera, 116, 118, 119, 127
Entoprocta, 5, 165, 168, 171, 178
Ephemeroptera, 116, 118, 119
Ephydatia fluviatilis, 15
Eucoelomata, 6
Eudendrium, 28, 30
Eudendrium racemosum, 30
Euplanaria, 39
Eurytoma aceculata, 129
Euscelis, 120
Euscelis plebejus, 133

Fasciola hepatica, 49
Filifera, 28
Filograna implexa, 67, 73
Flagellata, 9, 10
Formica rufa rufa-pratensis minor, 147

Gastropoda, 86, 88, 89
Gastrotricha, 5, 165, 171
Glomeris, 110
Grantia, 17, 18
Gryllus, 120, 124
Gymnolemata, 62, 64

Habrobracon, 122

Taxonomic index 241

Haliotus, 88
Hanseniella, 110
Haploembia solieri, 144
Haplotrips, 135
Helix aspersa, 102
Hemichordata, 6, 158
Hemimetabola, 116, 117, 118, 120, 122, 123, 125, 127, 156
Hemimysis, 152
Hemiptera, 148
Heteropeza, 144
Heteropeza pygmaea, 142, 143, 144
Heteroptera, 116, 119, 127
Hexacorallia, 35
Hexapoda, 6, 107, 108, 112, 155, 156
Hirudinea, 66, 69, 71, 74
Holometabola, 116, 120, 121, 122, 123, 124, 125, 127, 157
Holopedium, 151
Holothuria parvula, 84
Holothuroidea, 76, 77
Homoptera, 116, 127, 138, 148
Hyalospongia, 14
Hydra, 21, 22, 24, 25, 26, 27, 29, 31, 32, 185, 186, 187
H. attenuata, 24, 26
H. fusca, 32
H. littoralis, 32
H. pirardi, 32
H. viridis, 26, 32
Hydractinia, 27
Hydrozoa, 21, 22
Hymenolepis diminuta, 53
Hymenoptera, 116, 120, 121, 122, 125, 127, 128, 129, 136, 148, 157

Icerya purchasi, 133
Ilyanassa, 88, 91, 92, 93, 95, 96
Insecta, 112
Isopoda, 151
Isoptera, 116, 119
Isotoma cinerea, 115

Kalothermes, 120
Kinorhyncha, 5, 158

Lampyris noctiluca, 145
Lepidoptera, 116, 122, 125, 127, 138, 148
Lepisma, 114, 115
Leptinotarsa, 122, 123, 124, 126, 130, 136, 145, 146
Leptinotarsa decemlineata, 126
Leptomedusae, 27
Limax, 88, 95
Limax maximus, 92
Limnaea, 88, 89, 96, 100
Limnaea stagnalis, 69, 88, 89, 91, 95, 97, 102

Limnaea palustris, 97
Limnocnida, 26, 27, 28, 31
Limnodrilus, 73, 74
Limnomedusae, 26, 27, **30**
Limulus, 154
Lineus, 161
Liposcelis, 120
Litomosoides, 55
Littorina saxatilis, 102
Loligo, 98, 100
Loligo pealii, **99**, 100
Loxosomatidae, 166
Loxomella, 168
Lucilia, 122, 136, 137, 145, 147
Lumbricullus, 74

Machilis, 115
Malocostraca, 150, 151, 152
Mallophaga, 116
Mayetiola destructor, 142, 143
Mecoptera, 116, 128
Megaloptera, 116, 127
Mesoleius, 122
Mesozoa, 5, 162, 163, 171
Metazoa, 5, 9
Miastor, 130, 131, **141**, 142, 143, 144, 147
Microcyema vespa, 164
Modiolaria, 91
Mollusca, 6, 86, 97
Monogenea, 47
Monoplacophora, 86
Musca, 122
Mycetophilidae, 140, 174
Mycophila speyeri, 140
Myriapoda, 6, 104, 107, 108, 109, 111, 122, 155
Mytilus, 88, 91, 95

Naidiformes, 74
Narceus, 110
Nassarius, 91
Nautiloidea, 87, 97
Nebalia, 152
Nematoda, 5, 55
Nematomorpha, 5, 158
Nemertini, 5, 159, 171
Nephthys, 67
Nereis, 70
Nereis diversicolor, 68
Neuroptera, 116, 125, 128

Obelia, 26
Ocinebra, 91
Octocorallia, 35
Octopoda, 98
Octopus, 98, 100
Octopus vulgaris, 103

242 Taxonomic index

Odonata, 116, 117, 118, 119, 125
Oligarces paradoxus, 142, 147
Oligochaeta, 66, 71, **72**, 73
Onychophora, 6, 104, 106, 108, 110, 122, 155, 156
Opercularella, 26
Ophiuroidea, 76, 77
Opisthobranchia, 86
Orthocladiinae, 148
Orthoptera, 116, 117, 118, 119, 125, 127
Ostracoda, 150, 151
Ostria, 91

Palaeoptera, 116
Paludina, 88
Panorpa, 122, 125
Paracentrotus, **80**
Paraneoptera, 116, 117, 119
Parascaris, 55
Parazoa, 5
Patella, 88
Patella vulgata, 90, 96
Pauropoda, 109, 110, 112
Pauropus, 109, 110
Pecten, 91
Pedicellidae, 166
Pedicellina, 167
Pedicellina cornua, **166**
Pelecypoda, 87, 89
Peloscolex, 73
Pelmatohydra robusta, 32, 33
Pelmatozoa, 76
Pennaria, 22, 28, 30
Pennaria tiarella, 23, 30
Pentastomida, 6, 158
Perinereis cultrifera, 75
Perionyx, 74
Peripatopsis, 105, 106
Peripatopsis orientalis, 106
Pholas, 88, 92, 95
Phoronida, 6, 158
Phylactolemata, 62, 64
Phyllobothrium, 51
Phytomonadina, 9, 10
Pimpla, 122, 125, 138
Pimpla turionellae, 130
Planaria, 185
Planaria alpina, 42, 45
Planaria subtentaculata, 42, 45
Planaria vitta, 42, 45
Platycnemis, 120, 121, 124
Platyhelminthes, 5, 37, 178
Plecoptera, 116, 117, 148
Plumatella fungosa, **63**, 65
Podarke, **70**, 88
Podarke obscura, 69
Podocoryne carnea, 31

Pogonophora, 6, 158
Polycelis, 43, 45
Polycelis cornuta, 42, 45
Polycelis nigra, 41, 44, 45
Polychaeta, 66, 69, 73
Polycladida, 44
Polydora ciliata, 73
Polygordius, 70
Polyneoptera, 116
Polyphemus, 150, 151
Polyplacophora, 86, 88, 89
Pontonema vulgare, 56, 58
Porifera, 14
Priapulida, 6, 158
Protista, 5, 9, 13
Protostomia, 66
Prosobranchia, 86
Protura, 112
Psammocystus, 74
Pseudaulacaspis pentagona, 133, 143
Pseudocoelomata, 5
Psocoptera, 127
Pterygota, 6, 112, 116
Pulex, 140
Pulmonata, 86
Pyrilla, 120
Pyrilla perpusella, 133

Radiata, 5, 158, 171
Radiolaria, 10
Rathkea octopunctata, 30
Rhabditis, 55
Rhabdocoela, 44
Rhizopoda, 9, 10
Rotifera, 5, 163, 165, 171
Rynchocoela, 159

Sabellaria, 70, 71, 95
Salmacina dysteri, 73
Sarcodina, 9
Scaphopoda, 86, 89
Sciaridae, 140, 174
Sciara, 140, 142, 148
Sciara coprophila, 140, 142, 148
Sciara ocellaris, 148
Scolopendra sp., **111**
Scyphomedusae, 34
Scyphozoa, 21, 34, 35
Sepia, 103
Sepia officinalis, 100
Siphonoptera, 116, 128, 129, 157
Siphonophora, 22
Sipunculata, 158
Sipunculida, 6
Smittia, 122, 130, 184
Smittia parthenogenetica, 133
Solenobia triquetrella, 144

Spadella cephaloptera, 169, **170/1**
Sphaerium stricticum, 101
Spicula, 88, 92
Spirorbis, 68, 73, 74
Spirorbis pagenstecheri, 68
Spongilla lacustris, **15**, 16
Sporozoa, 9
Sternaspis, 70
Strepsiptera, 148
Strongylocentrotus, 82
Suberites domuncula, 15
Suberites massa, 18
Sycandra, **17**
Sycon, 17, 18
Sycon elegans, **17**
Sycon raphanus, **19**
Symphyla, 109, 100

Tachycines, 120, 121, 124
Tanaidacea, 151
Tardigrada, 6, 158
Tenebrio, 122
Tetrodontophora bielanensis, 113, 115
Thais, 88
Thermobia, 114, 115
Thermobia domestica, 115, **116**
Thysanura, 112, 113, 156
Thysanoptera, 116, 148
Tipula, 140
Tracheata, 109
Trematoda, 5, 46, 178

Trichoptera, 128
Tricladida, 45
Tubella pensylvanica, 16
Tubifex, 71, 73, 74
Tubifex tubifex, 68
Tubifex rivulorum, 72
Tubularia, 21, 22, 28, 29, 30
Tubularia crocea, 29
Turbatrix, 55, 58
Turbellaria, 5, 37, 39, 45, 46, 47, 178, 182

Urnatellidae, 166
Urosalpinx, 91

Viviparus, 102
Viviparus viviparus, 103
Volvocales, 9, 10
Volvox, 10, **11**, **12**, 16
Volvox aureus, **11**
Volvox rousseletti, 12
Volvox spermatosphaera, 12

Wachtliella, 122, 125, 139, 174, 185
Wachtliella persicariae, 130, 142, 143
Wyeomyia, 184

Xiphosura, 154, **156**
Xyleborus, 122

Zoothamnion, 10

Subject index

Entries dealing with major topics are in italics; entries which are met too frequently are marked as *not specified*.

abdomen (abdominal segments), 47, 67, 73, 74, 112–15, 118–22, 125, 127, 134, 135, 146, 153, 154 (post-), 157
actinomycin D, 157
acron, 67, 109
adult, *not specified*
agametes (agametic), 145, 162, 163
  see also gametes
aggregation (aggregates), 15, 18, 26, 42, 92, 94, 118, 133, 169
albumen sac, *see* yolk sac
albuminotrophic cells, 71, 72
alimentary tract, *see* digestive tract
alkaline phosphatase, 41, 129
alkylating agents, 27, 32
alternation (alternative development), 3, 9, 10, 13, 21, 29, 34, 49, 69, 88, 162, 175, 177, 180, 181
  see also switched into *and* reproduction/asexual and sexual
ambulacra (ambulacral grooves), 76, 77
ambulatory segments, 154, 156
amnion (amniotic), 82, 114, 115, 119, 121, 122, 155, 156
  cavity, 118, 119
  folds, 118
  pro-, 115
  –serosal space, 122
amoebocytes, 44, 73, 158
amoeboid activity/motility/movement, 14–16, 18, 23, 34, 40, 73, 123, 135
amphetamine, 46
amputation, 45, 68
  see also decapitation *and* transection
anaphase, 56, 59, 141, 153
anatrepsis, 119, 124
aneuploid, 40
androgenesis, 3
animal
  factor, 60
  pole, 16, 78, 79, 81, 89, 90, 96, 98, 154
  pole plasm, 59, 71, 89, 95
animalisation, 82
animalising agent, 82

antennae (antennal segment), 90, 104, 105, 109, 110–12, 148, 149, 151, 153
  pre-, 109, 110
  post-, 118, 119, 125
antennulae (antennular segment), 148, 149, 151, 153
anterior pole, 10, 12, 118–20, 126, 136–8
anus, 21, 55, 58, 63, 69, 70, 73, 76, 100, 104, 159, 163, 165–9
apical
  organ, 167
  tuft, 69, 70, 72, 79, 80, 82, 90, 95
apical–basal, *see* polarity/proximo-distal
apomictic parthenogenesis, 144
appendages (appendicular), 76–8, 87, 108, 109, 112, 114, 118, 119, 148, 151, 152
archaeocytes, 15, 16, 18–20, 175, 181
archenteron, 57, 79, 81, 82, 94, 96, 169
arms, 76–8, 80, 87, 98, 99
asexual form/generation/individual/state, 12, 13, 19–21, 29, 32, 44, 45, 49, 64, 68
  see also reproduction/asexual
association, 33, 41, 42, 59, 103, 106, 115, 128, 130–3, 137, 140, 151, 152, 174, 184, 187
  see also attachment
atrium (atrial cavity), 63, 165–7
attachment, 10, 22, 127, 128, 130, 131, 140, 163, 165, 167, 168
  see also association
autoradiography (autoradiographic), 135
autotomy, 67, 78
axial
  cells, 162, 163
    see also germinal cells
  organs, 83
axis, *see under* polarity
axoblasts, *see* axial cells
axocoel, 81

basal disc, 22–4
basophilic (basophilia), 24, 30, 38, 40, 52, 53, 120, 139, 142

'beta' cells, *see* neoblasts
bipotential, 46, 74, 75, 102
bisexual, 103, 144, 164
blastema, *see under* regeneration blastema
blastocoel (blastocoelic cavity), 64, 73, 79, 109, 154, 161
blastoderm, 100, 101, 105, 110, 113–15, 121, 122, 124–30, 173
   cellular, 117, 120, 121, 124, 126, 128, 130, 133, 154
   continuous, 129
   differentiated (differentiating), 117, 126, 127
   formation, 104, 126, 127, 133, 184
   pre-, 126, 136, 137
   pseudo-, 123
   syncytial, 117, 120, 121, 123, 124, 126, 128, 129, 133, 135, 138, 157
   uniform, 109, 114, 117, 120, 121, 123, 128
blastodisc, 98, 104
blastogenesis (blastogenetic), 27, 32, 64, 67, 68
blastokinesis, 113–15, 119, 121, 124, 127, 156
blastomeres, general, *not specified*
   animal, 79, 81, 83, 89, 161
   pyramidal, 109
   vegetal, 79, 81, 89, 161
blastopore (blastoporal), 18, 58, 64, 70, 161, 167, 169
   groove, 165
   lip, 73, 101
blastula, 17, 28, 69, 70, 72, 79, 80, 83, 88, 128, 149–51
   amphi-, 17
   coelo-, 27, 28, 34, 35, 64, 70, 128, 150, 154, 161, 167–9
   peri-, 113
   pseudo-, 64
   stereo-, 18, 27, 28, 70, 150, 161, 168
   stomo-, 16
   syncytial, 28
blood
   stream, 178
   vessels, 66, 68, 73, 159
body, *see* trunk
body cavity, *see* coelomic cavity
boundary, 98, 100, 138, 187
brain, *see* ganglia/cerebral
breakdown, 34, 69, 73, 130, 181
   see also differentiation/de-
bud formation, 3, 23–7, 29–32, 34, 35, 62–5, 68, 166, 167, 171, 178, 181
budding zone, 22, 30

calcium ion (Ca^{2+}), 83

calycocytes, *see* cuprophilic cells
calyx, 165–7
canal system, 14
carbon ^{14}C label, 38
caulus, *see* stalk
cauterisation, 138, 145
cell
   boundary, *see* boundary
   constancy, 55, 162, 163, 171
   division, general, *not specified*
      asynchronous, 98, 120
      equal, 16, 59, 92, 165
      synchronous, 13, 79, 81, 114, 117, 120
      unequal, 10, 13, 16, 48, 59, 92, 163
      see also cleavage *and* mitosis
   lineage, 45, 48, 57, 59, 60, 70, 87–9, 91, 98, 128, 164
   membrane, 97, 130, 131, 133, 175, 181, 188
   multiplication, 9, 25–7, 39, 40, 45, 49, 50, 52, 73, 102, 115, 118, 146, 162, 175
   -specific, *see under* specificity
   wall, *see* membrane
cellular inclusions, *see* organelles
'*cellules reproductrices*', *see* germ cells
centrifugation, 58, 60, 71, 82, 92, 94, 95, 138, 142, 147, 185
cephalic
   ganglia, *see* ganglia/cerebral
   influence, 68
   lobe, 110, 113, 114, 118, 121, 122, 134, 151, 154, 156, 159
cephalo-caudal, *see* polarity/cranio-caudal
cephalothorax, 152, 154
chambers, 14, 16, 19
cheliceres (cheliceral segment), 152, 154
chemotaxis, 175
chimaera (chimaeric), 24, 126
   sexual, 32, 126
choanocytes, 14–16, 18, 19, 185
chorion, 117, 120
chromatin, 4, 41, 43, 54, 56, 57, 59–61, 102, 140
   diminution, 4, 56, 57, 59–61, 140, 152, 157, 163, 171, 173
'chromatoid' bodies, *see* electron-dense bodies
'chromidium', 130
chromosome(s), 4, 9, 46, 56, 59, 60, 148, 152, 174
   E- or L-, 140, 143, 148, 174
   paternal, 147
   S-, 143
   X-, 138, 140, 147, 148
chromosome
   distribution, 148

246 Subject index

elimination, 4, 123, 139–44, 147, 148, 153, 157, 163, 173, 174, 181, 184, 185
pattern, 145
reduction, *see* elimination
cilia (ciliated), 10, 16, 34, 35, 47, 51, 62, 79, 90, 94, 158, 159, 161–3, 165, 167, 168
circulatory system, 37, 47, 55, 62, 66, 86, 87, 109, 112, 159, 168, 169
cleavage, general, *not specified*
  cycle, 113, 114, 120, 123, 129, 140–2, 148
  furrow, 98, 115
  pattern, 52, 56, 69–71, 81, 88, 96, 97, 108, 149, 150, 154, 159, 183
  plane, 56, 59, 71
  type
    bilateral, 149
    determinate, 44, 52, 58, 69, 71, 87, 88, 97, 159, 167, 171, 183
    dexiotropic, 69, 88
    discoidal, 87, 154
    equal, 16, 28, 59, 64, 71, 79, 88, 89, 91, 150, 169, 183
    equatorial, 56, 71, 79
    holoblastic, 16, 87
    intermediate, 150
    intralecithal, 35, 109, 113, 114, 117, 150, 154, 156
    irregular, 28
    laeotropic, 69, 88
    'Lepas', 150
    meridional, 56, 59, 69, 71, 79, 88
    meroblastic, 87, 98
    pseudo-, 123
    radial (radiate), 28, 56, 150
    spiral, 44, 56, 69–71, 87, 88, 106, 108, 149, 150, 155, 159, 164, 167, 168, 171, 183
    superficial, 87, 98, 113, 150
    tetrapolar, 60
    total, 27, 35, 47, 64, 71, 105, 113, 115, 128, 150, 154, 167
    transverse, *see* equatorial
clitellum, 66, 71
  see also gonadal region
closure (closing), 12, 58, 106, 119, 124, 167, 169
cnidoblasts, 21, 23–32, 34, 184
coelomic
  cavity, 62–4, 66, 75–7, 81, 82, 100, 104, 108, 118, 135, 152, 165 (pseudo-), 169
  complex, *see* peritoneal complex
  epithelium/wall/sac, *see* peritoneum
coelomoduct, 66, 154
colchicine, 26, 27, 92
colloidal carbon, 23

colonisation, 101, 126, 146
colony (colonial), 9–13, 21–3, 27, 36, 62, 64, 165, 174, 175, 178
compartments
  cortical/medullar, 102
  intranuclear, 143
complement (full) of developmental potentialities, 1, 2, 173, 176, 182, 186, 187
compression, 71, 92, 95
computer simulation, 69, 88
configuration (spatial), *see* pattern
connective tissue, *see* mesenchyme
constriction, 42, 78, 81, 84, 100, 142, 162, 165
  see also fission
contact between cells/layers, 44, 73, 89, 96, 100, 101, 187
contraction (contractility), 9, 10 (non-), 51, 79, 98, 124, 168
conversion (convertibility), 3, 3 (non-), 32, 104, 135
  see also transformation
copulation (copulatory), 46, 51, 66, 71, 112
cortex (cortical), 59, 60, 78, 92, 95–7, 100, 123, 138, 152
  granules, 78
  plasm, 115, 131
  sub-, 89, 91
cuprophilic cells, 135, 136, 176, 182
cuticle (cuticular), 10, 47, 50, 51, 55, 86, 104, 168
cystide, 62–5
cytochalasin B, 81, 92, 100
cytodiaeresis, *see* cytokinesis
cytokinesis, 35, 60, 150, 154
cytoplasm (cytoplasmic), general, *not specified*
  bodies, 102, 103
  composition, 2
  factors, 136
  inclusions, *see* organelles
  organelles, 9, 18, 24, 25, 41, 53, 56, 70, 92, 95, 124, 128, 138, 139, 187
  streaming, 123, 124, 135

de novo formation, 35, 39, 74, 129, 159, 162, 175, 177, 178, 181, 185
decapitation, 38, 39, 41, 42, 45, 74
  see also amputation *and* transection
defects, 100, 124, 125, 138
  see also elimination *and* removal
degeneration, 17, 27, 40, 49, 52, 58, 114, 128, 130, 135, 136, 140, 146, 148, 162, 164, 167
delamination, 27, 28, 35, 155
  see also gastrulation

Subject index

deletion, *see* destruction *and* elimination
'dense bodies', 92, 115
deoxyribonucleic acid (DNA), 41, 61, 130
  synthesis, 52, 95, 143
depression, *see* invagination
destruction, 16, 24, 26, 27, 40, 45, 72, 73, 84, 136–8, 145, 146
  see also elimination
detachment, 79, 139
detergents, 83
'*Determinanten*', 1, 173
determinate development, *see* embryonic development/mosaic
determination (determinancy), 10, 42, 44, 59, 60, 71, 75, 83, 88, 90, 95, 117, 136, 172, 173, 176, 182, 183
  pre-, 92, 128, 177
dictyosomes, 102
differential (division/distribution/fate/gene activation/transcription), 1, 2, 39, 54, 59, 60, 92, 135, 148, 152, 173, 174, 180, 181, 186
differentiation (differentiated)
  auto-, *see* self-
  cellular, general, *not specified*
  de-, 3, 26, 27, 32, 38–40, 45, 46, 52, 67–9, 78, 101, 162, 175, 177, 181, 182, 185, 186
  re-, 68, 181
  regional, *see* somatic differentiation
  self-, 75, 124
  un-, 9, 24, 29, 31, 32, 35, 38–46, 52–4, 62, 64, 75, 83, 90, 101, 103, 162, 168, 171, 174–7, 181, 182, 185
digestive canal/system/tract/tube, 21, 26, 37, 48, 55, 62, 63, 68, 69, 75–8, 85, 86, 104, 158, 159, 165, 166, 168, 169
dimorphism
  seasonal, 147
  sexual, 103, 147, 148, 164, 168
dioecious, 7, 12, 14, 32, 55, 66, 76, 77, 86, 87, 103, 104, 109, 112, 159, 165, 168
diploid, 13, 49, 50, 126, 144, 147, 148, 163
discs
  attachment, 167
  brain, 161
  cerebral, 161
  imaginal (including genital), 134
  proboscis, 161
disintegration, 38, 47, 59, 129
dispermy (dispermic), 60, 92
dispersion, 96, 102, 115, 135, 142
displacement, 23, 35, 56, 76, 77, 79, 89, 92, 130, 133–6, 138, 139, 142, 147, 148, 150, 152, 157, 169, 185
dissociation, 14, 24
distribution (distributed), 1, 2, 24, 32, 39, 46, 50, 54, 56, 59, 60, 71, 89, 102, 118, 123, 129, 131, 150, 152, 170–1, 173
dorsal organ, 113, 114, 119
dorsalisation, *see* symmetry/bilateral
'double gradient hypothesis', 82
  see also gradient
drones, 147
dwarf embryo, 82, 83, 100, 161

ectoderm (ectodermal) cells/layer, *not specified*
ecto-mesoderm/mesenchyme, 44, 62, 69, 70, 72, 158, 159, 161
ectopic pole cells/PGSs, 126, 136, 137, 139, 146, 147
'ectosomes', 89, 96, 152
egg (ovum), general, *not specified*
  membrane, 105, 175
  shell/membrane, 44, 51, 52, 56
  type
    centrolecithal, 113, 150
    'composite', 44
    isolecithal, 150
    mosaic, 125, 127
    regulative, 125, 127
    'simple', 44
    telolecithal, 150
electron-dense bodies/emissions lumps/material, 32, 33, 41–3, 46, 70, 92, 94, 130, 133 (rod-like), 140, 152
elimination, 4, 59, 61, 72, 74, 123, 139–44, 147, 148, 151, 174
elongation, 51, 70, 118, 122, 124, 127, 154, 155
EM pictures of germ cells and their precursors, 25, 33, 41, 43, 93, 94, 116, 131, 132, 170–1
embryogenesis, *see* embryonic development
embryonic
  anlage/area/primordium, *see* germ anlage
  cell, 129
  complex, 129
  covers, *see* extra-embryonic membranes
  state, 181
embryonic development
  systematic, *see* Contents, pp. vii–x
  type
    direct, 15, 18, 26, 29, 34, 64, 71, 72, 83, 89, 98, 118, 151, 154, 169
    epigenetic, 34, 59, 64, 71, 81, 83, 87, 96, 97, 172, 173, 180, 181, 183, 184
    epimorphic, 111, 117
    indirect, 16, 64, 69, 118, 161
    intra-uterine, 55
    mosaic, 59, 87, 96, 123, 161, 168, 183, 184

248 *Subject index*

post-, 55, 74, 112, 115, 127
preformistic, 61, 87, 97, 163, 180, 181, 183
regulative, 58, 59, 72, 81, 87, 97, 123, 138, 168, 183, 184
embryophore, 51, 167
endoderm (endodermal) cells/layers, *not specified*
endo-mesoderm, 44, 83, 98, 100, 159, 161, 167
endoplasmic reticulum/E.R., 24, 25, 41, 54, 71, 92, 102, 116
energids, 109, 113, 114, 117, 120, 123, 124, 126, 129, 137, 138, 147, 150
entocodon/bell nucleus, 29, 30, 35
environmental factors/conditions, 4, 13, 19, 31, 34, 35, 144, 175, 177, 183
epiboly, 18, 44, 47, 58, 70, 90, 119, 159, 160–1
epidermis (epidermal) cells/layers, *not specified*
epigenesis (epigenetic), *see under* embryonic development *and* germ cell formation (mode)
epithelio-muscular cells, 21, 24, 26, 32, 159, 185
epsilon granules, 130
  see also polar granules
equipotential, *see* isopotent
ergastoplasm, 41
esophagus, 62, 76, 77, 100, 104, 112
evagination, 29
evolutionary, *see* phylogenetic
excretory organs/tubules, 37, 53, 62, 76, 100, 102, 103, 109, 159, 162, 169
  see also nephridia
expansion, *see* extension
extension (extending), 29, 32, 34, 52, 62, 67, 74, 77, 81, 82, 98, 100, 105, 119, 124, 159, 176
extirpation, 84, 101, 145
  see also removal
extra-embryonic
  ectoderm, 98, 100, 114, 115, 119, 121, 122
    dorsal, 105, 106, 110, 111, 134, 149, 154, 156
    ventral, 105, 106, 110, 111, 154
  membranes, 113, 122, 124
extrusion, 17, 18, 49, 78, 113, 152
eye, 10 (spots), 37, 86, 87, 90, 93–5, 98–100, 104, 112 (facet), 153, 159, 168

fate (developmental), 60, 70, 133, 135
fate map, 27, 70, 71, 72, 88, 89, 105, 106, 108, 110, 114, 121, 122, 134, 149, 155, 156

genetic, 138, 146
feedback mechanism, 181
female, *not specified*
feminisation (feminising influence), 46, 75
fertile (fertility), 135, 138, 146, 165, 166
fertilisation (fertilised), 1, 13, 16, 49, 51, 56, 66 (cross), 78, 79, 83, 98, 117, 126, 130, 131, 144, 147, 161, 163
  (auto-)
  membrane, 78, 79
fibrils (fibrilous), 9, 14, 33
fibrillo-granular material, 33, 42, 43, 129, 133, 152, 169
  see also electron-dense bodies
field, *see* morphogenetic field
filopodia, *see* pseudopodia
fin, 169
fission, 10, 35, 67, 68, 78, 85, 181
flagella (flagellated), 10, 12–18, 175
follicle cells/epithelium, 89, 102, 117, 143, 145
foot, 86, 87, 90, 93–5, 165, 167
foregut, 110, 113, 134, 162
  see also stomodeum
'foreign' elements, 177, 187
fraction, 135, 139, 176, 182
  see also fragment (fragmentation)
fragment (fragmentation), 24, 34, 37, 42, 45, 58–60, 66, 78, 91, 129, 131, 133, 158, 161, 162
free-living/moving/swimming, 16, 17, 21, 22, 29, 34, 47, 48, 51, 52, 55, 64, 65, 69, 76, 83, 89, 151, 159, 161, 163, 167, 168, 174
frustula, 27
functional, 133, 136, 139, 152
  see also metabolic
funiculus, 62, 63
funnel, 99, 100, 166
furrows
  cephalic, 121, 134
  cleavage, 150, 154
  cross, 96
  secondary, 134
fuse (fusion), 29, 44, 58, 70, 89, 91–3, 100, 103, 112, 115, 119, 127, 129, 144, 145, 150, 154

gametes, 3, 7, 9, 13, 26, 27, 30, 35, 163, 185
gametogenesis, 7, 25, 32, 47, 143, 174, 183, 186
ganglia
  cerebral, 37, 46, 47, 58, 63, 66, 68, 69, 72, 74, 76, 81, 86, 87, 90, 100, 104, 108, 121, 134, 153, 159, 162, 168, 169

Subject index   249

ganglia–*cont.*
  circumenteric, 62, 109
  circumesophageal, 86
  segmental, *see* nerve cord
  subenteric, 165, 167
  supraesophageal, 68, 108
  ventral (trunk), 68, 168, 169
ganglionic complex, 86
gastral groove, *see* gastrulation groove
gastrocoelic cavity, 29, 30
gastrodermal cells/layer, 21, 23–7, 35, 39, 185
gastro-vascular system, 158
gastrula, 29, 57, 79, 80, 81, 126, 167
  stereo-, 18, 44, 47
gastrulation, 18, 27, 28, 44, 58, 70, 79, 90, 105, 106, 110, 113, 114, 118, 119, 122, 124, 127, 134, 151, 156, 159, 161, 165, 167, 169
  groove (ventral), 118, 119, 122, 124, 129, 154, 155
gemmule, 15, 16
gene, 1, 2, 4, 61, 173, 181
  activation (activity), 2, 173, 174, 180, 181, 186
  amplification, 4, 174
  derepression, 2, 3, 173
  inactivation, 2, 181
  repression, 173
genome, 4, 97
  activity/inactivity, 4
genotype, 126
generation/form, 1, 133, 176
  asexual, 21, 162
  sexual, 21, 162, 163
generative cells, *see* germinal and germ cells
genetic (genetically)
  constitution, 147
  control, 97, 123, 143
  evidence, 45
  fate mapping, *see under* fate map
  females, 152
  information, 2–4, 61, 157, 186
  marked, 126, 136
  material, 144
genital
  apparatus, 145
  atrium, 51
  cord, 77
  ring sinus, 84
  see further under gonad *and* gonadal
'genome reduction', 173, 174
  see also chromatin diminution *and* chromosome elimination
germ anlage/band/disc, 52, 110, 111, 113–15, 117–19, 122, 124, 127–9, 154, 155, 157

type
  'long', 121, 125
  'semi-long', 118, 121, 125
  'short', 118, 121, 125, 151
germ cells, general, *not specified*
  primordial (PGCs), *not specified*
  potential, 137, 139, 157, 183, 187
  presumptive (pPGCs), 7, 59, 74, 75, 136, 138, 150–2, 155, 176, 177, 179, 182
  (*germinal*) *determinant*, 2, 4, 18, 36, 101, 116, 128, 138, 139, 152, 157, 164, 169–71, 184, 185
  determination, 4, 137, 157, 176, 182, 184–6
    primary, 184
    secondary, 182, 186, 187
    see also formation (mode)
  development/differentiation/formation, xi, xii, 1–7, 9, 12, 13, 18, 29, 31–6, 45–7, 52, 54, 55, 60, 64, 65, 73, 83, 85, 101–3, 106, 111, 115, 126, 127, 137, 143, 151, 155, 159, 162, 164, 165, 168, 169, 174–7, 182, 184–7
  site of formation/origin, 6, 7, 127, 156
  formation (mode), xi, 1, 2, 4, 6, 7
    determinate, 4, 139, 157
    epigenetic, xii, 2, 4, 6, 13, 18, 34, 46, 50, 54, 59, 65, 75, 85, 103, 128, 157, 162, 168, 171, 175, 177–80, 182, 183, 186
    indeterminate, 128
    intermediate, 6, 85, 103, 157, 175, 177, 178, 180, 188
    preformistic, xii, 1, 4, 6, 13, 34, 50, 61, 85, 103, 157, 163, 165, 169, 171, 175–8, 180, 182–4, 186
  origin
    ectodermal, 31, 35, 102
    endodermal, 29, 35, 159, 171, 177
    mesodermal, 64, 65, 75, 85, 102, 106, 112, 115, 127, 156, 157, 162, 168, 171, 177
germ discs, *see* germ anlage
germ layer, 4, 17, 68, 69, 90, 124, 172, 173, 181
germ line, xi, 1, 2, 13, 19, 48, 49, 84, 85, 101, 143, 174–7
germ masses/balls, 49, 50
germ (germinal) (cyto-) plasm, 1, 4, 34, 36, 50, 54, 103, 120, 138, 157, 170–1, 177, 183, 185
  synthesis/formation, 34, 36
  see also pole plasm
germen, 1, 2, 3, 175, 176
germinal
  cells/layer, 10, 47–50, 52–5, 59, 102, 102 (non-), 162, 163, 171 (non-), 181, 181 (non-), 184

Subject index

epithelium, 83
granules, 32, 92, 139, 170–1, 184–7
  see also polar granules
  primordia/region, *see* germ anlage
  state, 59, 60
germinal vesicle, 69
germinative, *see* germinal
germinative–somatic cells, 53
gills (external), 77, 86, 87, 100
gland (glandular) cells, 26, 47, 145, 166, 167
  attachment, 165
  digestive, 76, 77
  egg shell, 44
  endodermal, 26, 39, 40
  epidermal, 52
  optic, 103
  penetration, 52
  pericardial, 103
  poison, 154
  salivary, 134
  shell, 90, 96, 98
  slime, 104, 105
  spinning, 154
  stalk, 163
glycogen, 52–4, 70, 92, 131, 152, 169
glycoprotein, 13
gnathal segments, 113, 114, 118
Golgi apparatus/complexes, 24, 33, 41, 54, 132
gonad (gonadal) anlage/bud/gland/organ, 7, 29, 30, 37, 45–7, 50, 51, 53, 55–7, 62, 63, 66, 73–8, 83–5, 87, 100, 101–3, 106, 111, 112, 115, 119, 120, 122, 127, 131, 135–8, 145–7, 151–3, 155, 157, 159, 164, 165, 167–9, 176, 177–82, 186–8
  indifferent, 131, 145, 146
gonadal
  area/region/zone, 22, 66, 135
  complex, 51, 77, 102
  differentiation, *see* sexual differentiation
  dissepiments/segments, 66, 73–5, 153, 156
  ducts, 53, 56, 77, 83, 84, 86, 101, 104, 112, 120, 148, 151, 153, 164–7
  interstitial cells, 147, 182
  sheath, 145, 147, 182
  tubules, 78
gonidia, 10–13, 175
gonoduct, *see* gonadal duct
gonophores, 22, 27, 29–31, 35
gradient
  animal–vegetal, 82, 89
  axial, 24
  cortical, 96
  cranio-caudal, 54, 67

field, *see* morphogenetic
morphogenetic, 46, 74, 96, 143, 173
sex-, 32
UV susceptibility, 60
vegetal–animal, 79, 82
grains/granular complex, *see* fibrillo-granular
granular bodies, *see* electron-dense bodies
granules (granular), 41, 43, 53, 54, 59, 70, 78, 120, 130, 140, 142, 152, 170–1
granulo-fibrillar, *see* fibrillo-granular
growth, 14, 16, 22, 25–7, 29, 35, 45, 67, 68, 71, 98, 119, 124, 129, 131, 145, 166
  over-, 72, 119
  phase, 13, 64, 130
  pattern, 23
  zone, 14, 22, 23, 110, 111, 114, 118, 121, 125, 151, 154–6, 166
gut, *see* intestine
gynandromorphs, 126, 138, 144, 146

haploid, 13, 47, 144, 147, 148, 163
hatching, 17, 45, 57, 75, 79, 102, 103, 110, 111, 117, 148
head, 34, 37, 39, 45, 55, 74, 86, 87, 90, 96, 98, 104, 105, 109, 112, 119, 148, 151, 152, 169
  lobe, *see* cephalic lobe
heart, 86, 87, 94, 95, 100, 102, 109
heredity (hereditary), 1, 46, 74, 133
hermaphroditism (hermaphrodite/hermaphroditic), 8, 11, 12, 14, 32, 37, 46–51, 55, 57, 62, 66, 71, 76, 77, 83, 86, 101–3, 144, 158, 165, 166, 169
  duct, *see* gonadal duct
heterochromatin, 59, 152
  see also chromatin
hexaploid, 40
hibernacula, 27, 62–4, 167
hind-gut, 58, 73, 102, 104, 110, 112, 113, 134, 162
  see also proctodaeum
histioblasts, 15, 16
holoblastic cleavage, *see* cleavage/total
hormones (hormonal), 44, 75, 152
  androgenic, 75, 152
  control, 103
  female, 152
  male, 152
hydrocoel, 81, 82
hydropore (hydroporal), 76, 81, 82, 83
  canal, 81, 82, 83
hydrostatic pressure, 136
hypostome, 22, 24

imago, 112, 174

immigration, *see* gastrulation *and* migration
implantation, *see* transplantation
in situ formation, 13, 45, 74, 125
indeterminate development, *see* embryonic development/regulative
induce (inducer/induction/inductive), 2, 4, 13, 30–2, 44, 46, 73 (mitogenic), 96, 97, 100, 101, 124, 139, 147, 172, 173, 176, 178, 183, 185
  see also interaction
infiltration, 27, 73
  see also gastrulation
information
  'blue print', 96, 97
  'executive', 96
  pre-programmed, 100
ingression, 28, 113, 161
  multiple (-polar), 28, 35
  unipolar, 27, 28, 34
  see also gastrulation
initial cells, 11–13
  see also gonidia
injection, 137, 185
interaction, 46, 74, 83, 96, 97, 172, 187
  interblastemic, 124
  nucleo-cytoplasmic, 124, 133, 173, 174
  see also induction
intercellular spaces, 187
internalisation, *see* invagination
interphase, 60
interradius, 76, 83, 84
intersexuality, 32
interstitial cells (*I cells*), 24–36, 39, 42, 175, 181, 184–7
  origin
    ectodermal, 26–9, 35
    endodermal, 26–9, 35
intestine (intestinal system/tract/tube), 37, 39, 40, 47, 50, 52, 66, 67, 69, 73, 77, 81, 86, 94, 95, 102, 112, 135, 149, 163, 165, 167, 168
  see also midgut
invagination, 12, 13, 18, 34, 35, 58, 64, 70, 72, 79–81, 90, 96, 105, 106, 109, 110, 113–15, 118, 119, 122, 124, 128, 133, 137, 147, 149, 155, 159–61, 165–7, 169
  see also gastrulation
invasion, 120, 133, 137, 154
inversion, 12, 13, 17, 18
irradiation
  UV-, 60, 100, 136, 137, 142, 145
    absorption spectrum, 92
    laser beam, 138
    micro-beam, 72, 74, 137
    sensitivity, 137
  X-, 24, 26, 27, 32, 38, 45, 67, 78, 84, 138
irreversible (changes), 126, 136, 144, 181, 186

isolation (isolates/isolated), 24, 34, 58, 82, 83, 94, 96, 102, 139, 161
isopotent, 2, 4, 126
isotopes, 38

jaws, 87, 104, 109

karyoplasm, 41
karyosomes, 60
katatrepsis, 119, 124
'*Keimplasma*' theory, 1, 173, 174
kidney, *see* excretory organs

labrum, 110, 150
larva, general, *not specified*
  auricularia, 82
  bipinnaria, 82
  cercaria, 48, 49
  cydippid, 159
  cysticercoid, 52
  ephyra, 34
  infusorigen, 163
  infusoriform, 163
  meta-cercaria, 48–50
  meta-trochophore, 73
  miracidium, 47–50
  nauplius, 151
  nematogen, 162–4
  oncosphere, 51, 52
  parenchymula, 18, 28
  pilidium, 161
  placula, 16
  planula, 27–9, 34, 35
  pluteus, 80, 82
  pre-oncosphere, 52
  pre-veliger, 89, 90, 103
  rediae, 47–50
  rhombogen, 162, 163
  scyphistoma, 34
  sporocyst, 47–50
  trochophore, 69, 73, 89, 102, 103, 167, 168
  veliger, 89
  vermiform, 162
legs, *see* limbs
life cycle, 9, 19, 21, 47–51, 55, 131, 162, 163, 175–7
ligation, *see* constriction
'*lignée germinale*', *see* germ line
limbs, 104, 109, 112, 148, 153, 155
limiting membrane, 33, 130, 140
lipid droplets, 33, 41, 70, 93, 131
lithium ion (Li^+), 82, 83, 91, 95, 96
locomotion (locomotory), 9, 10, 77, 86, 123, 124, 153, 167
longitudinal axis, *see* polarity/cranio-caudal and proximo-distal

Subject index

lophocytes, 14
lophophore, 62, 63
'lower', *see* 'primitive'
lymphocytes, 178
lysosomes, 38, 116

macromeres, 52, 79, 80, 88, 89, 96, 97, 154, 159–61, 167
madreporal interradius, 76, 82, 83
magnesium ion (Mg^{2+}), 83
male, *not specified*
malformations, 96, 124, 167
Malpighian tubules, 110, 112
mandible (mandibular segment), 109, 110, 112, 114, 148, 149, 151
 pre-, 109, 114
mantle, 86, 87, 98, 99
manubrium, 24, 29–31, 35
marking, 159
masculinisation (masculinising influence), 32, 46, 75
maturation (divisions), 49, 56, 89, 95, 117, 123, 136, 144
 see also meiosis
maxillae (maxillar segment), 109, 110, 112–14, 148, 151
 post-, 109, 110
maxillulae, 148, 151
medusa (medusoid), 21, 22, 24, 27, 29–32, 34, 35
 'ephyra', 34
meiosis, 9, 13, 16, 47, 49, 147, 148
merging, 117, 124, 129
 see also fusion
meridional canal, *see* radial canal
mesenchyme (mesenchymal cells), 15, 16, 18, 39, 47, 65, 67, 76, 103, 104, 158, 161, 162, 165–7
 primary, 79, 80, 172
 secondary, 79–81, 172
mesenteries, 76, 81, 169
mesentoblast, *see* endomesoderm
mesoderm (mesodermal cells/band/layer), *not specified*
mesoglea, 21, 22, 26, 31, 34, 35
mesomeres, 52, 79, 80
metabolism (metabolic), 41, 52, 54
metabolites, 131
metamere (metamery/metamerism), *see* segment (segmentation)
metamorphosis, 16, 27, 28, 47, 71, 73, 89, 112, 116, 135, 145, 161, 168, 175
metaphase, 56, 59, 117, 153
metaplasia, 69
 see also transformation
methyl-green pyronin, 44
microfilaments, 81, 91

micromeres, 44, 52, 69, 71, 79, 80, 82, 83, 88–90, 92, 94–7, 101, 154, 161, 167
 primary, 159–61
 secondary, 159–61
micropyle, 15, 78, 117
micropuncture, 74, 136, 137
microtubules, 92, 123
microvilli, 91
midgut, 67, 68, 71–3, 100, 104, 106, 112, 113, 115, 121, 133, 134–6, 139, 149, 151, 154, 155, 157, 176, 182
 anterior, 105, 119, 121, 122, 134, 149
 posterior, 105, 110, 119, 121, 122, 134, 147, 149, 155, 156
migration (migrated), 14, 18, 24, 26, 27, 29–31, 35, 38–40, 44–6, 67, 68, 73, 84, 101, 106, 109, 115, 117, 123, 127, 131, 133, 135, 137, 139, 146, 152, 155, 187
mitochondria, 24, 25, 33, 41–3, 53, 70, 71, 92, 93, 101, 102, 116, 130, 131, 138, 140, 152
mitosis (mitotic division), 23, 26, 27, 38–40, 53, 56, 57, 60, 78, 79, 120, 123, 124, 126, 129, 142, 147, 148, 181
 a-, 40, 129
 activity, 38
 apparatus/aster, 17, 123, 140
 rate, 39
 see also cleavage *and* cell division
moiety (animal and vegetal), 81, 82, 172, 173
molluscan cross arms, 89, 91, 96
monospermy (monospermic), 79
morphallaxis (morphallactic), 37
morphogens, 46, 68, 92
morphogenesis (morphogenetic), 24, 49, 71, 72, 74, 97, 98
 map, 100
 movements, 124
 pattern, 100
morphogenetic substance, *see* morphogens
morula, 52, 64, 105
mosaics, *see* chimaera
mouth, 21, 37, 47, 50, 55, 58, 62, 63, 70, 76, 77, 87, 98, 104, 105, 137, 155, 159, 162, 163, 165–9
movements, *see* displacements
moving inwards, *see* invagination
mucopolysaccharides, 41
multiploid, *see* polyploid
multipotential, *see* pluripotential
multivesicular bodies, 33, 152
muscle cells/fibrils/layer, 40–2, 45, 47, 50, 52–5, 66–8, 72, 73, 86, 104, 109, 112, 153, 158, 159, 161, 162, 165, 167, 168
mutants, 123, 123 (embryonic lethal), 133, 135–8, 146

Subject index   253

mycetocytes, 133
'mycetoma', 133
mycoplasm, 133
myoepithelial cells, *see* epithelio-muscular cells

nematocysts, *see* cnidoblasts
neoblasts, 38–42, 44–6, 65, 67–9, 73, 74, 162, 175, 181, 185
nephridia, 53, 62, 87
  meta-, 66, 86, 109
  proto-, 37, 42, 47, 55, 66, 69, 109, 159, 163, 165, 167
  see also excretory organs
nerve
  branches, 86
  cells, 26, 32, 41, 42, 72
  cords, 10, 37, 45, 47, 50, 55, 62, 66, 72, 75, 104, 121, 134, 153, 159, 168
  ring, 55, 76, 78
  strands, 76, 78
nervous system, 47, 50, 55, 66, 75, 76, 78, 104, 150, 153, 169
neurosecretory cells/activity, 68, 69, 75
nitrogen mustard, 26, 27
nuage material, 115, 116, 152
nuclear
  bodies, 130, 140
  compartments, 143
  emission, 33, 42, 43
  information, 123, 174
  lamellae, 143, 144
  membrane, 33, 41, 42, 59, 130, 131, 148
  pores, 33, 41, 42, 152
nucleic acid, 92
nucleo-cytoplasmic ratio, 53
nucleolus, 24, 25, 33, 40, 41, 54, 102, 115, 120, 123, 130, 144
nucleolus-like bodies, 115, 116
nucleus (nuclear/nucleated), *not specified*
  e-, 2, 3, 126
  intra-, 143
  mono-, 15
  para-, 129
  poly-, 15
  pro-, 3, 56, 117, 123, 128, 144, 152
nurse cells, 130, 131, 143, 148
nutrition, 16, 31, 89, 90, 104, 105, 128, 129, 148, 162, 187
nutritional cells, *see* yolk cells

ocelli, *see* eyes
ooecium, 63, 64
oocytes, 19, 31, 58, 63, 78, 82, 89, 101, 117, 130, 131, 136, 138, 144, 152, 161, 182

oogenesis, 16, 89, 130, 131, 138, 163, 176
oogonia, 7, 16, 18, 19, 30–3, 35, 46, 73, 83, 102, 131, 186
ooplasm, 118, 129, 139, 152
oosome, 115, 120, 129–30, 138
opisthosomal segments, 153–5
organ formation, *see* organogenesis
organelles, 9, 18, 24, 25, 41, 53, 56, 70, 92, 95, 124, 128, 138, 139, 187
organisation, 87, 88, 97
  adult, 22, 52
  centre, 46
  dis-, 15
  level, 87, 178
  polar, 60
  spatial, 9, 42, 158
  supracellular, 10
'organiser', 72, 88, 101
organogenesis, 98, 100, 101, 122, 125
orientation, *see* polarity
ostiole, 11, 12
outpocketings, *see* protrusions
ovarioles, *see* ovary
ovary, 7, 32, 37, 45, 46, 50, 51, 53, 64, 74, 75, 103, 104, 112, 133, 145, 158, 165, 166, 168, 169
overgrowth, *see* epiboly
oviduct, 105, 145, 166, 169
oviparous, 104
oviposition, 89, 95, 137
ovotestis, 7, 8, 102
ovo-viviparous, 104, 148, 154
oxygen supply, 128, 175

paedogenesis, 143, 144
pallium, *see* mantle
parabiosis, 32
parasite (parasitic), 10, 44, 47, 51, 55, 128, 174
  ecto-, 47, 50, 66
  endo-, 47, 162, 171, 174
parenchyma (parenchymal), 37–40, 43–7, 50, 51, 53, 54, 162, 171, 185
parietal/parietopleural, 62, 74, 75
parthenogenesis, 3, 49, 50, 87, 117, 143, 144, 164, 165
pathway, *see* germ cell/somatic differentiation
pattern
  basic, 72
  branching, 10
  formation, 125, 155
  hexagonal, 10
  mosaic, 95, 97
  movement, 23
  pre-, 100
  regional, 124, 126, 173

pattern – *cont.*
　streaming, 100
　see also under cleavage *and* chromosome
pedipalp (pedipalpal segment), 152, 154, 156
peduncle, *see* stalk
penis, 51, 145
peptides, 75
perchloric acid, 92
pericardium (pericardial cells), 101–3, 152
periplasm, 113, 114, 117, 120, 123–5, 129, 130, 137, 154
　'primary', 120
　'secondary', 120, 124
perisarc, 22, 23
peritoneum (peritoneal), 62, 66, 74, 75, 77, 83, 115, 153, 155, 157, 159, 168
　complex, 102, 103
phagocytose (phagocytosis), 15, 67, 73
pharynx, 35, 37, 42, 44, 46 (region), 47, 55, 62, 72, 77, 86, 87, 112, 158, 165
phospholipids, 92
phylogenetic relationship/origin/pattern, xi, 5, 48, 105, 106, 108, 150, 178, 180
pinacocytes, 14, 14 (endo-), 15
pinnules, 76, 77
pinocytosis, 89
placenta (placental)
　pseudo-, 148
　stalk/plate, 105, 106
planktonic, *see* free living
plasma membrane, *see* cell membrane
pluripotent (pluripotency/pluripotenciality), cells, 16, 26, 34–6, 101, 182
podia, 76, 77
polar
　body, 49, 97, 98, 105, 117, 129, 144
　cap, 129, 133, 135, 154
　differentiation, 60
　granules, 92, 94, 127–33, 136–40, 142, 157, 176, 185
　see also germinal granules
　lobe, 70, 71, 91–5, 95 (-less), 103
　region, 56, 60, 137
polarity, 10, 16, 18, 22, 52, 56, 59, 96, 150
　animal–vegetal, 16, 56, 60, 69, 80, 82, 83, 92, 95, 97, 100, 173
　antero-posterior, 10, 37, 42, 45, 46, 58, 77, 105, 110, 117, 120, 184
　cranio-caudal, 39, 47, 50, 54, 74, 119
　dorso-ventral, 37, 46, 50, 66, 69, 71, 77, 79, 83, 86, 89, 91, 93, 96, 97, 117, 120
　oral–aboral, 158, 159
　proximo-distal, 21–4, 32, 167
pole cell (formation), 33, 121, 126, 128–31, 133–9, 143–5, 147, 157, 176, 182–7

polyembryony, 49, 50, 87, 128, 163
　primary, 49
　secondary, 49
　'substitute', 128
polymorphism, 21, 22, 27
polyp, 21–7, 30, 31, 34, 35
　'scyphistoma', 34
polypide, 62–4
polyploid, 40, 45, 46, 143, 174
polysomes, 41, 116, 130–2, 143
polyspermy, 79, 117
post-naupliar segments, 149, 151
posterior pole, 118–20, 126, 127, 129–31, 133, 136–40, 145–7, 157, 185
　plasm, 120, 126, 133, 136–40, 142, 145, 147, 174, 176, 185
　see also germ plasm
post-trochal region, 90, 95
potassium ion (K⁺), 83
potentiality, 2, 3, 24, 32, 38, 75, 100, 159, 175, 176, 182, 186, 187
precheliceral lobes, 153, 154
　see also cephalic lobes
preformistic, *see under* embryonic development *and* germ cell formation (mode)
pre-trochal region, 90
pricking, *see* micropuncture
primary axis, *see* polarity/animal–vegetal
'primitive', 3, 87, 116, 175, 178, 180, 182, 187
proboscis, 51, 159, 161, 162, 168
proctodeum, 55, 58, 72, 105, 106, 110, 114, 115, 119, 121, 122, 134, 155, 156, 161, 165–7
　see also hindgut
progenitor cells, 57
progeny, 126, 136
proglottids, 51, 53
proliferation (proliferative), 27, 29, 30, 39, 40, 67, 102, 105, 110, 115, 117, 118, 121, 124, 155, 166
　zone, 51, 52, 54
propagation, *see* reproduction
propagatory cells, *see* germ cells
prophase, 152, 153
prosomal segments, 154
protandry (protandrous), 8, 12, 165
proteins (proteinaceous), 130, 140, 185
　basic, 60, 139, 187
　chromosomal, 60
　synthesis, 82, 95, 137
protocerebral, 149
protogyny (protogynous), 8, 165
prototroch, 69, 72, 90
protrusions, 129, 137
protuberances, *see* protrusions

Subject index

pseudopodia, 10, 79, 81
pulmonary sac, 86
pupa (pupal), 112, 116, 143
purine compounds, 92
pygidium (pygidial), 70, 73
  bud formation, 68
  regeneration, *see* regeneration/posterior
pyrimidine compounds, 92

quadrant (D), 88, 89, 93, 150
queens, 147

radialisation, *see* symmetry/radial
radial canal, 29, 30, 76, 81, 84, 158
radio-sensitive, 84
radio-resistant, 84
radula, 86, 87
reaggregation, 14, 24, 131–3
rearrangement, *see* reconstitution
recombination, 44, 82, 83, 96, 138 (somatic)
reconstitution (reconstituted), 14, 24, 37, 74, 85, 87, 95, 97
  see also redistribution *and* reorganisation
reconstruction, *see* reconstitution
redistribution, 32, 95, 129
  see also reconstitution *and* reorganisation
reduction (reduced), 9, 29, 40, 49, 52, 77, 86, 87, 89, 122, 135–8, 143, 148, 163, 174, 184, 185
  see also elimination
reduction bodies, 15
reduplication (reduplicated), 4, 184
reestablishment, *see* reconstitution
regenerate (*regeneration*) (*regenerative capacity/process*), 6, 14, 21, 23, 24–7, 34, 37–42, 45, 46, 50, 55, 66–9, 73–5, 78, 84, 85, 87, 101, 102, 158, 159, 161, 162, 167, 171, 175, 176, 179, 181, 185
  anterior, 67, 68
  blastema, 37–42, 46, 67–9, 78, 162, 185
  embryonic, 45
  field, 39
  multiple, 38, 40
  posterior, 67–9, 74
regional, *see under* pattern
regression, 29, 81, 136
regulative, *see under* embryonic development
reimplantation, 126, 136
release (released), 3, 7, 13, 45, 123, 167, 183
removal (removed), 74, 93, 94, 96, 97, 101, 125, 136, 139
reno-pericardial mesoderm, *see* peritoneal complex

reorganisation, 15, 112, 115
  see also reconstitution
repair, *see* replacement *and* regeneration
replacement (replacability), 3 (non-), 13, 25, 27, 44, 49, 55, 67, 68, 72, 89, 115, 116, 119, 128, 145, 173, 176, 177
reproduction
  asexual, 3, 6, 9, 10, 13, 15, 18, 21, 24, 27, 31, 32, 34, 35, 37, 42, 45, 49, 62, 65–9, 73–5, 78, 84, 85, 87, 158, 162, 166, 168, 171, 175, 177–9, 181, 182, 184–7
  sexual, 3, 9, 10, 13, 16, 18, 21, 27, 29, 31, 34, 35, 42, 64–7, 87, 167, 175, 177, 181, 185–7
reproductive organ/system, *see* gonad
'*reserve cells*' (*embryonic*), 16, 24, 26, 38, 67–9, 84 (PGCs), 175
respiratory organs/system, 37, 47, 55, 76–8, 104, 109, 112, 153
reticular cytoplasm, 117, 120, 154
retransformation, 46
reversible (reversibility), 2, 20, 22, 33, 34, 75, 136, 181, 183, 186, 187
rhynchocoel, 159, 161
ribonucleic acid (RNA), 40, 41, 59–61, 92 (mRNA), 95, 124 (synthesis), 130, 131 (mRNA), 137, 140, 169
ribosomes (ribosomal), 24 (free), 25, 40, 41 (free), 43, 53, 61 (cistrons), 92, 116, 130, 133, 137, 138, 143 (cistrons)
ring canal, 29, 30, 76, 84
rods, *see* skeleton/endo-
rotation, 10, 58, 119, 165, 167, 168

schizogenesis, *see* budding
scissoparity, 42, 68, 74, 162
sceroblasts, 14, 16
scolex, 51, 52
secretion (secretory cells), 14, 16, 22, 47, 52, 56, 68, 71, 86, 108, 113, 117, 166
segments, general, *not specified*
segmentation (segmented), 37, 53, 66, 68, 71, 72, 86, 104, 106, 109, 111–13, 119, 138, 154
  heteronomous, 66, 71, 108, 112
  homonomous, 66, 71, 104, 109
  primary, 154
  pseudo-, 159
  secondary, 154
  un-, 55, 66, 104, 128, 155, 159, 168
segregation, general, *not specified*
  cytoplasmic, 95, 96, 173
  intravitelline, 115
  ooplasmic, 69, 89, 90, 95
SEM pictures, 91

Subject index

seminal vesicle, 101
sense (sensory) organs, 17, 29
   cephalic, 37, 87, 158
septum, 34, 35, 84 (interradial), 152
sequences, 3, 5
   antero-posterior, 105, 110, 119
   non-repetitious, 61
   repetitious, 61, 144
   satellite, 143
serosa, 113–15, 115 (pro-), 119, 122, 124, 155, 156
sessile, 21, 29, 76, 163, 165 (non-), 165, 168
sex
   cells, *see* gametes *and* germ cells
   differentiation, *see* sexual differentiation
   chromosome, *see* X-chromosome
   determination, 102, 103, 145, 147
   inversion, *see* reversal
   reversal, 75, 152
sexual
   character (characteristics), *103, 152*
   differentiation, 7, 32, 45, 46, 68, 74, 75, 103, 140, 142, 145, 152, 162
   generation, 13, 19–21, 29, 31, 32, 42, 44, 45, 49, 50, 64, 147
   maturation, *see* differentiation
   state/form, *see* generation
   sexuality, *see* sexual generation
   shell (membrane), 86, 87, 93–5, 102, 164
   see also under gland
sinking inwards, *see* invagination
skeleton
   endo-, 14, 16, 76, 77, 79–81, 153
   exo-, 62, 108, 109, 112, 153
sodium ion (Na^+), 83
solitary, 21, 22, 165, 178
soma, 1–3, 175, 176
somatic cells/tissues, general, *not specified*
   cell line/type, *see* differentiation
   differentiation, 2, 3, 32, 54, 124, 126, 174–7, 181, 184–7
somatoblast, *see* teloblast
somatocoel, 81, 83, 85
somatogenesis, *see* somatic differentiation
somatopleure (somatopleural), 62–4, 75
somites, 72, 105, 106, 111, 113, 119, 120, 122, 151, 155, 156
somitic lobes
   appendicular, 105, 111, 113, 155
   dorso-lateral, 105, 111, 113
   medio-ventral, 105, 111, 113, 156
specialisation, *see* differentiation
specific (specificity)
   cell-type, 1, 3, 58, 157, 169, 176, 177, 181, 183, 187
   germ-layer, 68, 181
   inter-, 136

species, 13
spermatids, 63
spermatogenesis, 16, 152, 163, 176
spermatogonia, 7, 16, 18, 19, 30, 32, 35, 46, 73, 83, 102, 131, 186
spermiocyst, 16
sperm (spermatozoa), 3, 9, 12, 16, 29, 47, 64, 66, 78, 79, 117, 123, 144, 163, 175
spermioduct, 145
spicule, 15, 16, 80, 86
spindle (cleavage/nuclear), 59, 60, 71, 96, 140, 150, 152
splanchnopleure (splanchnopleuric wall), 73–5, 106, 119, 122, 127, 145
spongoblasts, 15
sporosac, 29, 49
stalk, 10, 22, 23, 29, 76 (-less), 78, 167
starvation, 38, 75
statoblast, *see* hibernaculum
stem, *see* stalk
stem cell, 26, 36, 56, 101
sterile (segment/region), 73, 74, 85, 136, 137, 146, 147, 165, 166
sterilisation, 137
sterility, 133, 135, 136, 145
stolon, 22, 23, 27, 165–7
stomach, 62, 69, 77, 112, 149
stomodaeum, 35, 37, 55, 58, 62, 67, 69, 70, 72, 80, 81, 90, 100, 105, 106, 110, 114, 115, 119, 121, 122, 134, 149, 150, 155, 156, 158, 161, 165–7
stone canal, 81, 82
stratification, 82, 95
'strobila', 51, 53
strobilation, 34
subhypostomal region, 23, 32, 35
subumbrellar cavity, 29, 30
sucker, 47, 51, 66, 87
sulphate ion (SO_4^{2-}), 83
suppression (suppressed), 29, 42, 45, 83, 88, 122, 136, 144, 146, 187
switched into, 31, 32, 42, 175, 177, 181, 182, 185–7
   see also alternation
symbiont-ball, 133
symmetry (symmetrisation), 52
   a-, 86
   bilateral, 37, 47, 50, 56, 71, 76, 77, 79, 81, 83, 86–8, 90, 96, 97, 108, 117, 145, 149, 168
   biradial, 158
   pentaradial, 76
   radial, 14, 21, 76, 77, 79, 83, 88, 93, 96
   radio-bilateral, 21
   tetraradial, 17
symplastic, 10
syncytium (syncytial), 10, 37 (semi-), 38,

40, 44 (yolk), 52, 98, 117, 120, 121, 123, 124, 128, 135, 185
syngamy, 117
synkaryon, 128, 144

tail, 37, 169
teloblasts
  ecto-, 69–73, 151, 168
  meso-, 64, 69–74, 151, 168
telophase, 56, 141, 153
telson, 110, 148, 155
temperature (influence), 27, 31, 32, 42, 74, 137, 147, 167
tentacles, 21–5, 29, 30, 62–4, 77, 86, 87, 104, 105, 165–8
testicular anlage/follicles/vesicles, *see* testis
testis, 7, 32, 37, 45, 46, 51, 53, 63, 64, 74, 103, 104, 112, 133, 145, 158, 166, 169
thicken (thickening), 81, 98, 113, 120, 166, 167
thocyanate, 82
thorax (thoracic segments), 67, 112 (pro-, meso-, meta-)–14, 118, 119, 121, 151, 152
thymidine ^{3}H label, 23, 24, 143
thymidine-kinase, 39
toluidine blue, 24
totipotent (totipotency), 1, 3, 20, 67–9, 101, 125, 172–5, 181, 182
totipotential (totipotentiality), 182, 183, 185
tracheae, 104, 109, 112, 134, 153
transection, 67, 74
  see also amputation *and* decapitation
transcription, 92
transfer, 16, 38, 40, 60, 122, 128, 133, 169
transition, 4, 19, 20, 113
transformation, 3, 13–16, 18, 19, 31, 34, 35, 47, 48, 54, 67, 73, 86, 112, 117, 133, 172
  see also conversion
transplantation, 2, 22, 32, 38, 46, 75, 125, 126 (nuclear), 136, 173 (nuclear)
trauma, 78
triploid, 40
trophamnion, 129, 148
trophic influence, 38, 40, 75
trophochorion, 148
trophocytes (trophocytic), 15, 130
trophoserosa, 148
trunk, 37, 51, 62, 104, 105, 109–11, 148, 151, 152, 168, 169
turns, *see* rotation

ultrasonic treatment, 136, 137, 147
upper lip, 112
uterus, 51, 56, 145

vagina, 51, 120, 145
vas deferens, 51, 169
vegetal
  body, 92–4
  pole, 60, 78, 79, 82, 83, 89, 90, 98, 113, 115, 169, 170–1
  pole plasm, 59, 60, 71, 89, 95
    see also germ plasm
vegetalisation, 82
vegetalising agent, 82
velum (velar plate), 24, 29, 30, 90, 94, 95
vesicles, 92 (multimembranous), 94, 115, 152, 166
vesiculum seminalis, 51, 145
vestibule, 165, 168
vitellarium, 37, 44, 47, 51
vitelline cells, *see* yolk cells
vitelline membrane, 117, 166, 167
vitellogenesis, 15, 16, 130, 131
vitellophages, 109, 113, 135, 154, 155, 157, 176, 182
  primary, 117, 120
  secondary, 117, 135, 182
  tertiary, 121
viviparous (viviparity), 104–6, 128, 148, 154, 164
  adenotrophic, 148
  haemocoelous, 148
  pseudo-placental, 148

water lungs, *see* repiratory organs
water-vascular system, 76, 77, 81
wings, 112, 145
  -bearing, 112
  covers, 112
  -less, 112
winter buds, *see* hibernacula
worker bees, 147
wound (wounding), 23, 39, 67, 68, 162
  area, 38–40, 68, 78
  reaction/stimulus, 38, 40, 41
  surface, 37, 38, 46, 69, 78

yolk, general, *not specified*
  cells, 18, 44, 51, 52, 105, 109, 164
  epithelium, 98
  formation, *see* vitellogenesis
  -free (blastomeres), 71, 117, 120, 149, 154
  gland, *see* vitellarium
  mass/system, 98, 105, 109, 113, 117, 119, 124, 128, 133, 150, 154, 155, 173
  platelets, 92
  -poor (eggs), 28, 34, 71, 88, 105, 128
  pyramids, 150, 154
  -rich (eggs), 28, 34, 35, 52, 64, 71, 89,

yolk – *cont.*
  104, 105, 109, 110, 113–15, 117, 150, 154, 167
 sac, 72, 89, 90, 98–100
 spheres, *see* cells
 syncytium, 98, 100, 101

zinc chloride ($ZnCl_2$), 75
zooecium (zooecide), 62, 64, 65
zooid, 10
zygote, 10, 13, 16, 47, 48, 50, 64, 105, 117, 144, 148, 163